Universities, Cities and Regions

Regions and cities are the natural loci where knowledge is created, and where it can be easily turned into a commercial product. Regions are territories where, under certain socio-economic conditions, a strong sense of belonging and mutual trust develops the ability to transform information and inventions into innovation and productivity increases, through cooperative or market interaction. Especially in contexts characterised by a plurality of agents – such as cities or industrial districts – knowledge is the result of cooperative learning processes, nourished by spatial proximity, network relations, interaction, creativity and recombination capability.

This book explains the logic behind these interactions and cooperative attitudes in regions and cities. One of the most significant channels comes from the presence of a university and its collaboration with firms and scientific research centres. These mutual relations between academic institutions and enterprises are of key importance.

The significance of universities in driving economic well-being and regional development has been well documented for some time now. Much of the research, however, has centred upon countries in Western Europe and the United States. Increasingly, and since the expansion of the European Union in 2004 in particular, themes of academic entrepreneurship, university–business links, knowledge and innovation have become important on a Europe-wide scale. This book draws together key thinkers from across the continent to analyse the importance of higher educational institutions in fostering development.

Roberta Capello is Full Professor of Regional Economics at the Polytechnic University of Milan, Italy.

Agnieszka Olechnicka is Assistant Professor at the Centre for European Regional and Local Studies at the University of Warsaw, Poland.

Grzegorz Gorzelak is Professor and Director of the Centre for European Regional and Local Studies at the University of Warsaw, Poland.

Routledge Studies in Global Competition
Edited by John Cantwell, Rutgers, the State University of New Jersey, USA and David Mowery, University of California, Berkeley, USA

1. **Japanese Firms in Europe**
   Edited by Frédérique Sachwald

2. **Technological Innovation, Multinational Corporations and New International Competitiveness**
   The case of intermediate countries
   Edited by José Molero

3. **Global Competition and the Labour Market**
   Nigel Driffield

4. **The Source of Capital Goods Innovation**
   The role of user firms in Japan and Korea
   Kong-Rae Lee

5. **Climates of Global Competition**
   Maria Bengtsson

6. **Multinational Enterprises and Technological Spillovers**
   Tommaso Perez

7. **Governance of International Strategic Alliances**
   Technology and transaction costs
   Joanne E. Oxley

8. **Strategy in Emerging Markets**
   Telecommunications establishments in Europe
   Anders Pehrsson

9. **Going Multinational**
   The Korean experience of direct investment
   Edited by Frédérique Sachwald

10. **Multinational Firms and Impacts on Employment, Trade and Technology**
    New perspectives for a new century
    Edited by Robert E. Lipsey and Jean-Louis Mucchielli

11. **Multinational Firms**
    The global-local dilemma
    Edited by John H. Dunning and Jean-Louis Mucchielli

12. **MIT and the Rise of Entrepreneurial Science**
    Henry Etzkowitz

13. **Technological Resources and the Logic of Corporate Diversification**
    Brian Silverman

14. **The Economics of Innovation, New Technologies and Structural Change**
    Cristiano Antonelli

15. **European Union Direct Investment in China**
    Characteristics, challenges and perspectives
    Daniel Van Den Bulcke, Haiyan Zhang and Maria do Céu Esteves

16. **Biotechnology in Comparative Perspective**
Edited by Gerhard Fuchs

17. **Technological Change and Economic Performance**
Albert L. Link and Donald S. Siegel

18. **Multinational Corporations and European Regional Systems of Innovation**
John Cantwell and Simona Iammarino

19. **Knowledge and Innovation in Regional Industry**
An entrepreneurial coalition
Roel Rutten

20. **Local Industrial Clusters**
Existence, emergence and evolution
Thomas Brenner

21. **The Emerging Industrial Structure of the Wider Europe**
Edited by Francis McGowen, Slavo Radosevic and Nick Von Tunzelmann

22. **Entrepreneurship**
A new perspective
Thomas Grebel

23. **Evaluating Public Research Institutions**
The U.S. Advanced Technology Program's intramural research initiative
Albert N. Link and John T. Scott

24. **Location and Competition**
Edited by Steven Brakman and Harry Garretsen

25. **Entrepreneurship and Dynamics in the Knowledge Economy**
Edited by Charlie Karlsson, Börje Johansson and Roger R. Stough

26. **Evolution and Design of Institutions**
Edited by Christian Schubert and Georg von Wangenheim

27. **The Changing Economic Geography of Globalization**
Reinventing space
Edited by Giovanna Vertova

28. **Economics of the Firm**
Analysis, evolution and history
Edited by Michael Dietrich,

29. **Innovation, Technology and Hypercompetition**
Hans Gottinger

30. **Mergers and Acquisitions in Asia**
A global perspective
Roger Y. W. Tang and Ali M. Metwalli

31. **Competitiveness of New Industries**
Institutional framework and learning in information technology in Japan, the U.S. and Germany
Edited Cornelia Storz and Andreas Moerke

32. **Entry and Post-Entry Performance of Newborn Firms**
Marco Vivarelli

33. **Changes in Regional Firm Founding Activities**
A theoretical explanation and empirical evidence
Dirk Fornahl

34. **Risk Appraisal and Venture Capital in High Technology New Ventures**
Gavin C. Reid and Julia A. Smith

35. **Competing for Knowledge**
Creating, connecting and growing
Robert Huggins and Hiro Izushi

36. **Corporate Governance, Finance and the Technological Advantage of Nations**
Andrew Tylecote and Francesca Visintin

37. **Dynamic Capabilities Between Firm Organisation and Local Systems of Production**
Edited by Riccardo Leoncini and Sandro Montresor

38. **Localised Technological Change**
Towards the economics of complexity
Cristiano Antonelli

39. **Knowledge Economies**
Innovation, organization and location
Wilfred Dolfsma

40. **Governance and Innovation**
Maria Brouwer

41. **Public Policy for Regional Development**
Edited by Jorge Martinez-Vazquez and François Vaillancourt

42. **Evolutionary Economic Geography**
Location of production and the European Union
Miroslav Jovanovic

43. **Broadband Economics**
Lessons from Japan
Takanori Ida

44. **Targeting Regional Economic Development**
Edited by Stephan J. Goetz, Steven C. Deller and Thomas R. Harris

45. **Innovation, Knowledge and Power**
Theodora Asimakou

46. **Creativity, Innovation and the Cultural Economy**
Edited by Andy C. Pratt and Paul Jeffcutt

47. **Co-opetition Strategy**
Giovanni Battista Dagnino and Elena Rocco

48. **Knowledge Intensive Entrepreneurship and Innovation Systems**
Evidence from Europe
Edited by Franco Malerba

49. **Innovation in Complex Social Systems**
Edited by Petra Ahrweiler

50. **Internationalization, Technological Change and the Theory of the Firm**
Edited by Nicola De Liso and Riccardo Leoncini

51. **Territory, Specialization and Globalization in European Manufacturing**
By Helena Marques and Francisco Puig

52. **Institutional Diversity and Innovation**
Continuing and emerging patterns in Japan and China
Cornelia Storz and Sebastian Schäfer

53. **Innovation and Economic Crisis**
Daniele Archibugi and Andrea Filippetti

54. **The Communications Industries in the Era of Convergence**
Catherine Mulligan

55. **Innovation, Technology and Knowledge**
Charlie Karlsson, Borje Johansson and Roger R. Stough

56. **Evolution of Competition Laws and their Enforcement**
Pradeep S. Mehta

57. **The Economics of Structural Change in Knowledge**
Francesco Quatraro

58. Economic Geography and the Unequal Development of Regions
Jean-Claude Prager and Jacques-François Thisse

59. Social Networks, Innovation and the Knowledge Economy
Edited by Isabel Salavisa and Margarida Fontes

60. The Economics of Creativity
Ideas, firms and markets
Edited by Thierry Burger-Helmchen

61. Epistemic Economics and Organization
Forms of rationality and governance for a discovery oriented economy
Anna Grandori

62. Universities, Cities and Regions
Loci for knowledge and innovation creation
Edited by Roberta Capello, Agnieszka Olechnicka and Grzegorz Gorzelak

Universities, Cities and Regions

Loci for knowledge and innovation creation

Edited by Roberta Capello, Agnieszka Olechnicka and Grzegorz Gorzelak

Routledge
Taylor & Francis Group
LONDON AND NEW YORK

First published 2013
by Routledge
2 Park Square, Milton Park, Abingdon, Oxfordshire OX14 4RN

Simultaneously published in the USA and Canada
by Routledge
711 Third Avenue, New York, NY 10017

First issued in paperback 2014

Routledge is an imprint of the Taylor and Francis Group, an informa business

British Library Cataloguing in Publication Data
A catalogue record for this book is available from the British Library

Library of Congress Cataloging in Publication Data
Universities, cities and regions : loci for knowledge and innovation creation / edited by Roberta Capello, Agnieszka Olechnicka and Grzegorz Gorzelak.
p. cm.
Includes bibliographical references and index.
1. Education, Higher--Economic aspects. 2. Universities and colleges--Economic aspects. 3. Regional economics. 4. Research--Economic aspects. 5. Diffusion of innovations. 6. Business and education. 7. Academic-industrial collaboration. I. Capello, Roberta, 1962- II. Olechnicka, Agnieszka. III. Gorzelak, Grzegorz.
LC67.6.U573 2012
338.4'3378--dc23
2012021774

ISBN 978-0-415-63108-2 (hbk)
ISBN 978-1-138-90375-3 (pbk)
ISBN 978-0-203-09714-4 (ebk)

Typeset in Times New Roman
by GreenGate Publishing Services, Tonbridge, Kent

Contents

Figures

Maps

Tables

Contributors

Sharmistha Bagchi-Sen is Professor and Chair, Department of Geography, State University of New York, Buffalo. Her research interests are in economic geography, regional development and international business.

Sascha Brinkhoff is a PhD candidate at the Department of Geography at the Humboldt-Universität zu Berlin. In his research he mainly focuses on the impact of knowledge management in territoral innovations systems and creative industries in metropolitan regions. Furthermore, he works as a policy consultant in regional economic development projects in Germany and internationally.

Vladislav Čadil specialises in evaluation of research and innovation policies, regional dimension of innovation processes, and science–industry cooperation. He holds a PhD in regional and political geography from the Charles University in Prague.

Roberta Capello is professor in Regional and urban economics at the Politecnico of Milan. Past-President of the Regional Science Association International (RSAI). Editor in chief of the Italian Journal of Regional Science and co-editor of Letters in Spatial and Resource Science (Springer Verlag). Author of many scientific papers and a textbook in Regional Economics, published in Italian and English.

Katalin Erdős is Assistant Lecturer at the University of Pecs, Hungary. Her main research interests are related to the role of universities in regional economic development, academic entrepreneurship, university spin-offs, and innovation in biotechnology.

Oliver Fiala is Research Assistant at the Chair of economics, esp. allocation theory, at the Technical University of Dresden, Germany. He graduated with a diploma in economics in 2011 with a thesis on the regional impact of a university medical centre. His main research interests are in the field of regional, environmental and health economics.

María Purificación García is Professor in the Department of Business and Private Law, at the University of León. She has a PhD in Labour Sciences and her topics of research include labour law, labour risk prevention and entrepreneurship education, as reflected in her multiple publications and international conferences.

John Goddard is Emeritus Professor of Regional Development Studies at Newcastle University UK. He founded and led the Centre for Urban and Regional Development Studies (CURDS) from 1977 to 1998 and was then Deputy Vice Chancellor of the University with responsibility for city and regional engagement until his retirement in 2008.

Grzegorz Gorzelak is Professor of Economics and director of EUROREG at the University of Warsaw. He has published many papers and books on regional development and on the financial crisis in Central and Eastern Europe. He is the president of the Polish section of the Regional Studies Association and Editor-in-Chief of the refereed journal Studia Regionalne i Lokalne.

Kevin Grant is a Lecturer and member of the Department of Business and Management at the University of Sussex, School of Business, Management and Economics. His main areas of research and teaching are in the field of strategic information systems, innovation, disruptive technologies and 'Green IT'. He has provided advice to a range of organisations on Strategic Information Management and Value Creation through appropriate and relevant technologies. Finally, he is a Fellow of British Computer Society and a Chartered IT Professional.

Dennis Häckl is Assistant Professor at the Chair of economics, esp. allocation theory, at the Technical University of Dresden, Germany. He graduated from Leipzig Graduate School of business with a PhD in economics in 2010. His main research interests are in the field of health economics, applied microeconomic theory and regulatory economics.

Mikołaj Herbst is assistant professor at the University of Warsaw, Centre for European Regional and Local Studies (EUROREG). He specialises in regional development, economics of education and public finance.

Louise Kempton is a Senior Research Associate at CURDS, focusing on the relationship between universities and regions and the nature of the 'Civic University'. She previously worked on economic strategy development at the regional development agency One North East.

Krzysztof Klincewicz, professor and Chair of the Department of Organization Theory and Methods at the Faculty of Management, University of Warsaw, specialising in the management of technology and innovation; Rector's Deputy for Intellectual Property Management at the University of Warsaw; member of the United Nations' Technology Executive Committee.

Elmar Kulke is a professor of economic geography at the Department of Geography and the Dean of the Natural Sciences Faculty at the Humboldt-Universität zu Berlin. Major research areas of Professor Kulke are the Geography of Services in a variety of sectors in Central Europe and South-East Asia besides innovation-driven case studies in Technology Parks. He has authored and edited many books; among recent publications are *Wirtschaftsgeographie* (2009, 4th edition) and *Wirtschaftsgeographie Deutschlands* (2010, 2nd edition).

Jari Kuusisto is director of the SC-Research Institute at the University of Vaasa since 2001. His current research activities focus on user-driven innovation, services innovation and intangibles management. In the policy related work he has extensive experience in services and user-driven innovation policy analysis, and design. He has provided policy advice to a range of government organisations in Australia, Sweden, Finland, Norway, Germany, Italy, UK, USA, South Africa, Estonia, France and Ireland. He has also worked with the OECD, European Commission, International Chamber of Commerce and World Intellectual Property Organisation. He is a member of IP Committee of the International Chamber of Commerce.

Ana Lanero is Assistant Professor in the Department of Business Management and Economy of the University of León. She has recently finished her PhD, with a thesis about entrepreneurship education in European university contexts. Her research lines are entrepreneurship education and corporate social responsibility.

Helen Lawton Smith is Professor of Entrepreneurship, Department of Management, Birkbeck, University of London. She is the Founder and Research Director of the Oxfordshire Economic Observatory, School of Geography and the Environment, Oxford University.

Martin Meyer is Head of Business and Management at the University of Sussex. His research focuses on knowledge exchange, academic patenting and emerging science-based technologies. For more than ten years, he has worked on various aspects of the Triple Helix, most notably science and technology indicators, the entrepreneurial university and global innovation environments. He has done extensive contract research and consultancy work in these areas. He also holds visiting appointments at the universities of Leuven, London and Vaasa.

Olga Mrinska holds a PhD in Economic and Social Geography. She has more than ten years experience in the area of regional policy, local and regional economic development and governance working for government institutions, think tanks, international projects and universities both in Ukraine and in the UK. She has a wide range of publications in her core areas of interest: regional policy in post-socialist economies and international comparative studies. She was formerly Research Director at the Institute of Public Policy Research North (UK). Currently she is an Associate with ILORI (Kyiv, Ukraine) and an independent consultant based in the UK.

Kazimierz Musiał is Assistant Professor for the Institute of Scandinavian Studies and Applied Linguistics at the University of Gdansk, Poland. He graduated in Scandinavian Studies at Poznań University, Poland and received a PhD in political science from Humboldt University in Berlin, Germany. His most recent research concerns two specific agendas, one relating to regional integration in the Baltic Sea region, and the other to the role of universities and reforms of higher education systems in Northern Europe. International experience includes longer research scholarships in Denmark, Norway and numerous shorter research visits in other Nordic countries, as well as work in Germany (research and teaching assistant at the Humboldt University Berlin 1997–2002 and research fellowship at Greifswald Institute for Advanced Study 2002–04).

Agnieszka Olechnicka, PhD is assistant professor at the University of Warsaw, Centre for European Regional and Local Studies (EUROREG). She has an academic background in economics and has been working on several projects focused mainly on regional development. Regional innovation, particularly the role of science institutions in regional development and the problems of knowledge absorption by enterprises are her current interest.

Małgorzata Runiewicz-Wardyn, PhD is an assistant professor in Kozminski University in Warsaw and a researcher in the Transformation, Integration and Globalization Economic Research Center (TIGER).

Jan Sadlak is President of IREG Observatory on Academic Ranking and Excellence. In the period September 2009 – January 2011, he was Professor and Vice-Rector for International Cooperation at the Warsaw School of Social Sciences and Humanities/SWPS, Poland. From 1999 to 2009, he was Director of UNESCO-European Centre for Higher Education (UNESCO-CEPES).

Suntje Schmidt is employed in the department Dynamics of Economic Spaces at the Leibniz Institute for Regional Development and Structural Planning in Erkner, Germany. Her research focuses on the spatial implications of the knowledge economy, volatile labour markets, and pre-conditions for the interregional transfer of practices and experiences. Her PhD research focused on the effects and spatial dimensions of knowledge spillovers.

Lech Suwala holds a PhD in economic geography, and is a junior lecturer at the Department of Geography at the Humboldt-Universität and at the University of Applied Sciences of Technology and Economics in Berlin. Furthermore, he worked as a Professional in a private policy consulting company for several years. His scientific interests comprise spatial aspects of creativity, innovation and entrepreneurship where he conducted manifold studies at the district, federal and European level.

Paul Vallance is a Research Associate at the Centre for Urban and Regional Development Studies (CURDS), Newcastle University. He is co-author of the forthcoming Routledge book *The University and the City*.

Marina van Geenhuizen is Professor of Innovation and Innovation Policy in Urban Economies, in the Faculty of Technology, Policy and Management, at Delft University of Technology, Delft, the Netherlands. She holds a PhD from Erasmus University in Rotterdam, Faculty of Economics. She is author of seventy reviewed journal articles and chief editor of six international volumes.

Attila Varga is Professor of Economics at the University of Pécs, Hungary. Research interest: regional economic development policy impact modelling, the economics of technological change, and the role of university research in regional economic growth.

José Luis Vázquez is Professor of Marketing in the Department of Business Management and Economy at the University of León. He has coordinated many institutional programmes on entrepreneurship, and currently he manages the Bancaja Chair of Young Entrepreneurship at the University of León.

Hans Wiesmeth holds the Chair of economics, esp. allocation theory, at the Technical University of Dresden, Germany, and is president of the Dresden International University in Dresden, Germany. His research interests include equilibrium theory in general and environmental economics in particular.

Stavros A. Zenios is professor at the University of Cyprus and Senior Fellow, the Wharton School, University of Pennsylvania. He served as the Rector of the University of Cyprus during 2002–10 and is serving a second four-year term as President of UNICA. He holds a BSc in Mathematics and BEng in Electrical Engineering. He received his PhD in Engineering-Management Systems from Princeton University in 1986. He has authored three books, edited sixteen books, and (co)authored over 160 scholarly articles in some of the premier journals in the field. His book on the Performance of Financial Institutions has been translated into Chinese. He has received the EURO Excellence in Research Award.

Pavla Žížalová is an economic development and innovation expert with over six years of experience. She specialises in innovation, competitiveness policy and strategy consulting for the public sector on regional and national levels. She works for PricewaterhouseCoopers CR in the Public Sector Advisory group.

Preface

Knowledge and innovation challenge knowledge and innovation

Stavros A. Zenios

There could not be a more timely moment for the University of Warsaw to bring us together in this International UNICA Conference to discuss the creation of knowledge and innovation. Let me explain.

We have been witnessing since 2008 the unraveling of the world economies in a scale not seen since the 1920s and in a very original development of events. We are observing at work several forces that leading scholars have talked about.

In the words of noted social and political scientist Anthony Giddens we are witnessing the complexities of an interconnected world of "manufactured uncertainty." The subprime mortgage market collapse in the United States was the result of unpredictable consequences of financial engineering of derivative securities. The collapse of the three major banks of Iceland—and the subsequent collapse of the country's economy in 2008—were a clear consequence of interconnectedness.

And while the financial world was led into self-destructive mode through forces of our own making that we could not fully understand when we unleashed them, at the same time we started witnessing severe adverse restrictions from environmental causes. The oft-quoted study of the Club of Rome *Limits to Growth* was republished in an updated form thirty years after it was first carried out in the 1970s. It verified that its earlier predictions were becoming a worrisome reality: Since the 1980s, forty-four nations, home to 12 percent of the world population, have seen declining per capita income. Many countries have seen the World Bank Adjusted Net Savings Index decline – both Poland where we meet today and my home country, Cyprus, have seen declining indices. (Cyprus is currently in the red.) The ecological footprint of planet earth is 1.5, that of Europe is 2, of the Mediterranean region is 3 and of my country is 6.5. As has been said by the publishers of the World Ecological Footprint report "we need 1.5 planet earths and of course we have only one."

And finally, we have Nobel laureate Joseph Stiglitz telling us that

> The world economy is undergoing seismic shifts. This is partly because of the success in increasing productivity in manufacturing, so that a small fraction of the population can produce all the toys, cars and TVs that even the most materialistic and profligate society might buy.

This has created huge trade imbalances between East and West and massive unemployment, especially of young people.

What do we gather from these facts? That the problems are of our own making. Innovation and automation are having adverse effects. They are responsible for manufacturing uncertainty and also for bringing us to the edge of the environmental boundaries.

But does this mean that innovation has been wrong for us? Of course one could give the pragmatic responses that "the genie is out of the bottle" and "there is no turning back." However, I do not think we should apologize—not yet—to the next generations for the innovations of current and past generations. For such an apology to be warranted we must analyze all the progress made owing to the innovations and weigh it against our current predicament.

However, we will indeed need to apologize if we fail to analyze the future prospects opening up from our current predicament and find ways forward. It will be inexcusable to have benefited from the innovations of the past, and leave the problems they created to the future. This would be tantamount to stealing your children's lunch money.

And the way forward, I believe, is through knowledge and innovation. It is through new knowledge and new innovations that we will discover where exactly things went wrong and how to fix them. In this way we can get ourselves out of the current predicament and unavoidably land ourselves into some other troubled spot in the future. This has been the price of progress since Prometheas stole fire from the gods. My main point is that we should strive to pay this price ourselves today for the benefits we have had in the past and the problems we have created for ourselves today—not leave the next generation to pay the price.

It is therefore important to reflect on how knowledge and innovation manufacture uncertainty so we can deconstruct uncertainty. It is important to understand how knowledge and innovation drive us to environmental limits so we can use knowledge and innovation to steer away from the brink. And the same applies to the organization of our economies.

A conference, therefore, that aims at understanding the interplay between cities and their universities in creating knowledge and innovations is bringing together the major actors of the drama. Those who make innovation happen—the Universities—with those who need it and those who pay the price—the cities and their citizens.

The fact that this event is organized within the Polish EU Presidency is a manifestation that in these difficult times Europe has not lost its capacity to reflect. This is the only way to make sure that in future editions of Jared Diamond's *Collapse: How Societies Choose to Fail or Succeed* there will not be a chapter on Europe. Many fear there will be, some hope there will be. Those of us who are gathered here today are working to prove them wrong.

References

Diamond, Jared M. (2005) *Collapse: How societies choose to fail or survive.* London: Allen Lane.

Giddens, Anthony (1994) *Beyond Left and Right: The future of radical politics.* Cambridge: Polity Press.

Meadows, Donella H., Randers, Jørgen and Meadows, Dennis L. (2004) *The Limits to Growth: The 30-year update.* White River Junction, VT: Chelsea Green Publishing Company.

Stiglitz, Joseph E. (2010) *Freefall: America, free markets, and the sinking of the world economy.* Pbk edn. New York: W. W. Norton & Co.

Introduction

Cities, regions and universities as knowledge and innovation creators

Roberta Capello, Agnieszka Olechnicka and Grzegorz Gorzelak

1 Cities and regions as sources of knowledge creation in modern societies

For some decades there has been a wide consensus among economists on the role played by non-material resources in economic growth. Knowledge has been especially pinpointed as the main discriminating element in economic performance. In the globalising economy, even regional competitiveness – and consequently regional growth – is no longer dependent on the traditional production resource endowment, capital and labour. The hyper-mobility that nowadays characterises these factors reduces their geographical concentration, and shifts the elements on which competitiveness rests from the availability of material resources to the presence of immobile local resources such as local culture, competence, innovative capacity; in general: knowledge.

Much research has been produced since the 1980s on the idea of a knowledge-based economy, and on the preconditions for knowledge creation. However, when one looks carefully into the existing literature, two striking aspects emerge. On the one hand, it appears evident that *the knowledge-based economy does not have a unique interpretative paradigm*, but has been defined on the basis of different approaches ranging from the earliest sectoral, through a more recent functional to the latest relation-based one (see Capello in this volume). As a consequence, the term is still vague and not precisely defined, and rather different policy suggestions have been highlighted.

On the other hand, it appears quite evident that the strategic aspect in the studies of knowledge and innovation creation has become the interpretation of the elements that allow the creation of knowledge and innovation. Different approaches to the concept share one common element, identifying as *central the role played by spatial elements in the creation and diffusion of knowledge*, both evidenced by empirical analyses or deductively derived from theoretical elements. Different reasons were given for the importance of space in the creation of a knowledge-based economy: externalities stemming from urban environment, knowledge spillovers subject to strong and visible distance-decay effects, collective learning based on a relational space where economic and social interactions take place and are embedded into geographical space.

This new approach links to the idea that knowledge develops and accumulates through slow individual and collective learning processes, and grows through information, interaction and local knowledge. Knowledge creation is therefore a local process, rooted in the historical development of the area, accumulated over time through experience, local culture in local labour markets and local context, and therefore difficult to transfer to somewhere else (Camagni 1991).

Regions and cities are the natural loci where knowledge is created, and where it is more easily turned into a commercial product. Regions are territories where, under certain socio-economic conditions, a strong sense of belonging and mutual trust develop a 'cognitive capability' (Foray 2000), intended as the ability to manage information in order to identify and solve problems, or, more precisely in the economic sphere, the ability to transform information and inventions into innovation and productivity increases, through cooperative or market interaction. The 'learning' region is identified as the place where such cognitive processes play a crucial role, combining existing but dispersed know-how, interpretations of market needs, information flows with intellectual artefacts such as theories and models, and allowing exchange of experiences and co-operation (Lundvall and Johnson 1994). Especially in contexts characterised by a plurality of agents – such as cities or industrial districts – knowledge evolution 'is not the result of individual efforts in R&D within individual firms, but rather the combination of complementary capacities and of widespread interactive learning processes, which involve many "customers" and "suppliers" along a well-defined filière or supply chain' (Cappellin 2003: 307).

In this approach, attention is focused mainly on the construction of knowledge through cooperative learning processes, nourished by spatial proximity ('atmosphere' effects), network relations (long-distance, selective relationships), interaction, creativity and recombination capability. A collective learning process of this kind was first hypothesised by the GREMI group (Camagni 1991; Perrin 1995) and subsequently widely adopted as a sound theoretical concept for the interpretation of knowledge-based development and innovation (Keeble and Wilkinson 1999 and 2000; Capello 1999; Cappellin 2003).

Within regions, cities are natural loci of knowledge creation; they host large and well-known scientific institutions, such as R&D centres and universities, and they have a high availability of high quality human capital. Urban space acts as a strong concentration mechanism of advanced facilities, which are mainly located inside large agglomerations or city-regions in order to benefit from scale effects in both input markets (human capital, private financial capital) and output markets (higher-education services, research services) (Scott 2001); secondly, urban space acts as drivers of knowledge spillover from R&D clusters, which are subject to strong and visible distance-decay effects. In both cases, space is treated in a widely abstract, indirect and stylised way: concentration and agglomeration of main R&D facilities are by and large assumed to be acceptable starting points for empirical analyses, while knowledge diffusion processes are analysed in terms of pure probability functions, decreasing with physical distance. Spatial spillover effects are considered as a black box, with no reference to real, territorialised, channels of direct knowledge interaction (Capello and Faggian 2005).

More recently, knowledge flows and information channels are investigated through the so-called cognitive approach, that highlights the role of the urban *milieu*: in this approach abstract space becomes real territory, a relational space where functional and hierarchical, economic and social interactions take place and are embedded into geographical space. The local *milieu* – a 'territory' identified by both geographical proximity (agglomeration economies, district economies) and cognitive proximity (shared behavioural codes, common culture, mutual trust and sense of belonging) – supplies the socio-economic and geographical substrate on which collective learning processes can be incorporated, mainly owing to three main processes (Camagni and Capello 2002):

- the huge mobility of professionals and skilled labour – between firms but internally to the local labour market defined by the district or the city, where this mobility is maximal);
- the intense cooperative relations among local actors, and in particular customer–supplier relationships in production, design, research and, finally, knowledge creation;
- spin-off from universities and large firms.

The *milieu* becomes therefore a 'cognitive engine' and possibly an innovation place: its characteristics enhance interaction and cooperation, reduce uncertainty (especially concerning the behaviour of competitors and partners), reduce information asymmetries (therefore reducing mutual suspicion among partners) and reduce probability of opportunistic behaviour under the threat of social sanctioning (Camagni 1991), all elements that are confirmed by many regional economics schools (Bellet *et al.* 1993; Rallet and Torre 1995; Cappellin 2003).

2 Universities and knowledge creation

The science sector is nowadays treated not only as a potential source of strategic possibilities and a set of instruments for problem solving, but as an engine of development, also at the regional level (Edqvist 2003). In the course of time, the perception of the functions of universities and their influence on the environment is subject to change (Uyarra 2008; Wissema 2009). We are at the stage of formation of a new type of university tailored to a developmental paradigm called the knowledge-based economy. Several fundamental issues determine our expectations of the contemporary university (cf. e.g. Kukliński 2001; Lawton Smith 2006; Youtie and Shapira 2008):

- increased demand of the economy for knowledge and results of research;
- increased demand for an educated labour force, especially in knowledge-intensive enterprises;

- changes towards greater interactivity, networking and openness of innovation processes;
- changes in the nature of regional/local development – universities constitute important elements of the local scene, are perceived as development stimuli in their areas of operation;
- competition from other educational and research institutions.

In consequence, the traditional functions of universities that include scientific research (knowledge production function) and education (human capital creation function) are broadened and defined anew. Moreover, they are supplemented with new functions, of which the most frequently mentioned is their capacity as regional actors (Boucher *et al.* 2003; Goldstein and Renault 2004). Discussion on the changes affecting the functions of the university constitutes an important element of numerous theoretical concepts, including the area of knowledge production methods, for example the concept of Mode 2 Knowledge Production, the triple helix model and the engaged university (Hessels and van Lente 2008), or the networking concept of the process of innovation, for example as regional innovation systems, an open model of innovation or the concept of clusters (Olechnicka 2012). Despite the differences among the above-mentioned approaches, all of them underline the increasing role of external relations in the system of knowledge production, which, in consequence, recognises the need to open up the scientific sector to the needs of the environments and circles in which it operates and to cooperation with a view to achieving development advantages.

From an innovation perspective, the most important influence of universities is the production of new knowledge: both that which is readily applied (applied research) and which can potentially bring results within a short period, as well as that which remains theoretical, the effects of which are of a long-term nature (basic research) (Drucker and Goldstein 2007). In this regard the mutual relations between academic institutions and enterprises are of key significance. There exists a long tradition of this form of cooperation, stretching as far back as the eighteenth century in Europe and the nineteenth century in the USA. The positive effects of cooperation between science and business have been confirmed in numerous examples, inter alia, in the economic history of Italy (Iammarino 2005), Scotland (Schwerin 2004), USA and Germany (Lawton Smith 2006). University-enterprise relations take on a variety of forms, the nature and intensity of cooperation depending on factors relating equally to the enterprise and the academic institution as well as to the researchers themselves (Perkmann and Walsh 2010; Bercovitz and Feldman 2006). A review of the literature confirms that cooperation with a university has a beneficial influence on the innovativeness of various kinds of enterprise (Kaufmann and Tödtling 2001; Arundel and Geuna 2004; Lindelöf and Löfsten 2004; Broström and Lööf 2006; Caloghirou *et al.* 2006; Vuola 2008).

The impact of the university on the regional enterprise sector is also evidenced by the arrival of firms from outside the region who wish to profit from the proximity of knowledge (Benko 1993; Westlund 2004). This indicates that the university itself can create a specific climate of enterprise and determine the attractiveness of the locality, by providing a range of benefits such as access to academic teams and scientific research, as well as supplying positive examples of firms established by graduates in the locality (Johnes-Evans 1997; Bramwell and Wolfe 2008). It is expected that the contemporary university will not only respond to the needs of its users, that is enterprises, for knowledge (user-driven commercialisation), but that it will above all create innovativeness based on its own potential (science-driven commercialisation) (Gulbrandsen and Slipersæter 2007). Universities are to be a source of new industry, in the form of so-called off-shoot firms or new firms established by graduates of a university, irrespective of its participation (Dahlstrand 1999).

At the same time the university is expected to be not only a source of knowledge, but equally to become an active participant in the process of transferring processed knowledge to the economy, for instance through creating and supporting various intermediary bodies (Etzkowitz and Leydersdorf 1997). These new expectations with regard to universities, particularly in terms of their engagement in knowledge spillover processes through a variety of formal and informal intermediary mechanisms, requires significant changes in the functioning of contemporary universities.

However, some analyses indicate that the role of the university as the main producer of knowledge for the needs of the region is in fact considerably more complex (Florax and Folmer 1992; Florida 1999; Mayer 2006; Meyer *et al.* in this volume). It is hard to distinguish the effects of the university's functioning from other parallel phenomena influencing the knowledge base of the region (e.g. the appearance of a major national or international investor). Moreover, the role of scientific institutions is often indirect and long term, and consists in catalysing and improving activity rather than initiating and driving new processes in the region (Colyvas *et al.* 2002; Wolfe 2009). It is also argued that innovative development in a region relies to a greater extent on the social capital, enterprise climate and inflow of knowledge along with direct international investment (Torre and Gilly 2000). At the same time, regions with strong scientific centres do not always display greater innovativeness, since making use of knowledge – resulting from a university's activity in the region – is variable and depends on many factors, including the absorption ability of the enterprise sector and on the growing significance of supra-regional (inter-regional and international) inflows of knowledge (Lawton Smith 2006; Capello in this volume; Olechnicka in this volume).

3 Aims and structure of the book

The effort of the present book is to present the most updated studies on the role of cities and universities in knowledge creation, and to understand in-depth the role of cities and universities in creating knowledge and supporting innovation, by highlighting the specific channels through which universities generate knowledge spillovers. In this way, universities play a major role in regional economic development.

The major novelty of the book is that on this subject much work has been developed, centred upon countries in Western Europe and the United States. Increasingly and since the expansion of the European Union in 2004 in particular, themes of academic entrepreneurship, university–business links, knowledge and innovation have become important on a Europe wide scale. In particular, this book brings together contributions from different Eastern European countries, that try to highlight the peculiarities that knowledge creation through universities and institutions have in post-soviet countries.

Scientific structures of the post-socialist Central European countries, in fact, have their roots in the continental European tradition, but they have inherited many features of the Soviet model as well. The more rigid Central European system thus seems to offer even fewer incentives for academic entrepreneurship, which suggests that the academic entrepreneur might not be a natural ingredient of the Central European university system (see Erdős and Varga in this volume). A legitimate question is whether the specific 'academic' drivers behind university spin-off firm formation are present at all within the Central European context. Empirical evidence on university spin-offs in Central and Eastern European countries is very limited so far; the literature pays even less attention to the particular 'academic' motivations behind the formation of university spin-offs in the Central European context (Erdős and Varga 2012).

This book has the aim to fill this gap, and goes a step further by presenting empirical evidence also from Western countries. The comparison between the Eastern and Western member countries in knowledge creation as driver of regional development can easily be made by a careful reader.

The book is structured into three parts, going from the most general issues to the most specific ones. The first part deals with the creation of knowledge and innovation by looking at what happens at the global scale. The second part is devoted to universities and entrepreneurship as key actors for an innovation-driven development. The third part is dedicated to the role played by the academic institutions in cities. With this structure the book provides a comprehensive picture of the process of knowledge and innovation as drivers of regional and urban development in two different development stages – namely, those of Eastern and Western Countries.

References

Arundel, A. and Geuna, A. (2004) 'Proximity and the use of public science by innovative European firms', *Economics of Innovation and New Technology*, 13(6), 559–580.

Bellet, M., Colletis, G. and Lung, Y. (1993) 'Introduction au Numéro Spécial sur Economie et Proximité', *Revue d'Economie Régionale et Urbaine*, 3, 357–361.

Benko G. (1993) *Geografia Technopolii.* Warszawa: PWN.

Bercovitz, J. and Feldman, M. (2006) 'Entrepreneurial universities and technology transfer: A conceptual framework for understanding knowledge-based economic development', *The Journal of Technology Transfer*, 31(1), 175–188.

Boucher, G., Conway, C. and Meer, E. V. D. (2003) 'Tiers of engagement by universities in their region's development', *Regional Studies*, 37(9), 887–897.

Bramwell, A., Wolfe, D. A. (2008) 'Universities and regional economic development: The entrepreneurial University of Waterloo', *Research Policy*, 37(8), 1175–1187.

Broström A. and Lööf, H. (2006) 'What do we know about firms' research collaboration with universities? New quantitative and qualitative evidence'. Stockholm: The Royal Institute of Technology Centre of Excellence for Science and Innovation Studies.

Caloghirou, Y., Constantelou, A. and Vonortas, N. S. (2006) 'Knowledge flows in European industry: An overview of evidence from surveys and case studies', in Y. Caloghirou, A. Constantelou and N. S. Vonortas (eds), *Knowledge Flows in European Industry*. London and New York: Routledge, pp. 183–209.

Camagni, R. (1991) 'Local milieu, uncertainty and innovation networks: Towards a new dynamic theory of economic space', in R. Camagni (ed.), *Innovation Networks: Spatial perspectives*. London: Belhaven-Pinter, pp. 121–144.

Camagni, R. and Capello, R. (2002) 'Milieux innovateurs and collective learning: From concepts to measurement', in Z. J. Acs, H. L. F. de Groot and P. Nijkamp (eds), *The Emergence of the Knowledge Economy*. Berlin: Springer, pp. 15–46.

Cappellin, R. (2003) 'Networks and technological change in regional clusters', in J. Bröcker, D. Dohse and R. Soltwedel (eds), *Innovation Clusters in Interregional Competition*. Berlin: Springer, pp. 53–78.

Capello, R. (1999) 'Spatial transfer of knowledge in high-technology milieux: Learning vs. collective learning processes', *Regional Studies*, 33(4), 353–365.

Capello, R. (2013) 'Science-based activities in European regions: The knowledge-innovation nexus', in A. Olechnicka, R. Capello and G. Gorzelak (eds), Chapter 2 in this volume.

Capello, R. and Faggian, A. (2005) 'Collective learning and relational capital in local innovation processes', *Regional Studies*, 39(1), 75–87.

Colyvas J., Crow, M., Gelijns, A., Mazzoleni, R., Nelson, R. R. and Rosenberg ,N. (2002) 'How do university inventions get into practice?' *Management Science*, 48(1), 61–72.

Dahlstrand, A. L. (1999) 'Technology-based SMEs in the Goteborg region: Their origin and interaction with universities and large firms', *Regional Studies*, 33(4), 379.

Drucker, J. and Goldstein, H. (2007) 'Assessing the regional economic development impacts of universities: A review of current approaches', *International Regional Science Review*, 30(1), 20–46.

Edqvist, O. (2003) 'Layered science and science policies', *Minerva*, 41(3), 207–221.

Etzkowitz, H. and Leydersdorf, L. (eds) (1997) *Universities in the Global Economy: A Triple Helix of University-Industry-Government Relations*. London: Cassell Academic.

Erdős K., Varga A. (2012) 'The academic entrepreneur: Myth or reality for increased regional growth in Europe?', in M. Geenhuizen and P. Nijkamp (eds), *Creative Knowledge Cities: Myths, Visions and Realities*. Cheltenham, UK; Northampton, MA: Edward Elgar, pp. 157–181.

Erdős, K. and Varga, A. (2013) 'Academic entrepreneurs in post-socialist Central European countries: Evidence from the Hungarian biotechnology sector', in A. Olechnicka, R. Capello and G. Gorzelak (eds), Chapter 11 in this volume.

Florax, R. and Folmer, H. (1992) 'Knowledge impacts of universities on industry: An aggregate simultaneous investment model', *Journal of Regional Science*, 32(4), 437–466.

Florida, R. (1999) 'The role of the university: Leveraging talent, not technology'. *Issues of Science and Technology*, from www.issues.org/15.4/florida.htm.

Foray D. (2000) *L'Economie de la Connaissance*. Paris : La Découverte.

Goldstein, H. A. and Renault, C. S. (2004) 'Contributions of universities to regional economic development: A quasi-experimental approach', *Regional Studies*, 38, 733–746.

Gulbrandsen, M. and Slipersæter, S. (2007) 'The third mission and the entrepreneurial university model' in A. Bonaccorsi and C. Daraio (eds), *Universities and Strategic Knowledge Creation. Specialization and Performance in Europe*. Cheltenham, UK, Northam, MA: Edward Elgar Publishing.

Hessels, L. K. and van Lente, H. (2008) 'Re-thinking new knowledge production: A literature review and a research agenda', *Research Policy*, 37(4), 740–760.

Iammarino, S. (2005) 'An evolutionary integrated view of regional systems of innovation: Concepts, measures and historical perspectives', *European Planning Studies*, 13(4), 497–519.

Johnes-Evans, D. (1997) 'Small firms, universities and technological development in a peripheral economy: The case of the Republic of Ireland', in B. Fynes and S. Ennis (eds), *Competing from Periphery*. Dublin: University College.

Kaufmann, A. and Tödtling, F. (2001) 'Science-industry interaction in the process of innovation: The importance of boundary-crossing between systems', *Research Policy*, 30(5), 791–804.

Keeble, D. and Wilkinson, F. (1999) 'Collective learning and knowledge development in the evolution of regional clusters of high-technology SMS in Europe', *Regional Studies*, 33(4), 295–303.

Keeble D. and Wilkinson F. (2000) *High Technology Clusters, Networking and Collective Learning in Europe*. Aldershot: Ashgate.

Kukliński, A. (2001) 'The role of universities in stimulating regional development and educating global elites', *Higher Education in Europe*, 26(3), 437–445.

Lawton Smith, H. (2006) *Universities, Innovation and the Economy*. London and New York: Routledge.

Lindelöf, P. and Löfsten, H. (2004) 'Proximity as a resource base for competitive advantage: University-industry links for technology transfer', *Journal of Technology Transfer*, 29(3/4), 311–326.

Lundvall, B. A. and Johnson, B. (1994), 'The Learning Economy', *Journal of Industry Studies*, 1(1), 23–42.

Mayer, H. (2006) 'What is the role of universities in high-tech economic development? The case of Portland, Oregon, and Washington, DC', *Local Economy*, 21(3), 292–315.

Meyer, M. (2013) 'The second coming of the Triple Helix and the emergence of hybrid innovation environments', in A. Olechnicka, R. Capello and G. Gorzelak (eds), Chapter 9 in this volume.

Olechnicka, A. (2012) *Potencjał nauki a innowacyjność regionu*. Warszawa: Wydawnictwo Naukowe Scholar.

Olechnicka, A. (2013) 'Regional cooperation or external links? Spatial proximity in science-business relations in Poland', in A. Olechnicka, R. Capello and G. Gorzelak (eds), Chapter 5 in this volume.

Perkmann, M. and Walsh, K. (2010) *The Two Faces of Collaboration: Impacts of university–industry relations on public research*. DRUID, Copenhagen Business School, Department of Industrial Economics and Strategy/Aalborg University, Department of Business Studies.

Perrin J.-C. (1995) 'Apprentissage collectif, territoire et milieu innovateur: un nouveau paradigme pour le développement', in J. Ferrão (ed.), *Políticas de Inovação e Desenvolvimento Regional et Local*, Edição do Instituto de Ciencias Sociais de Universidade de Lisboa, republished in R. Camagni and D. Maillat (eds) (2006), *Milieux Innovateurs*. Paris: Economica-Anthropos, pp. 99–128.

Rallet, A. and Torre, A. (eds) (1995) *Economie Industrielle et Economie Spatiale*. Paris: Economica.

Schwerin, J. (2004) 'The evolution of the Clyde region's shipbuilding innovation system in the second half of the nineteenth century', *Journal of Economic Geography*, 4(1), 83–101.

Scott, A. J. (ed.) (2001) *Global City-Regions: Trends, theory, policy*. Oxford: Oxford University Press.

Torre, A. and Gilly, J.-P. (2000) 'On the analytical dimension of proximity dynamics', *Regional Studies*, 34(2), 169–180.

Uyarra, E. (2008) 'The impact of universities on regional innovation: A critique and policy implications', Manchester Business School Working Paper, Vol. 564. Available from: www.mbs.ac.uk/research/workingpapers/.

Vuola, O. (2008) 'Big science as a catalyst of new business development', in E. Carayannis, D. Assimakopoulos and M. Kondo (eds), *Innovation Networks and Knowledge Clusters: Findings and insights from the US, EU and Japan*. New York: Palgrave McMillan, (pp. 110–132).

Westlund, H. (2004) 'Regional effects of universities and higher education: A knowledge overview of Swedish, Scandinavian and international experiences'. Paper presented at the ERSA conference papers. Available from: http://ideas.repec.org/p/wiw/wiwrsa/ersa04p154.html.

Wissema, J. G. (2009) *Uniwersytet trzeciej generacji. Uczelnia XXI w.* Zębice ZANTE.

Wolfe, D. A. (2009) 'Universities and knowledge transfer: Powering local economic and cluster development, in G. B. Doern and C. Stoney (eds), *Research and Innovation Policy: Changing federal government–university relations*. Toronto: University of Toronto Press.

Youtie, J. and Shapira, P. (2008) 'Building an innovation hub: A case study of the transformation of university roles in regional technological and economic development', *Research Policy*, 37(8), 1188–1204.

Part I

Universities

Local/regional institutions of knowledge creation and innovation in national and international contexts

1 Science-based activities in European regions

The knowledge-innovation nexus

Roberta Capello

1 Introduction[1]

The decisive globalization processes that have taken place in the last decade have increasingly put economic actors and policy makers under severe stress, in search for ad hoc strategies and policies to support competitiveness. At the European level, encouraged by the Lisbon Agenda (March 2000), all policy levels are contributing to the reinforcement of innovation and to the creation of the knowledge economy. Knowledge has in recent years become a key driver of growth in economic systems; access to knowledge is generally considered as a key condition for innovative activities in our modern economy. The idea of knowledge as the main discriminating element in economic and social performance was pointed out even before the Lisbon Agenda by some national governments, such as the United Kingdom, where in 1998 a white paper on 'Our Competitive Future, Building the Knowledge Driven Economy' was produced by the Department of Trade and Industry.

The well-known European strategy defined in the Lisbon and Luxembourg ministerial meetings (2000 and 2005) engages the Union to become the most competitive and dynamic knowledge-based economy in the world. A complex indicator for regional achievement of the Lisbon performance was circulated in the Luxembourg meeting, concentrating on private R&D investment and expenditure, educational levels of the labour force and productivity levels. An increasing flow of public resources into the scientific research system was requested up to the economic crisis (mid 2008) and is likely to be consented to by public authorities, giving rise to a huge scientific engagement into the measurement of the internal efficiency, productivity and impact of the research system itself (Okubo 1997; Joly 1997) (Table 1.1).

At the beginning of the discussion most initiatives and policy suggestions were focused on the context of European or national economies, but it soon became evident that the same reasoning should have been applied at a more territorially disaggregated level of analysis, such as the regional level. A bottom-up approach in the development of the knowledge economy was thought to be interesting, given the high spatial concentration that innovation and knowledge creation activities were subject to. Clusters of technologically advanced firms, such as Silicon Valley in California, 'Route 128' in the Boston area, Baden-Württenberg in the

Table 1.1 Average increase of R&D over GDP ratio over the period 1996–2007 in different countries

Countries	*Average increase R&D/GDP 1996–2007*
Australia	2.1
Brasil	2.6
China	8.2
India	1.9
Mexico	4.2
South Korea	3
Canada	1.8
France	1.6
Germany	−0.4
Italy	1.2
Japan	1.8

Source: World Bank and Eurostat data.

South of Germany, Jutland in Denmark, Småland in Sweden, Sophia-Antipolis close to Nice, to cite only some examples, testified the presence of some form of increasing returns on the concentration of innovative activity.

Different reasons were given for the importance of space in the creation of a knowledge economy: externalities stemming from the urban environment, knowledge spill-overs subject to strong and visible distance-decay effects, collective learning based on a relational space where economic and social interactions take place and are embedded into geographical space.

The new territorial approach links to the idea that knowledge develops and accumulates through slow individual and collective learning processes, and grows through information, interaction and local knowledge. Knowledge creation is therefore a local process, rooted in the historical development of the area, accumulated over time through experience, local culture in the local labour market and local context, and is therefore difficult to transfer elsewhere.

Based on these reflections, a large consensus was achieved in the scientific sphere about the fact that regional competitiveness – and consequently regional growth – was no longer dependent on the traditional production resource endowment, capital and labour. The hyper-mobility that nowadays characterizes these factors reduces their geographical concentration, and shifts the elements on which competitiveness rests from the availability of material resources to the presence of immobile local resources such as local culture, competence and innovative capacity; in general: knowledge.

In this spirit, even in recent times, in the Green Papers of the Territorial Cohesion and of the European Research Area, the European Commission has called for particular attention to the territorial dimension of innovation and knowledge creation. The diversity of innovation activities, of the spatial diffusion of innovation throughout the European territory, of the capacity of regions to create knowledge and to exploit knowledge from outside calls for an in-depth analysis of the territorial dimension of the knowledge economy, on which the so-called 'third generation of innovation policy at Community level' can rely.

As Danuta Hübner (2009: 2) claims:

> innovation policy at community level is now moving into its third generation. It is moving away from the approach of the first generation of innovation policy which focused on R&D through a linear process for the development of innovations, beginning with laboratory science moving through successive stages up to the inclusion of knowledge in commercial applications. Equally, it is building on the approach of the second generation which recognized the complexity of innovation systems (national, regional, local, sectoral), with many feedback loops between the different stages. In the approach of *this third generation*, innovation is not considered as a linear process that starts with research, eventually leading to development, translated later into growth in the territories that have more capabilities. Instead, it is the product of a policy mix, including several bodies and stakeholders in which *the territories, their specificities and conditions are paramount.*

Notwithstanding the general idea that a territorial approach is required for the study of innovation and knowledge creation, at the normative level the main policy action is summarized in the Agenda 2020 that calls for the achievement of 3 per cent of the EU's GDP (public and private) to be invested in R&D, in order to achieve smart growth.

The aim of the present chapter is to analyse the role played by science-based activities in generating a knowledge economy, and in increasing innovation capacities of European regions.[2] The chapter first presents the spatial trends of the 'knowledge economy' in Europe (section 3), measured on the basis of a definition of 'knowledge economy' (section 2). When associated to the spatial innovation adoption patterns (section 4), these trends show a striking discrepancy between knowledge and innovation at the regional level and witness that R&D (and formal knowledge in general) does not necessarily equal innovation. This main result allows to claim that the pathways towards innovation and modernization are differentiated among regions according to local specificities (section 5). From this statement, thematically focused innovation policies are suggested (section 6).

2 A multidimensional definition of the knowledge economy

Much research has been produced since the 1980s on the idea of a knowledge-based economy, and on the preconditions for knowledge creation. However, when we look carefully into the existing literature, two striking aspects emerge. On the one hand, it appears evident that the knowledge-based economy does not have a unique interpretative paradigm, but has been (and still can be) defined on the basis of different approaches ranging from the earliest sectoral, through a more recent functional to the latest relation-based approach. As a consequence, the term is still vague and not precisely defined, and thus rather different policy suggestions have been highlighted. On the other hand, it appears quite evident that the different approaches to the concept share one common element, that of the central role played by spatial elements in the creation and diffusion of knowledge, both evidenced by empirical analyses or deductively derived from theoretical elements.

Although early use of the term goes back to the work of Fritz Machlup (Machlup, 1962), only in recent years has the concept of the knowledge-based economy begun to spread in the scientific and political literature. This is mainly due to work sponsored by the OECD (David and Foray 1995; Foray and Lundvall 1996). The European Union set itself the goal in 2000 of becoming the most competitive and dynamic knowledge-based economy in the world. It subsequently confirmed that goal in 2005, submitting its Structural Fund resources to achieve it.

But what does this concept really mean? Vaguely, we know that research, human capital, creative utilization of scientific concepts and information should merge, giving rise to continuing innovation and advanced production. The OECD suggested using about sixty indicators - among which R&D and high technology activities play a dominant role – to measure the knowledge-based economy (OECD 2004; Van Oort and Raspe 2006).

If we want to adopt a historical approach to the interpretation of the concept, sector-based definitions and function-based definitions were successively proposed and held for long periods. While human capital has always been considered as a basic condition for any knowledge-based development, different factors were indicated as the driving forces of change. In an early stage, which can be located in the late 1970s and the 1980s, most attention was directed to 'science-based' (Pavitt 1984) or high-technology sectors; regions hosting these sectors were considered as 'advanced' regions leading the transformation of the economy. New jobs were expected mainly from these new sectors, while more traditional sectors were expected to restructure or even to relocate abroad, giving rise to serious tensions in the local labour markets.

It soon became evident that this dichotomy was too simplistic, and that many knowledge-based advances were possible and were actually introduced by 'traditional' sectors – such as textiles and car production – in their path towards rejuvenation. Furthermore, the growing complexity of technological *filière* inside the value chain increasingly underlined the relevance of advanced tertiary sectors. These supplied producer services mainly in the form of consultancy for process innovation (proper acquisition and use of advanced technologies, tailor-made software, systems integration in production, administration and logistic processes, organizational support) and for product innovation (marketing, design, testing, advertising, finance, distribution).

In the second stage, which developed mainly during the 1980s and 1990s, a function-based approach was preferred (even though it overlapped conceptually with the previous one), which stressed the importance of pervasive and horizontal functions such as R&D and higher education. 'Scientific' regions, hosting large and well-known scientific institutions, were studied deeply and relationships between these institutions and the industrial fabric were analysed, with some disappointment as far as an expected but not often visible direct linkage was concerned (MacDonald 1987; Monk *et al.* 1988; Massey *et al.* 1992; Storey and Tether 1998). Indicators of R&D inputs (e.g. public and private research investment and personnel) and, increasingly, indicators of R&D output (e.g. patenting activities) were used in order to understand the engagement of firms and territories on knowledge, intended as a necessary long-term precondition for continuing innovation (Dasgupta and Stiglitz 1980; Antonelli 1989; Griliches 1990). This approach, equating knowledge and scientific research, was the one re-launched by the European strategy defined in the Lisbon Agenda.

It is difficult to escape the impression that both the sector-based and the function-based approaches to the knowledge-based economy, both driven by the need to measure and quantify, result in a simplified picture of the complex nature of knowledge creation and its relation to inventive and innovative capability. The presence of advanced sectors and advanced functions such as R&D and higher education are special features of only some of the possible innovation paths and, though relevant, cannot be considered as necessary or sufficient preconditions for innovation. Furthermore, emphasizing the stock of human capital, advanced functions and sectors may risk overlooking the interactive process between the different actors of knowledge development, which is increasingly seen as the crucial element in knowledge creation and evolution. This element is typical of production contexts characterized by the presence of SMEs but also of the contexts where big firms develop their own internal knowledge, culture and know-how by enhancing internal interaction and boosting selective external interaction with industrial partners, schools, professionals and research centres. Therefore, a rather different approach should be utilized, a cognitive one, stressing the relational, cultural and psychological elements that define the preconditions for knowledge creation, development, transmission and diffusion.

The third stage of reflection, typical of the present day in which a relation-based approach is preferred, concentrates on the identification of a 'cognitive capability' (Foray 2000): the ability to manage information in order to identify and solve problems, or, more precisely in the economic sphere, the ability to transform information and inventions into innovation and productivity increases, through cooperative or market interaction. The 'learning' region is identified as the place where such cognitive processes play a crucial role, combining existing but dispersed know-how, interpretation of market needs, information flows with intellectual artefacts such as theories and models and allowing the exchange of experiences and cooperation (Lundvall and Johnson,1994). Especially in contexts characterized by a plurality of agents – such as cities or industrial districts – knowledge evolution 'is not the result of individual efforts in R&D within individual firms, but rather the combination of complementary capacities and of widespread interactive learning processes, which involve many "customers" and "suppliers" along a well-defined filière or supply chain' (Cappellin 2003: 307).

What is striking in all the approaches mentioned above is the central role played by *spatial elements* in creating new knowledge and in supporting inter-regional flows of knowledge, both evidenced by empirical analyses or deductively derived from theoretical elements. Each approach defines a type of innovative region (Table 1.2): with a sector-based approach, technologically advanced regions are highlighted; with a functional approach, scientific regions are evidenced; and with the relational approach, innovative networking regions are analysed.

Table 1.2. Complementary approaches to the knowledge economy

	Sector-based definitions (1970–80)	*Function-based definitions (1980–90)*	*Relation-based definitions (1990–2000)*
Driving forces of the knowledge-based economy	Science-based sectors, High-tech sectors	High education, R&D	Cognitive capability
Location regions	Advanced regions	Scientific regions	Innovative networking regions
Path towards innovation	Invention–innovation short circuit	Spin-offs, spatial spill-overs	Collective learning, local synergies
From innovation to performance	Radical innovation, Schumpeterian profits	Technological breakthrough, royalties on patents	Continuing innovation, productivity increases
Spatial context	High-tech clusters	Science parks, large city-regions	Innovative milieux, large cities
Role of space	Proximity economies, specialization advantages	Proximity and agglomeration economies	Uncertainty reduction, relational capital

Source: Camagni and Capello (2009).

All these kinds of region belong to a knowledge economy; they differ in terms of territorial elements that explain their innovative performance, as well as the spatial elements that support knowledge creation. A modern definition of the knowledge economy should embrace all these 'innovation models', taking into consideration that what is striking in all these approaches is the central role played by *spatial elements* in creating new knowledge and in supporting inter-regional flows of knowledge, both evidenced by empirical analyses or deductively derived from theoretical elements, as we highlight in the next section.

3 The knowledge economy in Europe

Following the definition of 'knowledge economy' outlined before, in the recent ESPON KIT project the identification of the knowledge economy in European regions has been one of the major empirical challenges. In particular, one aim of the project has been to measure the degree to which regions are 'knowledge economy' regions in Europe, by stressing the presence in each region of technologically advanced sectors, of scientific functions and of knowledge networking activities.

Technologically advanced regions (henceforth, TAR) have been identified as those regions that present simultaneous specialization in both medium and high tech manufacturing sectors and knowledge information service industries. Because of the deep technological content of both manufacturing as well as service activities, none of them is *ex ante* excluded from the definition, and regions are classified as technologically advanced if both technologically advanced manufacturing and services characterize the region. The definition of high tech industries is a broad one, encompassing industries with medium-high and high-tech content (henceforth, MHHT) in order to capture a wide range of industries characterized by consistent high-tech creation and deployment. High-tech industries are classified according to the OECD methodology (OECD 2005). Such industries include manufacturing of aircraft and spacecraft; pharmaceuticals; office, accounting and computing machinery; radio, TV and communications equipment; and medical, precision and optical instruments.[3] High-tech services follow the OECD classification too, labelled 'Knowledge-Intensive Service Activities'. Specialization is here defined with a location quotient calculated with respect to the EU27 average value; regional data include industry-specific employment in MHHT manufacturing and Knowledge Intensive Services (henceforth, KIS).[4] Specialization is calculated for two years (2002 and 2007) in order to identify time trends along with its spatial distribution.

The results of the location quotient on sectoral regional specialization in 2007 are shown in Map 1.1. Twenty-one regions identified as TAR with our methodology are German, thirteen British, eight French, five Belgian, four Swiss, three Swedish, two Finnish and Danish, and one each for Italy, Norway, Slovenia and Slovakia.

The geography of technology in Europe is indeed highly concentrated, although peripheral regions and regions with capital cities in new member states do play a major role. Over time (although the time span considered may be too short to draw safe conclusions), no region acquired or lost the status of technologically advanced regions. The productive fabric of Europe shows therefore a remarkable concentration of technology,[5] related either to advanced manufacturing or to services activities.

Scientific regions are defined as those regions showing at the same time a higher than average scientific activity and a higher than average quality of human capital. The level of human capital stock in a region is measured by means of the following indicators, as an average of a three-year span (2005–07):

- the percentage of population employed in the education sector;
- the share of population that has attained at least a university degree;
- funding per capita in the activities of the 5th Framework Programmes.

Similarly, the level of research activities is measured by an average of a two-year span 2006–07 of the following indicators:

- the R&D expenditures per capita;
- the percentage of employees in R&D;
- the number of patents per capita for all economic sectors;
- the number of patents per capita for the subsample of high-tech sectors.

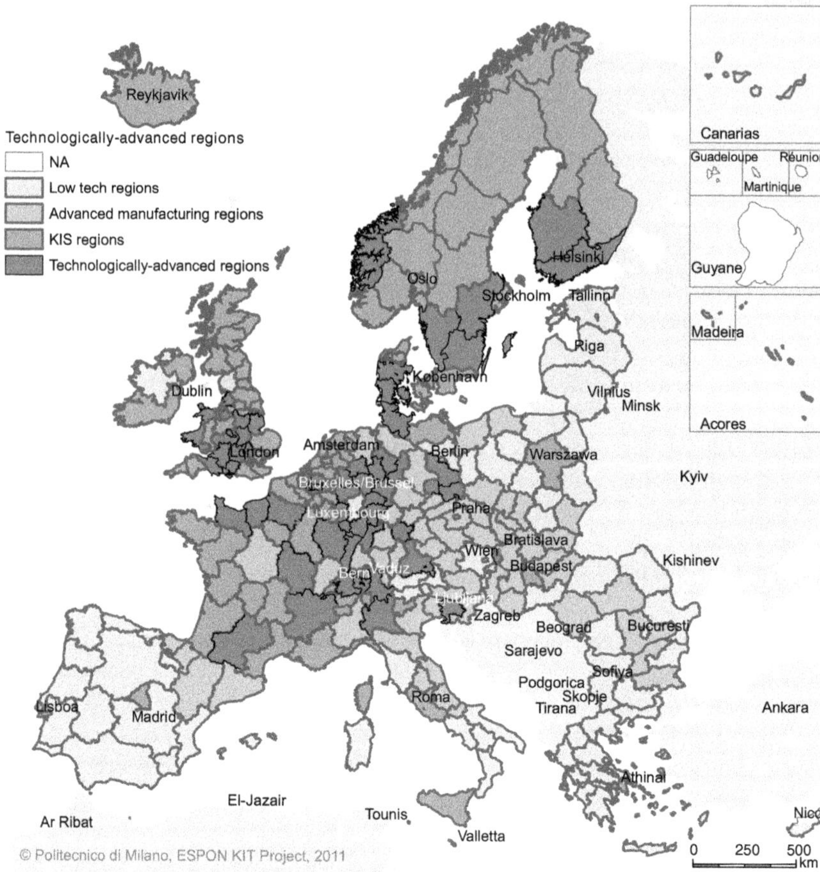

Map 1.1 Technologically advanced regions in Europe, 2007.

Source: Interim report KIT.

The results are shown in Map 1.2, where seventy-four scientific regions are depicted. But most regions, 126, are regions with no specialization in knowledge activities. Among the seventy-four scientific regions there are fifty-nine regions belonging to EU15 countries, three belonging to New Member States (NMS) and twelve belonging to Efta countries. Moreover, fifty-eight are competitive regions, three are convergence regions and only one is a transition region. Regions with no specialization in knowledge activities are mainly located on the peripheral territories of Europe and research-intensive regions are concentrated on territories characterized by a manufacturing specialization (i.e. Northern Italy, German regions). Finally, as expected, human capital-intensive regions are mainly in the north.

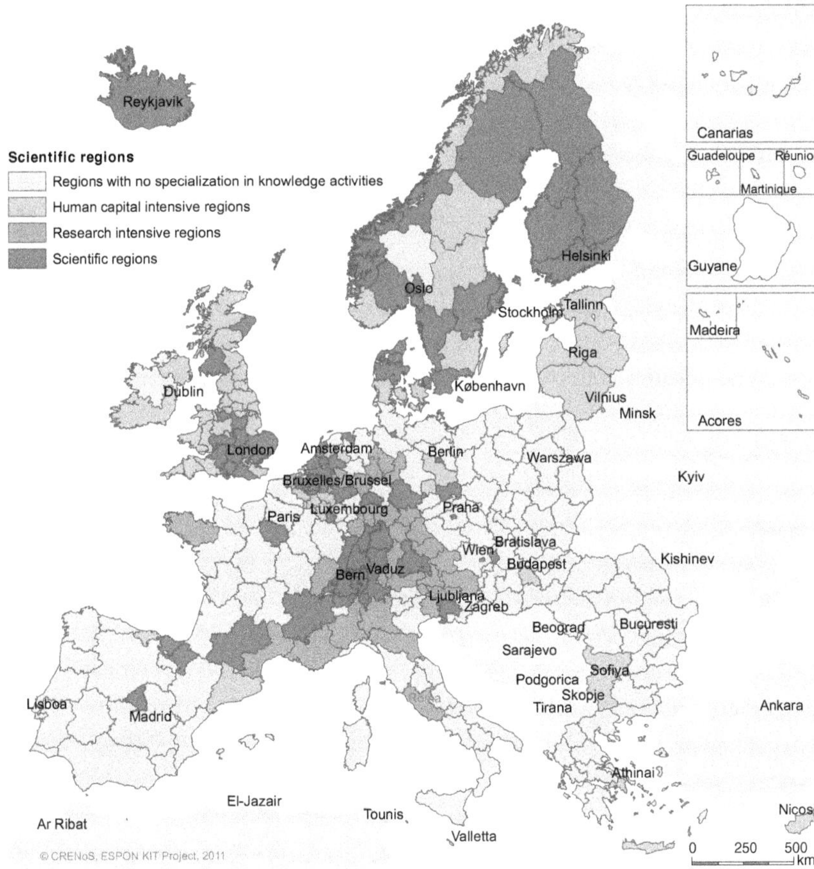

Map 1.2 Scientific regions in Europe, average data 2005–07.

Source: Interim report KIT.

The main idea behind the definition of *knowledge networking regions* is that knowledge is created within some crucial nodes (i.e. firms and universities) which tend to co-locate in specific places. Knowledge is then diffused and exchanged either through a diffusive pattern in which spatial proximity is essential or according to intentional relations based on intentional cooperation networks, for which spatial proximity is not the main explanation for cooperation to occur. Translating these ideas to the regional level, knowledge networking regions can be understood as regions that rely on external sources of knowledge and on facilitating interactive learning and interaction in innovation. This knowledge diffusion can take place through diffusive patterns based on spatial proximity (henceforth 'unintentional linkages') and/or through intentional relations based on intentional networks or non-spatially mediated mechanisms ('intentional linkages').

A combination of different measures is used to assess the degree of regional a-spatial linkages, namely:

- co-patents with other European regions: number of patents co-authored with inventors from outside the region;
- inflows: number of inflows of inventors from other regions (from where they bring knowledge, brain gain)[6];
- cross-regional patent citations: number of citations made to patents of other regions.

In spite of the advantages of citations as being a straightforward measure of knowledge flows, it could be argued that they represent the output of any form of knowledge transmission, instead of the exact mechanisms through which knowledge is transferred.

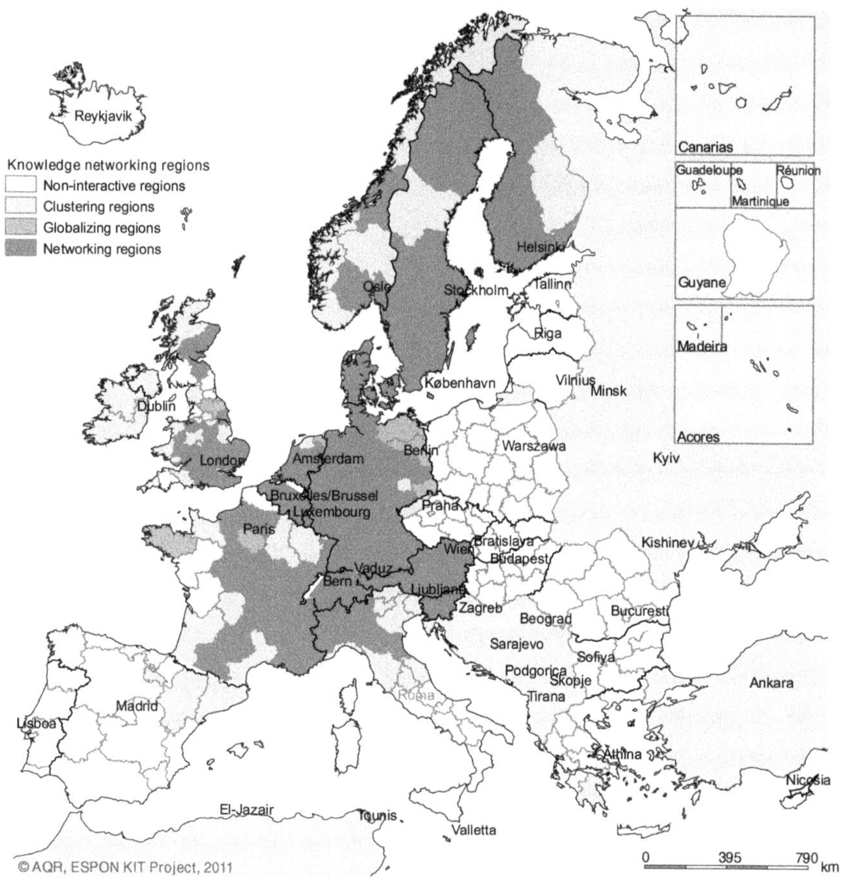

Map 1.3 Knowledge networking regions in Europe.

Source: Interim report KIT.

Map 1.3 shows the *knowledge networking regions*. They are concentrated in the centre of Europe as well as in the Scandinavian countries, whereas the *non-interactive regions* are mainly those belonging to the new member states and some specific regions in the South European countries (the whole of Portugal and Greece, most of Spain except the North-East area, and the South of Italy). This provides evidence of a clear core–periphery pattern in the geographical distribution of the regions that in one way or another rely on external sources of knowledge for the development of innovation.

As mentioned in the previous sections, a knowledge economy region can be identified, in the broadest sense, as either being specialized in high-tech sectors, or in high-tech functions, or in networking with other regions, or being the result of a combination of these definitions. Map 1.4 presents an integrated picture of the different typologies of knowledge economy regions.

Map 1.4 The knowledge economy in Europe.

Source: Interim report KIT.

The picture looks very fragmented, with quite a reasonable number of regions being only networking, and mainly in the central part of Europe. Only three technologically advanced regions host scientific functions (Dresden, Vestlandet and Bratislava), while most of the technologically advanced 'islands' that host knowledge are also networking regions. In general, scientific regions are also networking regions, witnessing that knowledge accumulation inside a region also requires networking activity, which allows for the acquisition of knowledge from outside. What is really impressive is that a very high number of European regions, mainly in Eastern countries and in the Southern peripheral countries, is below the EU average, which shows that *in most of European regions the knowledge economy is still in its infancy.*

An important step forward in the analysis of the knowledge economy is the degree of innovation that accompanies the presence of a knowledge economy. To what extent is knowledge applied for innovation purposes? Is innovation more dependent on a particular type of knowledge economy region, such as the scientific regions? This is the subject matter of the next section.

4 Regional innovation diffusion in Europe

A value added of the European KIT project is its estimation of regional innovation data, starting from CIS national data.[7] Data at regional level are made available by KIT on the share of firms having developed product innovation only; process innovation only; both product and process innovation; product and/or process innovation; marketing and organizational innovation.

Spatial concentration characterizes product innovation, as depicted in Map 1.5. This variable displays consistent concentration in strong countries, the core of product innovative activity in Europe being carried out in German, Scandinavian, Swiss and British regions, with a few notable exceptions outside these areas. EU15 regions tend on average to innovate more, and significantly so, than Eastern ones; the same applies to densely populated regions, while rural regions display a relatively lower product innovation rate. In general, in countries where product innovation is high, concentration seems pronounced. Spatial concentration of product innovation, on the contrary, strongly characterizes countries with low product innovation rates. This is the case in: Portugal, where Lisbon is the only area with some product innovation activity; Spain, with Madrid, Barcelona and a few Pyrenean regions; Greece; some NMS. Italy represents an exception to this pattern, since several regions in the northern and central part of the country display similar product innovation rates.

In general, process innovation shows a more dispersed pattern than product innovation (Map 1.6). Countries such as Portugal, Spain, France, Germany and the UK do not display a remarkable concentration of process innovation within their boundaries. The variance associated with this variable is much lower than the same measure associated with product innovation.

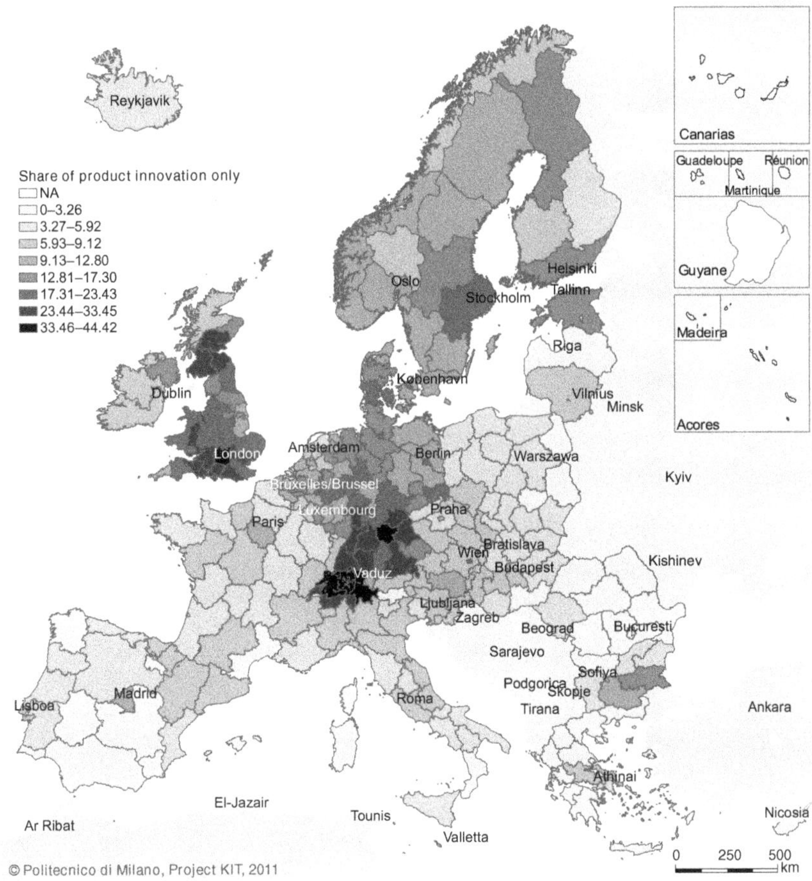

Map 1.5 Share of product innovation only.
Source: Interim report KIT.

A comparison between the knowledge economy and the innovation maps shows much more spatially dispersed patterns for both product and process innovation, indicating that innovation activities also take place in non-specialized knowledge economy regions. Table 1.3 presents the level of innovation in the different types of knowledge economy regions as well as in other regions which do not fall into this category. As expected, the greatest difference lies between knowledge economy regions and the other remaining regions. The former definitely show a higher innovation performance, whatever definition is adopted.

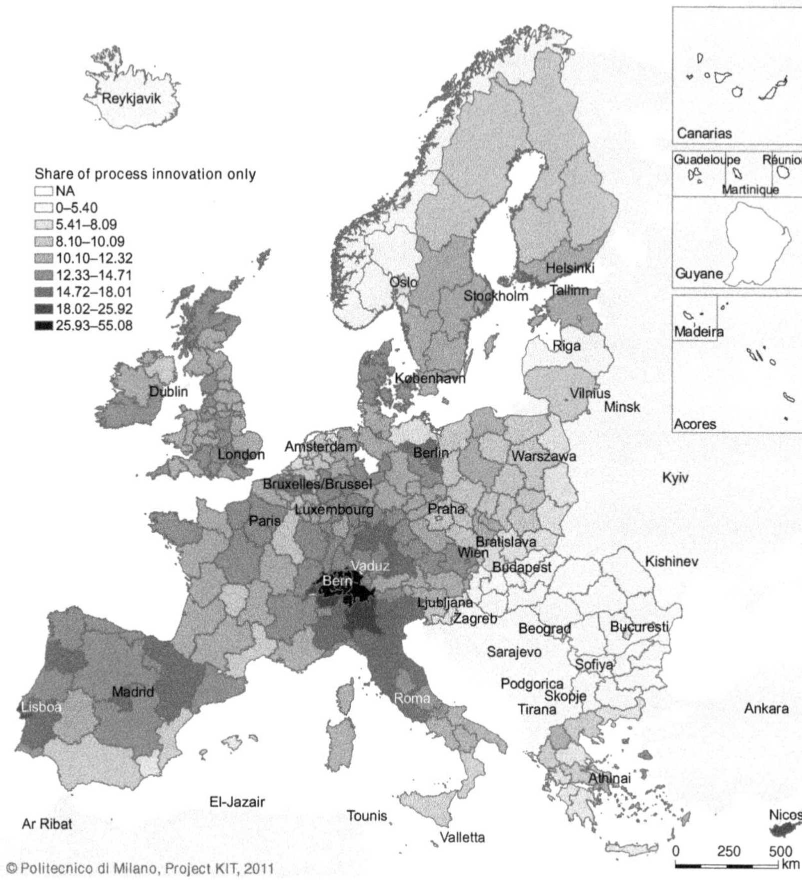

Map 1.6 Share of process innovation only.
Source: Interim report KIT.

What is striking in the results presented in Table 1.3 is that one would have expected the regions with the highest R&D and scientific activities in general to be the most innovative. Our empirical results show instead that scientific regions, although registering a high innovation rate, are not significantly more innovative than TAR or networking. Only a few regions show a pattern of innovation that goes from R&D to innovation, while most innovating regions seem to find other sources of knowledge. If this is true, a thematically/regionally neutral and generic orientation of R&D funding investments is not the right policy to increase EU competitiveness; a new conceptual approach has to be used to provide more ad hoc policy interventions, based on the idea that R&D does not represent the only source of knowledge and innovation.

Table 1.3 Share of innovation adoption by type of knowledge economy regions

Innovation typologies	*Share of product innovation only*	*Share of process innovation only*	*Share of product and/or process innovation*	*Share of marketing and organizational innovation*
Knowledge typologies				
Technologically advanced regions	17.42	13.76	43.66	32.75
Scientific regions	18.16	13.48	43.71	29.51
Networking regions	16.19	13.20	44.24	31.95
Other regions	6.34	9.88	27.40	20.58

Source: Capello *et al.*, 2012

5 Territorial patterns of innovation

Our empirical analysis suggests that knowledge, innovation and diffusion are not necessarily intertwined, especially at the local level. This can be explained by the fact that factors that enhance the implementation of new knowledge can be quite different from the factors that stimulate innovation. Firms and individuals that lead an invention are not necessarily also leaders in innovation or in the widespread diffusion of new technologies.

The history of technology and innovation is full of examples of this kind; the fax machine, first developed in Germany, was turned into a worldwide successful product by Japanese companies. Similarly, the anti-lock brake system (ABS) was invented by US car makers but became prominent primarily owing to German automotive suppliers (Licht 2009).

These reflections suggest that innovation can be the result of different patterns, with different modes of performing each phase of the innovation process. The variety of innovation patterns explains the failure of a 'one size fits all' policy for innovation, like the thematically/regionally neutral and generic R&D incentives, which are expected to develop a knowledge economy everywhere. On the contrary, innovation patterns typical of each specific area have to be identified, on the basis of which ad hoc and targeted innovation policies can be formulated.

The paradigmatic jump in interpreting regional innovation processes lies nowadays in the capacity to go beyond the single approaches developed in the scientific literature for the interpretation of knowledge and innovation, and to build a conceptual framework interpreting the different modes of performing each phase of the innovation process, highlighting the context conditions (internal and external to the region) that accompany each innovation pattern. In this way, it is possible to take into consideration alternative situations where innovation builds on internal knowledge, or where local creativity allows, even when there is a lack of local knowledge, an innovative application thanks to knowledge developed elsewhere and acquired via scientific linkages, or where innovation is made possible by an imitative process of innovations developed outside the region.

To this end, the concept of territorial patterns of innovation can be useful for interpreting reality. A territorial pattern of innovation is a combination of *context conditions* and of *specific modes of performing the different phases* of the innovation process. Among all possible combinations, the most interesting are the following, reflecting different knowledge and innovation aspects:

a) an endogenous innovation pattern in a scientific network, where the local conditions are all present to support the creation of knowledge, its local diffusion and transformation into innovation and its widespread local adoption so that higher growth rates can be achieved. Given the complex nature of knowledge nowadays, this pattern is expected to show a tight interplay in the creation of knowledge with other regions, and as part of an international scientific network. This pattern can easily be built from the conceptual point of view on all the literature dealing with knowledge and innovation creation and knowledge diffusion (Figure 1.1a);
b) a creative application pattern, characterized by the presence of creative actors interested and curious enough to look for knowledge, lacking inside the region, in the external world, and creative enough to apply external knowledge to local innovation needs. This approach is conceptually built on the literature on regional innovation creation (Figure 1.1b);
c) an imitative innovation pattern, where the actors base their innovation capacity on imitative processes that can take place with different degrees of creativity in the adaptation of an already existing innovation. This pattern is based on the literature dealing with innovation diffusion (Figure 1.1c).

Conceptually speaking, these three patterns represent the different ways in which knowledge and innovation can take place in a regional economy. Each of them represents a different way of innovating, and calls for different policy styles to support innovation. An R&D incentive policy can be extremely useful for the first kind of innovation pattern; incentives for co-inventing applications (the typical Schumpeterian profits), enhancing the ability of regions to change rapidly in response to external stimuli (such as the emergence of a new technology) and to promote 'shifting' from old to new uses, is a good policy aim for the second pattern. A maximum return to imitation is the right policy aim for the third innovation pattern, and this is achieved through the creative adaptation of already existing innovation to meet local needs.

Empirical analysis indicates the existence of a wider variety of possible innovation patterns than the ones conceptually envisaged, namely (Capello and Lenzi, 2012):

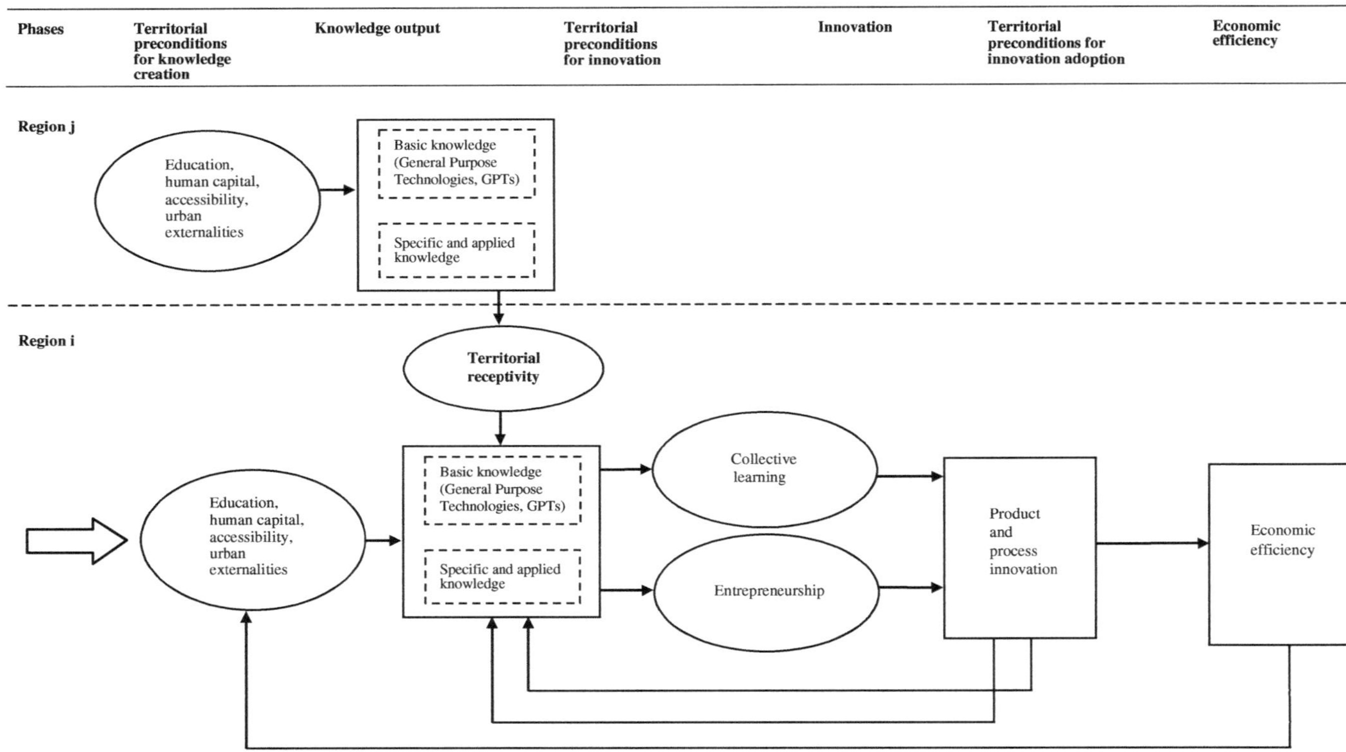

Figure 1.1a Endogenous innovative pattern in a scientific network.
Source: Capello (2011).

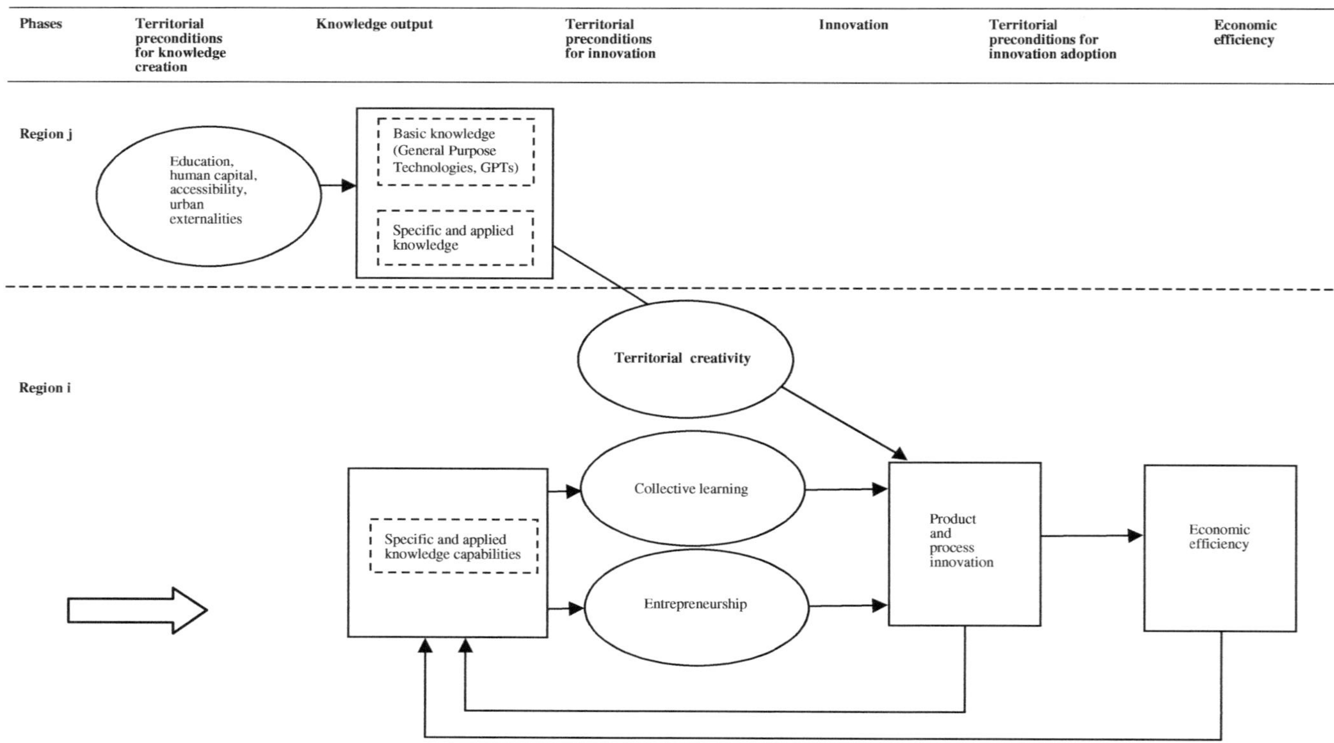

Figure 1.1b A creative application pattern.
Source: Capello (2011).

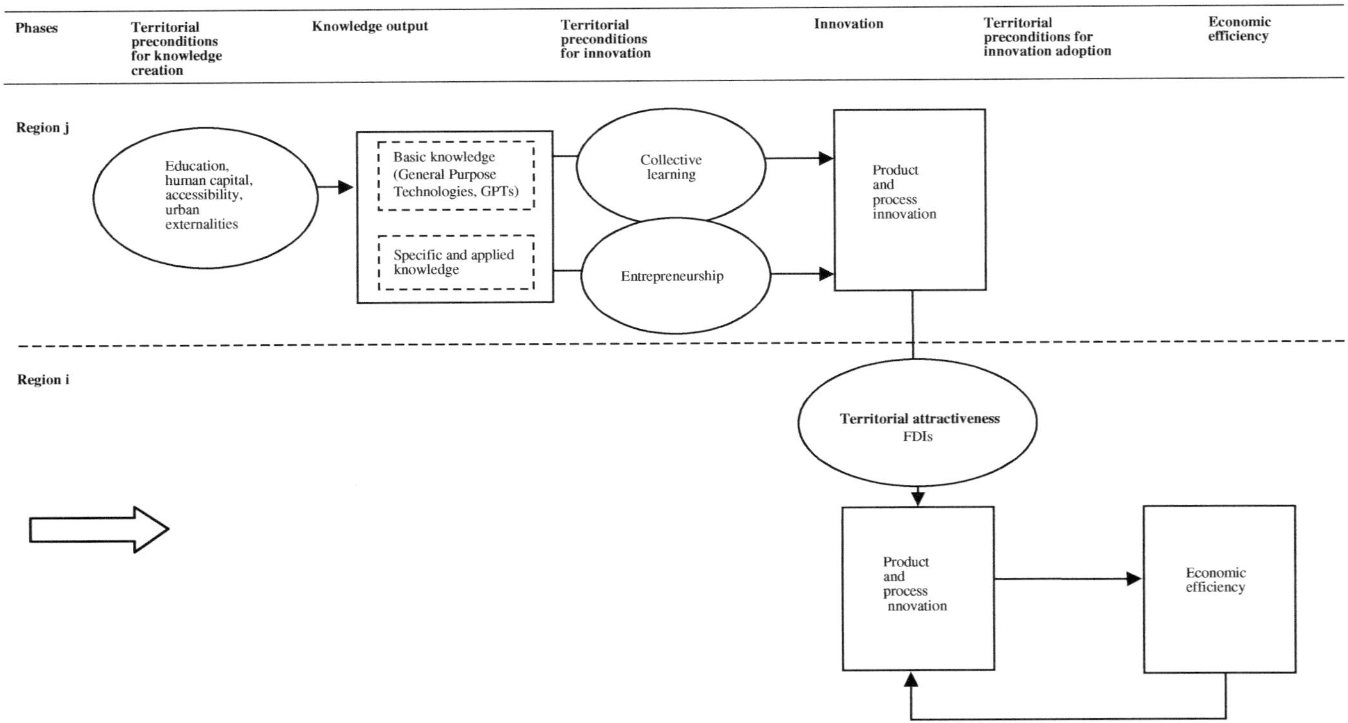

Figure 1.1c An imitative innovation pattern.
Source: Capello (2011).

A European science-based area

This cluster of regions is composed of regions that are the most knowledge and innovation intensive. Their innovative attitude is well above the EU average across all dimensions (i.e. product, process, marketing and/or organizational innovation). This is coupled with a very strong knowledge orientation that is more directed to general purpose technologies (GPT) than in the other cases (and above the EU average) both in terms of the amount of knowledge developed as well as in terms of specialization profile. Interestingly, this knowledge tends to be of greater generality and originality, that is, of greater technological value and more radical than the EU average. The regions in this cluster are also well endowed with those preconditions frequently associated with a greater endogenous capacity for knowledge creation, namely the presence of a highly educated population and, more importantly, the presence of scientific human capital, here measured by the share of inventors in the total population. Their accessibility is also the greatest, indicating that these regions probably cover more urban and metropolitan settings (as confirmed by the variable accounting for the number of agglomerated regions), which are traditionally more open and fertile environments for the generation of new ideas (Carlino *et al.* 2007).

All in all, these observations suggest that these regions show a strong knowledge and innovation orientation that is primarily linked to their endogenous capacity to create new knowledge and to efficiently translate it into new products and processes as well as into managerial and/or organizational changes. This marked orientation suggests that these regions potentially constitute a *European science-based area* and be part of what has been defined as the European Research Area (Foray *et al.* 2009; Pontikakis *et al.* 2009). These regions are mostly located in Germany, with the addition of Wien, Bruxelles, and Syddanmark in Denmark (Capello and Lenzi 2012).

An applied science area

This group includes a wider number of regions that share similar characteristics with regions in cluster 5, although most of the variables show lower mean values, particularly the share of EU total patents, which is almost halved, as well as the share of scientific human capital. Interestingly, the relevance of GPTs is lower both in terms of the share of GPT patents developed as well as in terms of specialization profile. Importantly, these regions appear more entrepreneurial, creative, attractive and with greater capabilities potential than regions in the previous cluster, albeit less than the EU average. These regions thus maintain a fairly strong knowledge and innovation intensity, that is they form a knowledge area but, in contrast to the previous cluster, they are less focused on GPTs, and, accordingly, more technologically diversified. Together with the 'European science-based area', these regions could constitute a *European Research Area*.

These regions are mostly agglomerated and located in central and northern Europe, namely in Austria, Belgium, Luxembourg, France (i.e. Paris), Germany, Ireland (i.e. Dublin) Denmark, Finland and Sweden with some notable exceptions in the east such as Praha, Cyprus and Estonia and in the south such as Lisboa and Attiki (Capello and Lenzi 2012).

Here we face strong knowledge producing regions that distinguish themselves from the European science-based area by their diversified knowledge production profile. From the normative point of view, these regions have the chance to strengthen their position by specializing in the production of applied knowledge, making use of the basic knowledge produced from the science-based area. If this is the case, this group can become an '*applied science area*' of Europe.

A smart technological application area

Regions in this cluster look quite different from regions belonging to the science-based area. They are comparable to regions in the applied science area in terms of size of knowledge base and its characteristics (i.e. relevance of GPTs, generality and originality), show greater endowment of embedded knowledge in human capital (i.e. capabilities) but are different in terms of innovation profile. In particular, they have a stronger orientation towards product innovation, are somehow weaker in terms of innovation processes (even though they are more innovative than the EU average in this respect) and are among the weakest performers in terms of marketing and/or organizational innovation.

These regions have the greatest advantage in terms of product innovation, accompanied by a high degree of knowledge potential flows and internal preconditions to translate external knowledge into innovation, thanks to high creativity. These results suggest that these regions are able to efficiently translate internal and external knowledge into new specific commercial applications. This cluster can easily represent our conceptual pattern 2, the creative application pattern, where co-invention of application is the result of internal creativity and external basic knowledge. It includes mostly agglomerated regions in the EU15, such as the northern part of Spain and Madrid, Northern Italy, the French Alpine regions, the Netherlands, the Czech Republic, Sweden and the UK (Capello and Lenzi 2012). Normative interventions should strengthen these peculiarities and push this group into becoming a '*smart technological application area*' of Europe.

A smart and creative diversification area

This group of regions shows some distinctive traits that clearly distinguish it from the others. In particular, the knowledge and innovation variables show smaller values than the EU average while the capabilities indicator shows the highest mean value. This suggests that the not negligible innovation activities carried out in regions belonging to this cluster mainly rely upon tacit knowledge embedded into human capital. Moreover, regions in this cluster appear highly entrepreneurial (this variable takes the highest mean value in this cluster) and, importantly, are strongly endowed with those characteristics such as creativity and attractiveness that help to absorb and to adopt innovations developed elsewhere. Additionally, whereas the knowledge potential does not look prominent, the capabilities and innovation potentials are well above the EU average. Thus, the key advantages of these regions reside in their embedded human capital and the entrepreneurial and creative attitudes that can be wisely exploited in the pursuit of upgrading innovative strategies. These regions are mainly located in Mediterranean countries (i.e. most of the Spanish regions, Central Italy, Greece, Portugal), in EU12 agglomerated regions (Slovakia and Slovenia, Poland and the Czech Republic), and a few regions in Northern Europe (Finland and the UK) (Capello and Lenzi 2012).

In these regions, a different type of pattern 2 emerges with respect to the previous cluster. In these regions, the type of knowledge that is acquired from outside is neither basic nor applied formal knowledge; these regions benefit greatly from external knowledge that is embedded in technical and organizational capabilities, in technicians and SMEs managers (Cooke 2009); thanks to the high degree of creativity present in the area, these regions are able to take advantage of the specific capabilities present in regions with similar sectoral profiles, and innovate in different products and in different industries. Normative interventions should strengthen this innovative attitude and push these regions into becoming a 'smart and creative diversification area' in Europe.

An imitative innovation area

Finally, the last group could be associated to pattern 3. In fact, it is composed of regions that have a rather narrow knowledge and innovation profile and are the poorest performers in both respects. However, some key traits characterize this cluster. In particular, entrepreneurship, creativity, attractiveness, capabilities and innovation potentials show greater values than the EU average. Attractiveness is particularly strong compared to other clusters. These dimensions can be enhanced and supported to creatively embrace new adoption, imitation and innovation strategies. For this reason, this group of regions can form an '*imitative innovation area*' in Europe. Most of these regions are in the EU12 such as all regions in Bulgaria and Hungary, Latvia, Malta, several regions in Poland, Romania, and Slovakia, but also in Southern Italy.

The high level of creativity, entrepreneurship and collective learning present in this cluster provide potential assets to turn this area, in an evolutionary perspective, into a smart upgrading diversification area, through normative intervention to help exploit creativity and entrepreneurship and thus increase indigenous innovation activities, and not only imitative innovation.

The variety of innovation patterns that exists in reality explains the failure of thematically/regionally neutral and generic R&D incentives; the latter are not suitable for the widespread development of a knowledge economy. On the contrary, innovation patterns typical of each specific area have to be identified, on the basis of which ad hoc and targeted innovation policies can be formulated.

6 Thematically/regionally focused innovation policies

Our empirical analysis has shown the existence of territorial innovation patterns that differ from one another in terms of the different modes of combining knowledge and innovation, owing to different territorial (context) conditions that support the creation/diffusion of knowledge and innovation. It is on these different territorial innovation patterns that thematically/regionally focused, ad hoc, innovation policies have to be built. Table 1.4 presents the policy suggestions for each territorial pattern of innovation.

Three different policy goals characterize the different patterns. The 'European science-based' and the 'applied science' patterns, characterized by high R&D endowments, share the aim to achieve a maximum return to R&D investments. By contrast, the latter are not a natural innovation policy aim for the 'smart technological application' pattern and the 'smart and creative diversification' patterns, where local R&D endowment is relatively low and inefficient in generating innovation and higher regional growth; the innovation policy aim in these patterns can be found in a maximum return to co-inventing applications (the typical Schumpeterian profits), which deeply depends on the ability of regions to change rapidly in response to external stimuli (such as the emergence of a new technology), promoting 'shifts' from old to new uses. Finally, in the 'imitative innovation' pattern, the aim must be to achieve a maximum return to imitation, through the creative adaptation of already existing innovation to meet local needs.

The methods of achieving common aims are differentiated among patterns. R&D spending is a typical policy aim for the 'European science-based' and the 'applied science' patterns; the differentiation lies in the scientific fields in which R&D funding are spent. In the 'European science-based' area the maximum return of R&D spending is obtained if R&D is spent in general purpose technologies. Applied scientific fields of research should instead absorb much of the R&D funds in the 'applied science' area. Support of creative application, shifting capacity from old to new uses, and deepening or improving productivity in existing uses is the right policy tool for maximizing the return to co-inventing applications. Incentives for technological projects are suggested for the reinforcement of the 'smart technological application area', while support devoted to the identification of international best practice, and support to entrepreneurial creativity are more suited to an innovation pattern such as the 'smart and creative diversification area'. The maximum return to imitation is supported by incentives to local firms for the development of creative projects with MNEs.

Table 1.4. Policy suggestions by territorial innovation pattern

Territorial patterns of innovation *Policy aspects*	*European science-based area*	*Applied science area*	*Smart technological application area*	*Smart and creative diversification area*	*Creative imitation area*
Club policy goals	Maximum return to R&D investments		Maximum return to co-inventing application		Maximum return to imitation
Club policies:					
Policy actions for local knowledge	R&D support in:		Support of creative application, shifting capacity from old to new uses, deepening or improving productivity in existing uses		Support to local firms for creative projects with MNE
	General Purpose Technologies	Specialized technological fields	Incentives for technological projects	Support the identification of international best practices Support entrepreneurial creativity	
Policy actions for knowledge spillover exploitation	Incentives for attracting inventors				Incentives for attracting MNE
	Support of research cooperation in:		Incentives for creative applications through:		
	GPT and trans-territorial projects (ERA)	specific technologies and trans-territorial projects (ERA), especially in related sectors Encourage labour mobility in related sectors	cooperative research activities among related sectors	participation of local actors in specialized international fairs attraction of "stars" even for short periods a work experience in best practice knowledge creation firms of the same sectors	

Policy style	Provide a critical mass of funds Stimulate knowledge transfer mechanisms in related sectors Thematically/ regionally orientation of R&D funding:		Stimulate a bottom up identification of industrial vocations, by raising awareness of local capabilities		Stimulate cooperation projects between MNE and local firms
	given to a critical mass of researchers and of laboratories in general purpose technologies	in specific fields of technological specialization of the area Targeted choice of joint research activities with partners external to the area	Stimulate cooperation with strong knowledge creation external partners in the specialized sectors of the area	Support local actors' participation to international fairs in their specialized sectors Support targeted work experience in best practice knowledge creation firms	
Beneficiaries	University, research centres, large local firms		Local firms	Local entrepreneurs	Local firms

Source: draft final report, KIT, ESPON

Combined with the previous policy recommendations aiming to foster the creation of local knowledge, policy interventions can be suggested for knowledge acquisition from outside the region. In the two patterns where external science-based knowledge is merged with local knowledge, the policy tools to attract new knowledge from outside are: in the 'European science-based area' – incentives for attracting inventors and support of research cooperation in GPT and trans-territorial projects; in the 'applied science area' – support of research cooperation in specific fields of technological specialization. The latter suggestions are in line with the creation of the European Research Area (ERA) suggested by the European Commission, an area composed of all research and development activities, programmes and policies in Europe that involve a transnational perspective. The 'applied science area' could also benefit from the encouragement of labour mobility between related sectors, enabling skills and experience to move across sectors and regions (Asheim *et al.* 2011).

Policy tools for knowledge acquisition in the 'smart technological application area' and in the 'smart and creative diversification area' are incentives for creative applications through cooperative research activities in related sectors in those regions where a small applied science base exists, and through participation of local actors in specialized international fairs, through the attraction of 'stars' even for short periods of time, or through work experience in best practice knowledge creation firms of related sectors for the 'smart and creative diversification area' whose innovation capacity lies in the aptitude of their entrepreneurs to find outside the area the right applied science on which to innovate.

The traditional incentives to attract MNEs remain the most efficient tool for attracting new knowledge in areas with very limited knowledge – formal or informal, scientific or technical.

The policies suggested require some care in their design in order to be efficient. R&D funding should be provided with a critical mass, and should also be accompanied by stimuli to knowledge transfer mechanisms in related sectors, in order to guarantee the permeability of new knowledge. Moreover, the efficiency of R&D funding is dependent on the identification of thematic and regional scientific specializations. Where support of creative application is concerned, its efficiency is guaranteed through the bottom-up identification of local industrial vocations, raising awareness of local capabilities. Where imitation is the key, the efficiency of the policies is dependent on the stimulation of cooperation projects between MNEs and local firms.

Beneficiaries of these policy recommendations differ among patterns; universities, research centres and large R&D laboratories of private firms are the natural beneficiaries of the 'European science-based area' and the 'applied science area'. Local firms are the natural recipients of 'smart technological application area' policies; entrepreneurs and small firms are the natural recipients of policies devoted to the 'smart and creative diversification area'; Local firms are thought to be the right subjects for the 'imitative innovation area' policies.

7 Conclusions

The knowledge economy shows a very differentiated and fragmented spatial pattern in Europe, and for many European regions it is still in its infancy. The achievement of smart growth – which requires 3 per cent of the EU's GDP (public and private) to be invested in R&D/innovation – is still a long way off. In 2007, thirty-three regions had achieved 3 per cent of R&D expenditures on GDP (11 per cent of NUTS2, representing 16 per cent of EU GDP) and these were concentrated in a few countries in Northern Europe. Moreover, a very high number of regions belong to the lowest class, with R&D on GDP lower than 0.5 per cent (representing 5 per cent of GDP).

Moreover, the so-called 'scientific regions', where R&D and human capital exceed the European average, although registering a high innovation rate, innovate only slightly more than other knowledge economy regions in Europe. We have to depart from the idea that R&D equals knowledge and that knowledge equals innovation. The distinction between the process of invention in general-purpose, basic technology pervading horizontally different sectors once an invention is turned into an innovation, and the process of inventing an application of basic knowledge in a specific sector, innovating in new products and new market niches is vital in understanding the present patterns of innovation. This becomes even more important if we consider that the factors that stimulate new knowledge, invention, innovation and innovation diffusion differ; invention and innovation are not necessarily intertwined and this gives rise, even at the local level, to very different and multifaceted situations; some regions have the capacity to go through all phases of the 'linear model', from knowledge creation to innovation and growth, with all feedback that can be foreseen from growth to knowledge and innovation. Other regions reinforce this 'linear model', exchanging knowledge with other regions, gaining complementary assets through a scientific network. There is, however, a completely different situation in which regions innovate by combining their creative thinking with basic knowledge accumulated in other regions, developing co-inventing applications.

These results are some of the main messages that come out of the empirical analysis, justifying doubts that R&D funding is the most appropriate innovation policy for all regions in Europe. Do we really take advantage from an innovation policy with a common aim for all countries/regions? The reply is straightforward: no. In this sense, we are in line with the 'smart specialization debate' that stresses the importance of 'smart specialization R&D activities' at regional level to exploit the advantages stemming from specialized R&D concentration (Foray *et al.* 2009).

However, thanks to our conceptual and empirical analysis, we have the necessary arguments to go two steps further than the smart specialization debate. First of all, we claim that *R&D specialization is not a 'simple' matter of core–periphery dichotomy*, interpreting core regions as natural places for general purpose technologies, and peripheral regions for co-inventions of applications. The 'geography' of innovation in Europe is much more complex, as our territorial patterns of innovation suggest and show. Second, the presence of advanced sectors and advanced functions, such as R&D and higher education, are special features of only *some of the possible innovation paths.*

Each territorial innovation pattern calls for specific ad hoc innovation policy goals: a maximum return to R&D investment may be the right goal for a region specialized in knowledge creation, but may not be the right policy goal for regions that innovate by exploiting external knowledge, or for regions that imitate innovation processes. For the former, the ad hoc policy goal is a maximum return to co-inventing applications, which happens when the region promotes changes in response to external stimuli (such as the emergence of a new technology). A maximum return to imitation, pushing towards creative imitation, is the right policy aim for regions that rely on external innovation processes. Each region has to succeed in discovering its territorial innovation pattern, and only through awareness of its original and unique territorial innovation pattern can a region hope to excel in exploiting its innovation efficiency (Capello 2012).

A single overall strategy is likely to be unfit to provide the right stimuli and incentives in the different contexts; in order to increase its regional innovation capacity, Europe needs normative interventions towards thematically/regionally focused innovation policies, based on specific territorial innovation patterns.

Notes

1 This chapter draws on part of the results of the ESPON KIT project, coordinated by the author at the Politecnico of Milan. The report of the project is available on the ESPON website: www.espon.eu/main/Menu_Projects/Menu_AppliedResearch/kit.html.

2 For details in the construction of the regional taxonomies, see the interim report of KIT available at: www.espon.eu/main/Menu_Projects/Menu_AppliedResearch/kit.html

3 Medium-high and high-tech manufacturing industries correspond to employment in chemicals (NACE24), machinery (NACE29), office equipment (NACE30), electrical equipment (NACE31), telecommunications and related equipment (NACE32), precision instruments (NACE33), automobiles (NACE34) and aerospace and other transport (NACE35); KIS include water transport (NACE 61), air transport (NACE 62), post and telecommunications (NACE64), financial intermediation (NACE 65), insurance and pension funding (NACE 66), activities auxiliary to financial intermediation (NACE 67), real estate activities (NACE 70), renting of machinery and equipment (NACE 71), computer and related activities (NACE72), research and development (NACE73) and other business activities (NACE 74).

4 Source of the data is EUROSTAT.

5 Moran's I index, measuring the degree of spatial autocorrelation among regions and calculated on the basis of a rook contiguity matrix of second order, is equal to 0.18, and significant at all conventional levels, for the categorical variable 'technologically advanced region' depicted in Map 1.1.

6 As some authors have pointed out (Saxenian 2006) the dichotomy brain drain/brain gain might be overcome by the concept of brain circulation. Thus, it is perfectly possible that regions exporting talent may benefit as well from knowledge inputs from outside the region because of enduring social relationships between the talented individual and his/her former colleagues. This is the reason why only the inflows as an indicator for a-spatial linkages has been used, and not the net flow indicator.

7 For the methodology applied, see Capello *et al.* (2012).

References

Antonelli, C. (1989) 'A failure inducement model of research and development expenditure: Italian evidence from the early 1980s', *Journal of Economic Behaviour and Organization*, 12(2), 159–180.

Asheim, B. T., Boschma, R. and Cooke, P. (2011) 'Constructing regional advantage: Platform policies based on elated variety and differentiated knowledge bases', *Regional Studies*, 45(7), 893–904.

Camagni, R. and Capello, R. (2009) 'Knowledge-based economy and knowledge creation: The role of space', in U. Fratesi and L. Senn (eds), *Growth and Innovation of Competitive Regions: the role of internal and external connections*. Berlin: Springer, pp. 145–166.

Cappellin, R. (2003) 'Territorial knowledge management: Towards a metrics of the cognitive dimension of agglomeration economies', *International Journal of Technology Management*, 26(2/4), 303–325.

Capello, R. (2012) 'Territorial patterns of innovation', in Cappellin, R., Ferlaino, F. and Rizzi, P. (eds), *La città nell'economia della conoscenza*, Milan: Franco Angeli, pp. 51–79.

Capello, R. and Lenzi, C. (2012) 'Territorial patterns of innovation: A taxonomy of innovative regions in Europe', forthcoming in the Annals of Regional Science.

Capello, R., Caragliu, A. and Lenzi, C. (2012) 'Is innovation in cities a matter of knowledge intensive services? An empirical investigation', *Innovation – The European Journal for Social Science Research*, 25(2): pp. 147–170.

Carlino G., Chatterjee, S. and Hunt R.M. (2007) 'Urban density and the rate of invention', *Journal of Urban Economics*, 61(3): 389–419.

Cooke, Ph. (2009) 'The knowledge economy, spillovers, proxmity and specialization', JRC, European Commission, Directoral General for Research, Brussels, pp. 29–39.

Dasgupta, P. and Stiglitz, J. (1980) 'Uncertainty, industrial structure and the speed of R&D', *Bell Journal of Economics*, 11(1), 1–28.

David P. and Foray D. (1995), 'Accessing and expanding the science and technology knowledge base', *STI Review*, n. 16, pp. 16–38.

Foray, D. (2000) *L'Economie de la Connaissance*. Paris, La Découverte.

Foray D. and Lundvall B.A. (eds) (1996), *Employment and Growth in the Knowledge-Based Economy*, Paris: OECD.

Foray, D., David, P. and Hall, B. (2009) 'Smart specialisation: The concept', *Knowledge Economists Policy Brief*, no. 9.

Griliches, Z. (1990) 'Patent statistics as economic indicators: A survey', *Journal of Economic Literature*, December, pp. 1661–1707

Hübner, D. (2009) 'Towards third generation of regional innovation policy', paper presented at the Conference on 'Innovation, Research and Development in the Framework of the European Regional Policy', held in Athens, 26 February.

Joly, P-B. (1997) 'Chercheurs et laboratoires dans la nouvelle economie de la science', *Revue d'Economie Industrielle*, 79(1), 77–94.

Licht, G. (2009) 'How to better diffuse technologies in Europe', in Research Commissioner Janez Potočnich's Extert Group (ed.), *Knowledge for Growth*, http://ec.europa.eu/invest-in-research/pdf/download_en/selected_papers_en.pdf, pp. 30–34.

Lundvall, B. A. and Johnson, B. (1994) 'The learning economy', *Journal of Industry Studies*, 1(1), 23–42.

MacDonald, S. (1987) 'British science parks: Reflections on the politics of high technology', *R&D Management*, 17(1), 25–37.

Machlup, F. (1962) *The Production and Distribution of Knowledge in the United States.* New York: Princeton University Press.

Massey, D., Quintas, P. and Wield, D. (1992) *High Tech Fantasies: Science parks in society, science and space.* London: Routledge.

Monk, C. S. P., Porter, R. B., Quintas, P., Storey, D. and Wynarczyk, P. (1988) *Science Parks and the Growth of High Technology Firms.* London: Croom Helm.

OECD (2004) *Global Knowledge Flows and Economic Development.* Paris: OECD.

OECD (2005) *Science, technology and industry scoreboard*, Paris: OECD.

Okubo, Y. (1997) 'Bibliometric indicators and analysis of research systems: Methods and examples', *STI Working Papers* 1/97. Paris: OECD.

Pavitt, K. (1984) 'Sectoral patterns of technical change: Towards a taxonomy and a theory', *Research Policy*, 13(4), 343–373.

Pontikakis, D., Chorafakis, G. and Kyriakou, D. (2009) 'R&D specialization in Europe: From stylized observations to evidence-based policy', in D. Pontikakis, D. Kyriakou and R. van Bavel (eds), *The Question of R&D Specialisation.* Brussels: JRC, European Commission, Directoral General for Research, pp. 71–84.

Saxenian, A.-L. (2006), *The New Argonauts: Regional advantage in a global economy.* Cambridge, MA: Harvard University Press.

Storey D. J. and Tether, B. S. (1998) 'Public policy measures to support new technology-based firms in the European Union', *Research Policy*, 26(9), 1037–1057.

Torre, A. and Rallet, A. (2005) 'Proximity and localization', *Regional Studies*, 39(1), 47–59.

Van Oort, F. and Raspe, O. (2006) 'Economic growth and the urban knowledge economy', Utrecht University Papers in Evolutionary Economic Geography (PEEG), no. 06/07.

2 The civic university

Connecting the global and the local

John Goddard, Louise Kempton and Paul Vallance

Introduction

The contemporary meaning of being 'an urban university rather than just a university built in a city' (Goodall 1970: 48) is ambiguous. Is the university *in* the city or part *of* the city? This is a question often asked by policy makers. In this chapter we make the case for the civic university working with others in the leadership of the city in order to ensure that its universities are both globally competitive and locally engaged.

The ambiguity about the university and the city is reflected in the existence of two separate knowledge communities. The first backs into the university from a city and regional development perspective and the second backs into the city from a focus on the university as an institution. Each of these has its own 'community of practice' linked to the knowledge base and enshrined in two separate domains of government – higher education and territorial development. From an urban and regional policy perspective the university is sometimes seen as providing the answer to all manner of urban ills, from a shortage of jobs through to the inclusion of marginalised communities into the socio-economic mainstream. From a higher education policy perspective engagement with a city can provide an outward and visible sign of the university's contribution to civil society. But there are potential tensions here between internationally acknowledged academic excellence and societal accessibility to knowledge that urban engagement implies (Calhoun 2006).

We suggest here that the concept of the civic university can provide a bridge between the internal and external drivers on higher education. However, building this bridge poses leadership challenges, both within the city and within the university, and between the two. We illustrate these challenges with reference to experience in the UK and the US, our own large-scale survey of how academics in six universities in three English cities view the impact of their own research, and the design of a universities and civic leadership development programme based on interviews with key actors in these cities.

The civic university in the UK and the land-grant university in the US

Both the UK and the US have long established institutions that historically combine academic excellence and public service. Many of the UK's great universities were created in the nineteenth century to meet the needs of growing manufacturing cities. Local entrepreneurs responded to the needs for scientific knowledge and a healthy and skilled workforce by funding universities. During the second half of the twentieth century, central government increased control of higher education through public funding, cities deindustrialised, and many of these earlier foundations turned their backs on the cities in which they were based. Increased public funding for research followed narrowly defined academic criteria and higher education was rolled out across the nation to fill in the map of teaching and learning with a diverse set of institutions. Now nearly all UK cities have one or more universities.

Elsewhere we have argued that now is the time to reinvent the notion of the broadly based civic university that served the UK so well in the nineteenth century, but is now set in the context of a more globalised economy and society (Goddard 2009). We suggest that such a university should:

- provide opportunities for the society of which it is part (individual learners, businesses, public institutions);
- engage as a whole not piecemeal with its surroundings;
- partner with other local universities and colleges;
- be managed in a way that facilitates institutional wide engagement with the city and region of which it forms part;
- operate on a global scale but use its location to form its identity.

The following statements made by the Vice Chancellors of Birmingham and Newcastle Universities, two of the UK's great 'civic' institutions, demonstrate this desire to 'reconnect' with their local roots.

> The vision of the founders of much of our higher education system, who sought to create institutions to enable 'the advancement of learning and ennoblement of life', still provides us with a significant challenge today. These are aspirations which are enshrined in the charters of universities in many of our towns and cities and provide us with a benchmark for assessing the extent to which today's institutions match these ideals. These founders were particularly interested in universities' civilising influence and how they could boost economies and transform people within their communities and beyond.
>
> (Eastwood 2009: 3)

> The combination of being globally competitive and regionally rooted underpins Newcastle University's vision for the future. We see ourselves not only as doing high-quality academic work … but also as choosing to work in areas responsive to large-scale societal needs and demands, particularly those manifested in our own city and region.
>
> (Newcastle University 2009: 5)

In pursuit of its mission to be a 'World Class Civic University', Newcastle University is seeking to mobilise its intellectual resources around a number of grand societal challenge themes that have global and local resonance. In partnership with the City it has established the Newcastle Institute for Research on Sustainability which aims 'to bring people together throughout the university and the wider community to develop sustainable responses to the great challenge of our age: ensuring everyone has access to a fair share of the world's resources in perpetuity'. Key themes are: urban living; low carbon energy and transport; food security; water management and clean manufacturing. Another challenge theme is that of an ageing population. The University's Institute for Ageing and Health brings together basic, clinical, social and computer scientists to address: how and why we age; the treatment of age associated diseases and disabilities; the support of through-life health, well-being and independence. It uses research, training, public engagement and commercialisation to fulfil its mission.

While Newcastle University badges itself as a 'World Class Civic University', University College London sees itself as 'London's Global University'. Its Provost has identified four grand societal challenge themes that the University will address – sustainable cities, human well-being, global health and inter-cultural interactions. These are informed by UCL's commitment to contribute to 'The Wisdom Agenda', defined as 'the application of knowledge for the good of humanity'. In each of these cases it is clear that the universities are moving beyond research for its own sake and beginning to address the public value issues that should underpin such endeavours.

Notwithstanding differences in scale, a similar narrative of local engagement, disengagement and re-engagement exists with respect to American land-grant universities. These were established under the principles laid down in the Morrill Act of 1862 which enabled the creation of universities to serve the agricultural and subsequently industrial development needs of individual states. But, as in the UK during the latter part of the twentieth century, many land-grant universities lost sight of their roots in the quest for academic status. Thus, the president of Arizona State University, Michael Crow, has noted:

> Institutional inertia is nowhere more evident than in the academic valorization of increasingly specialised knowledge. In our effort to produce abstract knowledge without regard for its impact, many universities have lost sight of the fact that they are also institutions with the capacity to create products and processes and ideas with entrepreneurial potential. … Through some strange elitist logic, the concept of entrepreneurship has been eradicated from institutions of higher education in this nation.
>
> (Crow 2008: 16)

In a similar vein the senior vice president of Oklahoma State University Robert Sternberg has written:

> Land-grant institutions, contrary to some popular beliefs, are not merely about agricultural development, but rather, about changing the world in a positive, meaningful and enduring way. ... What is important in a land-grant institution is developing future ethical leaders who will enrich their communities and

> their societies ... these institutions are about admitting people who will make the difference to the state and the society.
>
> (Sternberg 2010)

Sternberg contrasts the land-grant universities to the elite institution where there is a:

> kind of curious disconnection between the university and society. In a land-grant institution traditional scholarly endeavour still matters, but work that gives back to society receives especial plaudits. It thus becomes easier for state legislatures and the people of a state to see why research is important to them, not merely to the advancement of individual researchers' scholarly careers.
>
> (Sternberg 2010)

These views of institutional leaders are reflected in the pronouncements of representative bodies such as The Association of Public and Land-grant Universities (A-P-L-U) whose Kellogg Commission report on the future of these institutions is entitled *Returning to our Roots: The Engaged Institution*. The Commission argues that the engaged institution must accomplish at least three things:

> 1 It must be organized to respond to the needs of today's students and tomorrow's, not yesterday's.
> 2 It must enrich student's experiences by bringing research and engagement into the curriculum and offering practical opportunities for students to prepare for the world they will enter.
> 3 It must put its critical resources (knowledge and expertise) to work on the problems the communities it serves face.
>
> (Kellogg Commission on the Future of State and Land-Grant Universities 1999: 10)

The Association's Council on Engagement and Outreach argues that the publicly engaged institution should be fully committed to direct, two-way interactions with communities and other external constituencies through the development, exchange, and application of knowledge, information and expertise for mutual benefit. This has been expressed by the A-P-L-U as 'Stepping Forward as the Stewards of Place'. The dimensions of this 'stewardship' are defined and described in the following box.

Place-Related. While the demands of the economy and society have forced institutions to be nationally and globally aware, the fact remains that state colleges and universities are inextricably linked with the communities and regions in which they are located. Exercising "stewardship of place" does not mean limiting the institution's worldview; rather, it means pursuing that

Place-Related. While the demands of the economy and society have forced institutions to be nationally and globally aware, the fact remains that state colleges and universities are inextricably linked with the communities and regions in which they are located. Exercising "stewardship of place" does not mean limiting the institution's worldview; rather, it means pursuing that worldview in a way that has meaning to the institution's neighbours, who can be its most consistent and reliable advocates.

Interactive. The etymology of the word "engage" speaks to the intertwining or meshing of entities. In this context, engagement refers to a spirit of give and take by the university and its partners. For institutions, this means occupying the role of learner as well as teacher. For community and regional partners, this means looking to the university as a resource, not necessarily as "the answer."

Mutually Beneficial. Engagement should inure to the benefit of both parties involved. These initiatives should expand the learning and discovery functions of the institutions while enhancing community capacity to address and resolve the issues they confront. The work of the engaged institution is to be responsive to public needs in ways that are appropriate to the institution's mission and academic strengths. Engagement initiatives should also build greater public understanding of and support for the role of the campus as a knowledge asset and resource.

Integrated. At a campus level, engagement must permeate all levels of the institution, and be integrated into its policies, incentive structures, and priorities. At a departmental level, engagement cuts across the imperatives of teaching and scholarship to bring unparalleled opportunities for the entire campus community— faculty, staff, and students.

Source: AASCU (2002: 9)

Goddard (2009: 32–34) argues that Michigan State University, the first land-grant university to be established, exemplifies many of these principles. Here, the University Office of Outreach and Engagement is the central resource dedicated to developing connections and partnerships with external audiences. Often, this is helping staff to develop collaborative community-based applied research and evaluation, or providing technical assistance and consulting. Many staff are supported to extend their teaching to engage with non-traditional students at off-campus sites or by technology-delivered distance education, in many forms of continuing professional development. Lots of academics provide clinical services. Many teachers use community-based learning experiences as part of their courses. Others have developed and managed learning environments and exhibitions. Currently, MSU has around seventy community-based projects and has 170 partnerships in more than fifty countries. It also collaborates with two other regional universities in a research corridor. There are outreach and engagement projects in many civic areas: research and practical initiatives in urban regeneration and redesigning communities; cleaning polluted groundwater with schools; developing literacy; meeting the nursing shortage and helping to rebuild Rwanda. The Office also plays a central role in regional economic partnerships, such as Leap Inc, providing companies with easy access to the range of Michigan's capital, locations, people, university and industry partnerships, in tandem with services to accelerate business opportunities/development. This helps diversify the regional economy, attract investment and create jobs.

Reuniting the city and the university: the territorial development drivers and barriers

From a policy perspective OECD has argued that reuniting cities and regions with their universities requires a better understanding of the drivers and barriers to engagement operating in each domain (OECD 2007; Goddard and Puukka 2008). In the academic literature, universities have come into play in the context of a discourse centred on building knowledge economies at the local level (Harloe and Perry 2004; Harding *et al.* 2007; Goddard and Vallance 2010). Knowledge that originates from universities and 'spills over' to nearby firms is seen as a key factor in the development of urban or regional clusters in high-technology sectors (Cooke 2002). Within both academic and policy circles, attention has shifted to supporting the institutional base of a territory that can support collective learning and innovation in the economy as a whole (Lundvall 1992; Storper 1997; Cooke and Morgan 1998). This focus has been reinforced by the strengthening of a parallel policy discourse around the notion that cities and regions themselves compete with each other as relatively coherent economic units (Begg 1999; Sheppard 2000; Bristow 2005), and that the basis of this competitiveness in advanced economies is increasingly related to the capacity of supporting ongoing innovation (Amin 1999). So, in practical terms, many city and regional development authorities are looking to universities to act as global gateways for attracting inward investment, to generate new business via spin-outs, to act as a source of advice for established businesses and to enhance human capital through graduate retention and professional updating of employees; in short, they are going beyond considering universities just as major businesses and anchor institutions in local economies (European Commission 2011). However there are barriers to realising this potential. These include: the fact that national higher education policy is of a rule 'spatially blind' and not within the domain of often fragmented local governments; the limited absorptive capacity of local businesses, especially SMEs; how the economic development function of universities is funded especially when the impact arises in areas not directly linked to the core functions of teaching and research, and last but not least, the limited capacity of institutional leaders to reward and manage this activity.

The potential contribution of universities to city or regional development is, however, not limited to this economic development sphere, but may also encompass their role in other societal areas such as public health, sustainable development, and arts and culture (Goddard and Vallance in press). These forms of contribution have not yet been discussed in as much detail within the regional development literature (Goddard *et al.* 2011), but a wider possible context for this exists through recent arguments that local innovation is not just economic or technology based, but may also be social in nature. The concept of 'social innovation' in the urban and regional development literature is focused on the problems of social exclusion in specific geographical contexts, and the 'innovation' in question largely takes the form of the development of new forms of community-based social relation and organisation from within the sphere of civil society (as opposed to the market or the state) (see Moulaert and Mehmood 2011).

The notion of social innovation is, however, beginning to develop a wider currency in policy circles, albeit typically with a much looser and non-territorial definition. For instance, in European policy terms it is linked to the idea of 'Smart, Sustainable and Inclusive Growth'. According to the Board of European Policy Advisors:

> Social innovations are innovations that are social in both their ends and their means. Specifically, we define social innovations as new ideas (products, services and models) that simultaneously meet social needs (more effectively than alternatives) and create new social relationships or collaborations. They are innovations that are not only good for society but also enhance society's capacity to act. The *process of social interactions* between individuals undertaken to reach certain outcomes is participative, involves a number of actors and stakeholders who have a vested interest in solving a social problem, and empowers the beneficiaries.
>
> (BEPA 2010: 9–10; emphasis in original)

As we have pointed out elsewhere, the social dimension of university engagement with the city is less well conceptualised than its involvement with economic development (Goddard *et al.* 2011). However, Hazelkorn (2010) argues that the concept of social innovation opens up a way of thinking about the engaged civic university. She suggests that:

> rather than seeing [the university contribution to] innovation as purely a discovery process that is commercialized, it is viewed as a complex iterative process involving an array of stakeholders and (end) users – from the private/public sector and/or wider civil society – coupled with feedback loops and market linkages.
>
> (Hazelkorn 2010: 74)

This points to the need to move beyond the 'triple helix' model of business/university/government relations to a 'quadruple helix' which embraces civil society actors. According to Arnkil *et al.* (2010):

> the quadruple helix, with its emphasis on broad cooperation in innovation, represents a shift towards systemic, open and user-centric innovation policy. An era of linear, top-down, expert driven development, production and services is giving way to different forms and levels of co-production with consumers, customers and citizens.
>
> (Arnkil *et al.* 2010: 6)

So, according to Hazelkorn, universities need to be involved in the establishment of 'Think & Do' fora which:

> bring together actors from civil society, the state and state agencies, and higher education to mobilize and harness knowledge, talent and investment in order to address a diverse range of problems and need through co-ordinated action. … [S]ustained, embedded and reciprocal engagement is defined as learning beyond the campus walls, discovery which is useful beyond the academic community and service that directly benefits the public.
>
> (Hazelkorn 2010: 69)

Reuniting the university and the city: higher education drivers and barriers

Recent thinking on the university as an institution in civil society has suggested that there is scope to move this broader civic role more centrally to its mission and identity (Goddard 2009). For instance, in his 2001 book entitled *Challenging Knowledge: The University and the Knowledge Society*, Delanty argues that widely discussed collaborative notions of knowledge production (e.g. Gibbons *et al.* 1994) are overly based on a 'client model of the user' (Delanty 2001: 102), and that these societal transformations present opportunities for universities to engage a broader constituency of civil society in the knowledge production process (also see Nowotny *et al.* 2001). He suggests that:

> The crucial issue … [is] whether the embracing of the user will allow technological innovation to be shaped by the demands of citizenship. For this to be possible, the university will have to be a forum for users drawn from not only industry but from other domains in society. The university is the institution in society most capable of linking the requirements of industry, technology and market forces with the demands of citizenship. Given the enormous dependence of these forces on university based experts, the university is in fact in a position of strength, not of weakness. While it is true that the new production of knowledge is dominated by the instrumentalisation of knowledge and that as a result the traditional role of the university has been undermined, it is now in a position to serve social goals more fully than previously when other goals were more prominent.
>
> (Delanty 2001: 113)

Against the largely optimistic view of the opportunities for civic engagement, there are equally institutional and cultural barriers within higher education systems that can prevent universities from being oriented towards these social goals, especially as they pertain specifically to local development. Calhoun (2006) outlines these issues in reference to major structural tensions that universities in advanced economies now face. He argues that recent transformations have pushed universities to a point where they struggle to accommodate their diverging institutional responsibilities of producing knowledge that is 'both applicable in deterministic ways and valuable for informing personal and public choice', whilst also maintaining both excellence (associated with exclusivity and the specialisation of research) and public accessibility in the production and dissemination of this knowledge (Calhoun 2006: 18–19). He argues the current situation is one in which the balance is too much in favour of the 'excellence' side of the dichotomy with accessibility, primarily because 'research and the rewards for research are deeply tied up with the production of an academic hierarchy – not just with the advance of knowledge for all' (Calhoun 2006: 31).

The promise and the practice – what motivates academics?

How do these tensions play out in terms of the actual behaviour of academics? In the summer of 2010 we undertook an online questionnaire survey that explored the ways in which academics across different institutions and disciplines understand their research to have an 'impact' in the broadest possible sense (see Vallance *et al.* 2011). Our aim was to contribute to a better understanding of the range of both academic and non-academic impacts that result from different forms of research and their possible relationships to activities such as teaching, consultancy, professional practice, commercialisation and public engagement, and thereby set societal impact in the context of the total activity of academics and the environments in which they work. Invitations to participate in the survey were sent to a random sample of one third of all academic staff (including research staff) from the six universities in three large English cities – Bristol, Newcastle upon Tyne, and Sheffield. In total 711 responses were received.

Questions posed in the first half of the survey enquired about the broad areas in which participants thought their research was having an impact, the groups or organisations that are beneficiaries of the research, and the mechanisms they use to deliver these research impacts. For these three questions, we introduced a distinction between 'primary' and 'secondary' research impacts or beneficiaries to differentiate between the more and less direct ways in which research can have an impact. This distinction was defined for participants in the survey as such: 'by a primary impact or beneficiary we mean the main areas of groups for which your research is designed to directly and intentionally result in benefits. By a secondary impact or beneficiary we mean the other areas or groups that your research may indirectly benefit, even if this impact is not one of the main aims of the research'. Figures 2.1 and 2.2 give the responses to the areas of research impact and beneficiaries of research impact questions by the levels of all participants who responded that the given categories were either a primary or secondary area/beneficiary of their research. The other response options in the questions were 'no impact' and 'not applicable to me'.

As Figure 2.1 shows, the majority of respondents saw 'contribution to scientific/academic knowledge or method' as the primary impact of their research. While impact on 'education and the development of people' (related to teaching) was the next most commonly identified impact of research, it is interesting to note that the majority of academics who identified this impact saw it as secondary. While almost a quarter of all respondents believed their work to have a direct impact on 'informing public policy', fewer (20 per cent or less in each case) felt their work impacts directly on the other socio-economic issues. However, if we consider the intended secondary impacts of their research, significant numbers of respondents do mention 'public good' activities such as informing public policy,

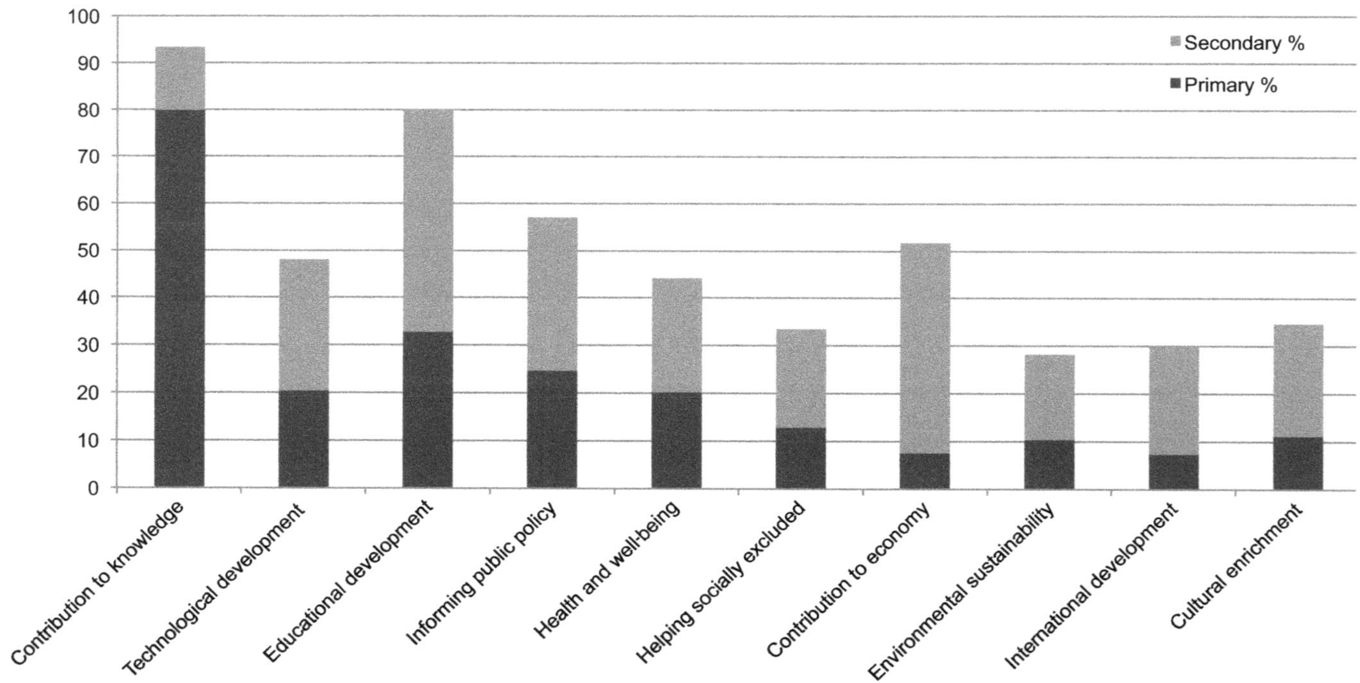

Figure 2.1 In which of the following areas do you think your research is having either a primary or secondary impact? – Overall (n=711).

contributing to health and well-being, environmental sustainability and cultural enrichment and helping the socially excluded, while over 40 per cent saw contributing to the economy as an area where their research had an indirect impact. In many areas this secondary impact outweighed the permitted category of 'no impact'.

With these points in mind it is not surprising that the majority of respondents identified their fellow academics as the primary beneficiaries of their research (Figure 2.2). Students were the next biggest beneficiary group, with almost all respondents seeing them as primary or secondary beneficiaries. Again, less than 20 per cent saw categories relating to the non-academic world (government – local and national, firms – large and small, the third sector and the general public) as primary beneficiaries of their research. For both of these questions, however, this level is much higher in some categories in the response from different academic disciplines (e.g. the medical sciences for health-related categories, engineering for economic-related categories, the social sciences for government-related categories) (see Vallance *et al.* 2011). However when we consider groups where the secondary impacts of academics' research might arise, then a number of 'publics' have a significantly greater score including national and local government, international government organisations, the health care sector, the professions and last but not least the general public

The second half of the survey consisted of questions about the personal, institutional, and wider environmental factors that encouraged or supported participants' research and its intended impact, and the factors that they had experienced as barriers to their research and its intended impact. Table 2.1 gives the response for all participants to the first of these questions, which asked them to rate the importance of the given factors in motivating their research and its intended impacts.

The vast majority saw making a contribution to knowledge and their own intellectual curiosity or interest as the greatest motivators (almost 87 per cent ranking both as 'high importance'). Far fewer were motivated by ambitions of advancement or enhancing status, whether their own career or that of a wider organisation, with less than 40 per cent of respondents ranking these as highly motivating. In terms of being motivated by the public (i.e. non-academic) benefits of their work, national or international impacts were considerably more motivating than local or regional. Almost 40 per cent saw the latter as being of little or no motivation, while more than 40 per cent were motivated by the potential for national or international impacts. Nevertheless, the public good aspects of research are highly or moderately important for the majority of academics.

As these results show, while the leadership of the university may aspire to connect it more firmly to its locale, and be seen to be responding to local demands for support in tackling socio-economic problems, most academics see this activity as coming a very distant second to their core function as researchers and educators. Universities as institutions and their public, private and third sector partners also need to acknowledge what it is that motivates academics if they are to successfully engage them in local and regional development. They must also recognise that the socio-economic impact of academic work may arise in an indirect manner and not necessarily focused on the locality and as a result mechanisms both within the university and the city need to put in place to facilitate this, a leadership challenge to which we now turn.

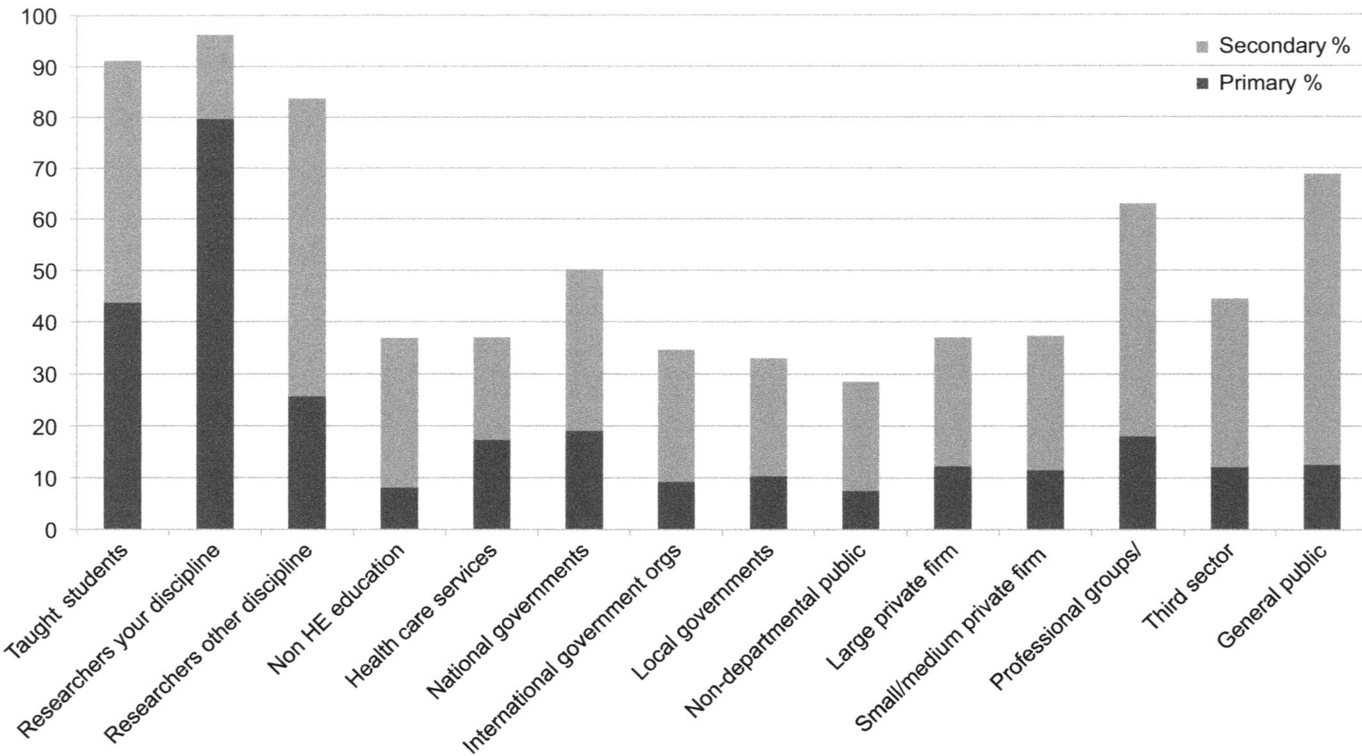

Figure 2.2 Which of the following types of group or organisation do you think are either primary or secondary beneficiaries of your research? Overall (n=711).

Table 2.1 How important are the following factors in personally motivating your research and its intended impacts? – Overall (n=711)

Personal factors motivating research and intended impacts	*High %*	*Moderate %*	*Low %*	*No %*	*Skipped %*
Making a contribution to scientific/academic knowledge	86.9	10.4	1.1	0.7	*0.9*
It supports your teaching	26.4	42.3	21.9	7.7	*1.7*
Your intellectual curiosity or personal interest in the subject	86.9	10.7	0.7	0.1	*1.6*
Advancement of your career	36.1	45.1	14.3	2.5	*2*
Your political or ethical beliefs and values	30.7	31.5	18	17.7	*2.1*
Boosting status of your department/school/research centre or institute	30.5	48.7	16.3	2.8	*1.7*
The public benefits (social, economic, or other) for the home city or region of your university	26.2	34.2	24.9	13.4	*1.3*
The public benefits (social, economic, or other) nationally or internationally	44	32.3	14.2	7.6	*1.9*

The leadership challenge

The ongoing programme of OECD reviews of higher education in city and regional development have highlighted the leadership challenge of working across the boundary between universities and civil society. All too often city leaders do not understand the drivers behind higher education, particularly the tension between academic excellence and public engagement. For them the university can be a 'black box' with mysterious ways of working. Equally, for many in the academy, the drivers behind city development and the responsibilities of many public authorities are opaque. However, common ground may emerge around the idea that the university and the public and private sectors can come together around 'the leadership of place'. There is a growing body of academic literature on the role of leadership in shaping the integrated development of places – cities and neighbourhoods within cities. According to Gibney *et al.* (2009) this 'new' leadership of place is concerned with:

> facilitating interdisciplinarity across institutional boundaries, technology themes, sub-territories and professional cultures in order to promote the development of innovation across the public and private sector domain …

> [and] needs to ensure the comprehensive engagement of local communities so that they can both contribute to and fully benefit from the outcomes – thus avoiding the danger of … creating deeper forms of social polarisation.
>
> (Gibney *et al.* 2009: 10)

Collinge and Gibney (2010) suggest that new complexities are being encountered by leaders working outside their own organisation:

> leaders find themselves representing places rather than organisations; there are more uncertainties to be accommodated as outcomes are difficult to pin down and there are more unknowns; leaders are increasingly required to lead initiatives without formal power but with responsibility; they must accommodate the views of organisations, groups and communities historically excluded (consider for example, the engagement of social enterprises with the knowledge-based economy).
>
> (Collinge and Gibney 2010: 386).

Where do universities and their relationship to place fit into this picture? Hambleton has argued that 'universities, provided they see themselves as "civic" or "engaged" universities, can make a significant contribution not just to the promotion of innovation (defined broadly) in their area, but also in assisting with the development of place-based leadership' (Hambleton 2009: 1). However, there are barriers in the way to universities realising this role. The rest of this section is adapted from a research project and scoping exercise for a proposed leadership development programme that would focus on improving partnership working between universities and cities. The research involved interviews with university and various civic leaders in the three cities – Bristol, Newcastle upon Tyne, and Sheffield (see Goddard *et al.* 2010). The next two sub-sections describe the obstacles and challenges to boundary-spanning collaboration experienced by two groups – those faced by the universities in 'reaching out' to civic society, and those faced by civic society in 'reaching in' to universities. The third sub-section outlines some of the findings and conclusions of this project as regards the role of leadership and best ways of developing this.

Obstacles and challenges faced by universities working with civil society

'Civic partnership' is often not seen as part of the universities' 'core business', but only a means to other ends, and hence there are few people within universities whose main role and responsibilities are to support these relationships. Many of the activities that fall underneath the label civic partnerships are cross-subsidised from other funding sources, and therefore may not be financially sustainable in times of reduced resources. Coupled with many universities' already stretched resources this places limits on the degree to which they can get involved in a range of external projects. Even where the university is involved in a range of 'civic' partnerships it is sometimes difficult to know whether their activities are having an impact, particularly in the long-term.

Universities are not institutions located directly within the local political sphere, and are therefore unable to exert significant influence here. Furthermore, the multitude of organisations that are involved in the political and economic governance of cities and regions in the UK creates challenges of understanding the 'local political–organisational map' and knowing who are the most important partners with which universities need to work. The way this varies across geographic and administrative boundaries can be a source of further complexity, while instability and changes in the leadership of local politics can make it hard for university leaders to build strong relationships with city councils.

The burdens of incentives and targets in the university system occupy gifted academics, meaning they do not have the spare time outside their main responsibilities to pursue other external engagement activities or interests. For some academics these systemic constraints will discourage them from taking risks by seeking to work across the boundaries of academia.

Obstacles and challenges faced by civic leaders working with universities

The university is not a homogeneous body: its size and diversity can make it hard for civic partners to get a single view from people there. Many civic partners do not know who to work with in the university below the executive level. This applies to academics, meaning that large potential sources of expertise remain untapped, but also to people in the administrative or support services, where the civic leaders may not always be clear of who is responsible in areas such as finance or estates.

Many people outside universities do not understand their organisational structures and procedures well enough to be able to interact with them properly. In particular they may not be familiar with the terminology used by people within the university system to describe these structures and procedures: for instance, in a UK context, the differences between Vice-Chancellors, Deputy Vice-Chancellors, Pro Vice-Chancellors, Deans etc. The language or jargon used by academics more generally can also be a barrier to their effective external engagement.

Civic partners sometimes find that universities work slowly in comparison to them and often do not have the same level of urgency in responding to opportunities or following up on agreements. Some feel that university procedures can also be overly-bureaucratic on occasions. External organisations in the private and third (voluntary) sectors may have a poor perception of universities as being unreliable, inefficient, or overly self-interested, and are therefore discouraged from working with them.

Alternatively, many external bodies may still perceive universities as solely inward-focused teaching and research organisations, and therefore are not aware of the opportunities working with them offers. Even when there is recognition of the 'value added' interaction with a university can bring, there may be a lack of demand or absorptive capacity for the knowledge that universities could supply. For instance, only a small proportion of SMEs would actually benefit from academic research outputs.

How to develop leadership that promotes civic partnership

It is possible to identify two relatively distinct forms of leadership that are required to overcome these obstacles: first, the internal leadership of large organisations (principally universities) so that they can become more externally engaged; second, leadership within city partnerships that require the collaboration of multiple organisational stakeholders. These two forms of leadership have different challenges attached to them and require different approaches. University and non-university leaders have a role in promoting enterprising or boundary-crossing behaviour within their organisations by recognising the potential of outstanding individuals with the ability to make wider connections, and then supporting, protecting and valuing them. People with the personal attributes and sense of purpose to assume a leading or mobilising role can emerge from levels further down the organisational hierarchy, but they may need to be supported by upper management and their endeavours aligned with the strategic priorities of the institution to have a full impact. Hence, Vice-Chancellors, chief executives and other equivalent organisational leaders have a vital role in clearly setting out and promoting the civic agenda within their organisation. One Vice-Chancellor described this in terms of 'articulating that you're interested where the city's going [and] permeating that sense of availability and openness down the organisation'. This function of leadership is of particular importance in universities in helping to overcome the disconnect between strategic and operational levels. The creation of new positions dedicated to civic engagement is one option, but this may have limited impact if they are not linked to wider cultural or systemic change within the institutions. Efforts should rather focus on building the principle of valuing outreach or engagement or knowledge exchange activities into core university structures such as promotion pathways or workload models. However, these external engagement activities should not be compulsory for all academics, as when a person unsuited to this type of role is pushed into a leadership position it is more likely to have a damaging effect on the institution's external relationships and reputation.

Turning to inter-organisational partnerships these are distinguished by people being as committed to the mutual benefits it will bring to the city as they are to the interests of their own organisation. These civic partnerships should be relatively independent of the transactional relationships that exist between organisations, (for instance between city councils and universities on estate matters), so that tensions or disagreements that inevitably arise on these fronts do not spill over to negatively affect overall relationships and be detrimental to the city as a whole. This should, therefore, allow leadership of these partnerships to focus on the joint benefits they can bring to their city through real change, instead of preserving the status quo relationships and 'not falling out'. As with leadership within organisations, good civic leadership, whether from the city council or other possible spheres, requires being able to effectively articulate the future direction of the city's development. A process of agreeing and clearly setting out this vision in strategic plan documents, so that all stakeholders are aligned behind the partnership, can see their role, and will be committed to delivering is required. This also indicates the mutual responsibility and trust that these civic relationships entail: if individual or group leaders fail to deliver themselves, it can destroy confidence and trust within the wider partnership. The form of collaborative or distributed leadership that characterises good civic partnerships does not just involve key individuals, but also works through intermediary partnership organisations. Because these organisations normally have only very limited resources themselves, their style of leadership must necessarily be facilitative and understated, concentrating on mobilising and aligning key public and private organisations within their city to achieve their goals.

The analysis above points to the need for developing leaders from the city and its universities to enhance their skills in working together on key challenges facing the city – to lead the city not just to lead in the city. To kick-start this process leaders from the university and the city need to come together and identify a key challenge such as removing barriers to social mobility or developing a sustainable city and then hand over to an operational group of future leaders from the university and the city. A key task for the university would be to mobilise its global knowledge around the chosen theme and translate this so that it has meaning for the city. Such intellectual leadership should embrace the political, managerial and community dimensions of civic leadership. Leadership development would involve building networks between key actors, locally, nationally and internationally, and developing skills in partnership working through joint projects and benchmarking against best practise elsewhere.

Leadership development programmes with their emphasis on the interpersonal skills of individuals are a necessary but not sufficient condition for creating the civic university. Formal structures within organisations and regulatory regimes can be barriers that even enterprising individuals cannot circumnavigate. Within Europe there is much discussion of the need to 'modernise' the management and governance of universities to meet the 'Grand Challenges' facing the European Union (e.g. The Lund Declaration 2009). Achieving major structural changes in universities is much easier when there is some external shock to the higher education system in the form of a major change in government policy towards the sector and/or institutional mergers lubricated by substantial additional funding. A recent example of this has been the reform of the Finnish higher education system introduced in 2010. This among other things facilitated the merger of three Helsinki universities – the Universities of Technology and Art and Design and the Helsinki Business School to form the Alto University. This new institution has objectives and values that clearly fit many of our criteria for a civic university:

> [To be a] world class university combining science and art, technology, business and industrial design to stimulate innovations … [and to] educate

> responsible and broad-minded experts with wide perspective to act as future visionaries in the society. … [An institution with the values of] passion to explore boundaries; freedom to be creative and critical; courage to influence and excel; duty to care, accept and inspire; high ethics, openness and equality.
> (Pursula 2010)

Conclusion

Universities are quintessentially urban institutions. According to the Carnegie foundation 46 per cent of US universities are to be found in large or mid-sized cities (see Moore *et al.* 2010: 5). The world's leading city regions account for the lion's share of academic publications and citations (Matthiessen *et al.* 2010). In terms of spatial organisation, cities are often claimed to be national and international hubs in the knowledge economy. Citing the analysis by Matthiessen *et al.* (2010), the prestigious scientific journal *Nature* asks the question 'why do so many scientists ignore the needs of our cities? … Researchers who benefit from opportunities in cities should ask what they can give back' (Nature 2010: 883–884). This suggests that there are deep rooted forces that result in a disconnection between universities and the cities where they are located such that the presence of a university is *not* a guarantee of local economic success or a vibrant and inclusive urban community. Mobilising universities in support of city development in the round therefore needs robust partnerships between universities and local civil society. And achieving this needs effective and distributed leadership on both sides. While some universities and city authorities are devoting resources to enhancing *internal* leadership the challenge now is to develop capacity in leadership across the boundaries between organisations. The role of universities in the 'leadership of place' can provide a powerful focus for such people development programmes.

We have suggested here that promoting the 'Civic University' as a model can contribute to breaking down the barriers between universities and cities. Such universities should not be confined to working with their immediate city but rather use the 'local' as a crucible in which to forge a more open institution able to address multi-scalar challenges such as sustainable development. Such institutions would not only mobilise their research around grand challenges but also organise their teaching with a view to producing future citizens whose decisions as consumers, workers or entrepreneurs will bring about societal innovation in the broader public interest. In doing this, the civic university could be regarded as a social innovator, behaving as a multi-level actor linking the global, local and national domains; working across the silos of the disciplines and of the private and public sectors and linking with both business and the community; developing the boundary spanning and social entrepreneurship skills of the professionals it trains and testing its research in local 'living labs'.

Acknowledgement

The last section of the chapter draws on a study sponsored by the Higher Education Funding Council for England (HEFCE) undertaken by two of the authors, together with Lynne Howlett of Newcastle University and Tom Kennie of the Leadership Foundation for Higher Education whose contribution is gratefully acknowledged.

References

AASCU (2002) *Stepping Forward as Stewards of Place*. Washington: American Association of State Colleges and Universities.

Amin, A. (1999) 'An institutionalist perspective on regional economic development', *International Journal of Urban and Regional Research*, 23: 365–378.

Arnkil, R., Järvensivu, A., Koski, P. and Piirainen, T. (2010) 'Exploring Quadruple Helix: Outlining user-oriented innovation models. Final report on Quadruple Helix Research for the CLIQ project', *Työraportteja 85/2010 Working Papers.*

Begg, I. (1999) 'Cities and competitiveness', *Urban Studies* 36: 795–809.

BEPA (2010) *Empowering People, Driving Change: Social innovation in the European Union*, Brussels: European Commission.

Bristow, G. (2005) 'Everyone's a 'winner': Problematising the discourse of regional competitiveness', *Journal of Economic Geography*, 5: 285–304.

Calhoun, C. (2006) 'The university and the public good', *Thesis Eleven*, 84: 7–43.

Collinge, C. and Gibney, J. (2010) 'Connecting place, policy and leadership', *Policy Studies*, 31: 379–391.

Cooke, P. (2002) *Knowledge Economies: Clusters, learning and cooperative advantage*. London: Routledge.

Cooke, P. and Morgan, K. (1998) *The Associational Economy: Firms, Regions and Innovation*. Oxford: Oxford University Press.

Crow, M. (2008) 'Building an entrepreneurial university', in *The Future of the Research University: Meeting the global challenges of the 21st century*. Kansas, KS: Kauffman Foundation.

Delanty, G. (2001) *Challenging Knowledge: The university in the knowledge society*. Buckingham, Open University Press.

Eastwood, D. (2009) 'Introduction', in *Urban Regeneration: Making a difference*. Newcastle upon Tyne: Northumbria University.

European Commission (2011) *Connecting Universities to Regional Growth: A practical guide*. Brussels: European Commission.

Gibbons, M., Limoges, C., Nowotny, H., Schwartzman, S., Scott, P. and Trow, M. (1994) *The New Production of Knowledge: The dynamics of science and research in contemporary societies*. London: Sage.

Gibney, J., Copeland, S. and Murie, A. (2009) 'Towards a "new" strategic leadership of place for the knowledge-based economy', *Leadership*, 5: 5–23.

Goddard, J. (2009) *Re-inventing the Civic University*, London: NESTA.

Goddard, J. and Puukka, J. (2008) 'The engagement of higher education institutions in regional development: An overview of the opportunities and challenges', *Higher Education Management and Policy*, 20: 11–41.

Goddard, J. and Vallance, P. (2010) 'Universities and regional development', in A. Pike, A. Rodríguez-Pose and J. Tomaney (eds), *Handbook of Local and Regional Development*. London: Routledge.

Goddard, J. and Vallance, P. (in press) *The University and the City*. London: Routledge

Goddard, J. Vallance, P. and Puukka, J. (2011) 'Experience of engagement between universities and cities: Drivers and barriers in three European cities', *Built Environment*, 37: 299–316.

Goddard, J., Howlett, L., Vallance, P. and Kennie, T. (2010) *Researching and Scoping a Higher Education and Civic Leadership Development Programme*. London, LFHE.

Goodall, L. E. (1970) 'The urban university: Is there such a thing?', *The Journal of Higher Education*, 41: 44–54.

Hambleton, R. (2009) 'Place-based leadership and public service innovation', *Think-piece*. Available from: www.urbananswers.co.uk/html/current_comments.html.

Harding, A., Scott, A., Laske, S. and Burtscher, C. (eds) (2007) *Bright Satanic Mills: Universities, regional development and the knowledge econom*. Aldershot: Ashgate.

Harloe, M. and Perry, B. (2004) 'Universities, localities and regional development: The emergence of the "mode 2" university?', *International Journal of Urban and Regional Research*, 28: 212–223.

Hazelkorn, E. (2010) 'Community engagement as social innovation', in L. E. Weber and J. J. Duderstadt (eds), *University Research for Innovation*. London: Economica Ltd.

Kellogg Commission on the Future of State and Land-Grant Universities (1999) *Returning to Our Roots: The Engaged Institution*. New York: National Association of State Universities and Land Grant Colleges.

The Lund Declaration (2009) *The Lund Declaration: Europe must focus on the grand challenges of our time*. Available from: www.se2009.eu/en/meetings_news/2009/7/8/declaration_from_the_research_conference_in_lund_european_research_must_focus_on_the_grand_challenges.

Lundvall, B-Å. (ed.) (1992) *National Systems of Innovation: Towards a theory of innovation and interactive learning*. London: Pinter.

Matthiessen, C. W., Schwarz, A. W. and Find, S. (2010) 'World cities of scientific knowledge systems: networks and potential dynamics: An analysis based on bibliometric indicators', *Urban Studies*, 47: 1879–1897.

Moore, B., Hughes, A. and Ulrichsen, T. (2010) *The Higher Education Knowledge Exchange System in the United States*. Cambridge: Public and Corporate Economic Consultants.

Moulaert, F. and Mehmood, A. (2011) 'Spaces of social innovation', in A. Pike, A. Rodríguez-Pose and J. Tomaney (eds), *Handbook of Local and Regional Development*. London: Routledge.

Nature (2010) 'Save our cities', *Nature*, 467: 883–884.

Newcastle University (2009) *Vision 2021: A world-class civic university*. Newcastle upon Tyne: Newcastle University.

Nowotny, H., Scott, P. and Gibbons, M. (2001) *Re-Thinking Science: Knowledge and the public in an age of uncertainty*. Cambridge: Polity Press.

OECD (2007) *Higher Education and Regions: Globally competitive, locally engaged*. Paris: OECD.

Pursula, M. (2010) 'Aalto University as an example of strategic university reform', *PowerPoint Presentation at The Dean Conference, Barcelona 16.11.10*.

Sheppard, E. (2000) 'Competition in space and between places', in E. Sheppard and T. Barnes (eds), *A Companion to Economic Geography*. Oxford: Blackwell.

Sternberg, R. (2010) 'Defining a great university', *Inside Higher Education*, November 29, 2010.

Storper, M. (1997) *The Regional World: Territorial development in a global economy*. London: The Guildford Press.

Vallance, P., Goddard, J. and Kempton, L. (2011) *Research Impacts – The Perspective of the Academic Community: Summary of the main findings from an online survey*. Available fom: www.ncl.ac.uk/curds/publications/publications/2011.

3 The innovation process of European regions

From absorption to knowledge creation capacities of European regions

Małgorzata Runiewicz-Wardyn

1 Introduction

Nowadays, it seems increasingly clear that economic development cannot be managed at the macroeconomic level alone. Critical elements of the successful transition to the knowledge-based economy depends on the capability of local level to upgrade its productive structures, generate, diffuse and apply knowledge and technological excellence into new and existing products, services and organizations.

Moreover, the process of innovation does not follow a straightforward 'linear model' path from the laboratory directly to the marketplace. The innovative process is circular rather than sequential. Schematically, university research feeds private business research, whereas private business research which is also fuelled by diverse types of expertise coming from customers and supplies, feeds back university research (Kline and Rosenberg 1987). Furthermore, the productivity and efficiency of this infrastructure heavily depends on the region's characteristics in terms of its absorption capacity; thus, as the qualification level of human capital, the willingness to cooperate and the openness and degree of insertion into the world economy.

This chapter attempts to draw some conclusions on the respective contributions of university R&D and business R&D to the knowledge creation at the European regional level; determine what are relations between university R&D and private business R&D in the knowledge creation process in European regions; define the role of human qualifications within the process of knowledge creation as well as identify relative efficiencies of regional innovation systems of the EU15 and EU10 member states.

The remainder of the chapter is organized as follows. The first section gives an overview of the subject related literature. The second section provides some useful facts and figures that highlight the heterogeneity of the European regional landscape regarding wealth and knowledge creation capacities. The third section presents the methodological aspects of the model applied and discusses the role of variables taken into consideration. The estimation results and model output is presented in the explained in section fourth section. The section five highlights some outcome for the EU15 and EU10 regional innovation processes. The chapter ends up with relative summarizes and conclusions of the most important findings.

2 The knowledge production function – overview of closely related literature

There exists a wide range of literature focusing on particular aspects of this process. Jaffe (1989) was the first to use the Griliches (1979) production function framework for a sample of twenty-nine US states over a disconnected period of eight years. Jaffe's findings suggest that both industry and university research are an important source of innovation for the business sector; universities induce the location of industry R&D spending nearby. However, he did not find evidence of a significant impact of industrial R&D on university R&D.

Inspired by the work of Jaffe (1989), Feldman and Florida (1994) investigated a very similar model but instead of corporate patents, they considered commercial product-innovation citations as a measure of innovative output. They found evidence that both industrial and university R&D positively and significantly influence private business innovation. Anselin *et al.* (1997) broadened the cross-section database used by the former authors and tested the system for a sample of 125 MSA. A major innovation with respect to other studies in this field is the introduction of spatially lagged explanatory variables. They found a positive impact of university R&D and industrial R&D on innovative output. Moreover, university R&D significantly influences private business R&D.

Although the research of Griliches (1979), Jaffe (1989) and Anselin *et al.* (1997) concerned the US regions, their methodology has been recently tested and improved by many European authors. A knowledge production factor analysis by Dunnewijk *et al.* (2008) resulted in a factor which indicated that at regional level high scores on patents and business R&D was associated with a large share of high-tech manufacturing. Grimpe and Patuelli (2011) analysed German regions from 2000 to 2004 and proved that both public and private R&D are relevant for nano-material patenting in an additive but also an interactive way. This finding suggests that co-located R&D provides opportunities for knowledge spillovers and collaboration between the actors.

Inspired by the above-mentioned research on the EU regions and the earlier results achieved by the US researchers, the chapter focuses on the European regional landscape of EU25 regions[1] for which this kind of investigation has never been undertaken. Particular attention is drawn onto the qualification level of human capital since the latter has been shown to largely determine a region's absorption capacity.

3 Knowledge absorption determinants of the EU regions

Knowledge creation requires an efficient innovation system interconnecting universities, private enterprises and the government. In some regions the innovation system and the underlying technological infrastructure are not sufficiently developed. It is therefore essential for these regions to develop their knowledge absorption (Capron 2002; Greunz 2003) or social capability skills (Abramovitz 1986). Social capability is considered as a precondition for potential knowledge creation and spillover capacities which are examined next. According to Capron (2002), the capacity to create knowledge requires first that the knowledge absorption and knowledge transfer capacities are developed. This knowledge transfer capacity depends on the local qualification level of the working age population and the employment concentration in the knowledge-intensive sector. The importance of these variables for the knowledge production was also emphasized by Abramovitz (1986), who referred to a region's 'social capability' as to factors that facilitate not only innovation but also technological imitation and the implementation of technology spillovers.

The share of human resources involved in science and technology (S&T) of the working age population and the number of students in tertiary education are proxies of a region's endowment of human capital and, thus, the 'social capability'. Knowledge spillovers occur if the regions have the capacity to integrate them. The latter depends on the region's social capacity, which is dependent on the stock of already accumulated knowledge.

Table 3.1 gives an overview of the current absorption capacities of European regions and indicates the most important trends. Absorption capacity is particularly strong in the northern part of Europe. Without exception, all Swedish, Finnish and UK regions, but also German regions, Denmark, along with two capital regions of Poland and the Czech Republic combine a high share of S&T population and above European average levels of educational attainment. In general, EU15 regions have better developed absorption capacities, with a few exceptions in the Estonia, Bratislava region. Swedish, Finnish and Dutch regions, followed by a few Spanish and French regions have one of the highest shares of S&T human capital with a relatively high level of tertiary education ratio. Most Spanish, French, Austrian and a few Italian regions have moderate absorption capacities. The absorption capacity of Italian, Greek and Portuguese regions is generally low. A common feature of Polish regions is their low S&T human capital and the high number of tertiary education students. A similar situation concerns other EU10 regions, mainly Hungarian and Slovakian regions, Latvia and Lithuania, for which the general weakness is low S&T capital. Moreover, compared to the French case, all Austrian regions have higher participation rates of lifelong learning. Belgian regions are in an intermediate position both in terms of participation in lifelong learning and educational attainment.

4 Knowledge transfer capacity: university, private business and government R&D

As previously stated, knowledge creation and innovation may be the outcome of the region's own R&D efforts. However, for regions where the innovation system is not in place, knowledge creation hardly occurs. These regions are generally characterized by relatively low business R&D intensities which are necessary for efficient technology diffusion. For these regions, absorption capacity is mainly embodied in university labs and government research centres. Since the direct measures of knowledge transfer among institutional sectors such as higher education,

Table 3.1 Knowledge absorption capacities of the EU regions

		Students in tertiary education by NUTS2 regions (2008)		
		High (S>111877,67)	*Medium (41513,67<S <111877,67)*	*Low (S<41513,67)*
Human resources in S&T by NUTS2 regions % of economically active population in 2008	High (H>30,13)	Vlaams Gewest (33,6)	Région de Bruxelles (39,2)	Principado de Asturias (31)
		Région wallonne (30,6)	Brandenburg (35,9)	Comunidad Foral de Navarra (34,9)
		Praha (44)	Hamburg (37,4)	Cyprus (32,5)
		Denmark (36,7)	Mecklenburg-Vorpommern (32,4)	Groningen (35,5)
		Baden-Württemberg (35,5)	Sachsen-Anhalt (32)	Friesland (NL) (30,9)
		Bayern (33,2)	Thüringen (35,9)	Drenthe (34,3)
		Berlin (38,1)	Estonia (34,6)	Overijssel (33,8)
		Hessen (32,6)	País Vasco (38,9)	Flevoland (37,6)
		Sachsen (36,2)	Aragón (31,2)	Zeeland (30,6)
		Ireland (32,2)	Alsace (32)	Limburg (NL) (31,1)
		Comunidad de Madrid (40,6)	Midi-Pyrénées (33,3)	Pohjois-Suomi (31,4)
		Île de France (42,4)	Auvergne (32,9)	Åland (30,8)
		Rhône-Alpes (31,4)	Gelderland (37)	Mellersta Norrland (30,7)
		Közép-Magyarország (31,6)	Utrecht (44,3)	
		Zuid-Holland (37,8)	Noord-Holland (44,2)	
		Wien (33,8)	Noord-Brabant (34,9)	
		Mazowieckie (30,8)	Bratislavský kraj (42,3)	
		Manner-Suomi (36,6)	Länsi-Suomi (32,3)	
		Etelä-Suomi (41,9)	Stockholm (47,9)	
		East of England (31,9)	Östra Mellansverige (33,1)	
		London (42,6)	Sydsverige (36,2)	
		South East (UK) (36,7)		
		South West (UK) (32,8)		
		Scotland (32,6)		

Continued

Table 3.1 (continued)

		Students in tertiary education by NUTS2 regions (2008)		
		High (S>111877,67)	*Medium (41513,67<S <111877,67)*	*Low (S<41513,*
Human resources in S&T by NUTS2 regions % of economically active population in 2008	Medium	Niedersachsen (28,4)	Jihovýchod (27,2)	Strední Cechy (25,6)
	(22,90<A <30,13)	Nordrhein-Westfalen (29,8)	Moravskoslezsko (23)	Jihozápad (24,9)
		Cataluña (28)	Schleswig-Holstein (28,8)	Severovýchod (22,9)
		Comunidad Valenciana (25,4)	Galicia (27,7)	Bremen (28,2)
		Andalucía (23,2)	Castilla y León (29)	Saarland (28,5)
		Nord - Pas-de-Calais (27)	Región de Murcia (24,7)	La Rioja (28,9)
		Provence-Alpes-Côte d'Azur (28,8)	Picardie (24,9)	Basse-Normandie (26,9)
		Lombardia (24,3)	Haute-Normandie (24,1)	Bourgogne (26,8)
		Emilia-Romagna (25)	Lorraine (25,2)	Franche-Comté (28,9)
		Lazio (24,6)	Pays de la Loire (28)	Limousin (27)
		Latvia (28,1)	Bretagne (29,9)	Liguria (25)
		Lithuania (28,8)	Poitou-Charentes (24)	Provincia Autonoma Trento (24,3)
		Slaskie (22,9)	Aquitaine (29,8)	Niederösterreich (27,5)
		Pomorskie (23,2)	Languedoc-Roussillon (28,6)	Kärnten (23,9)
		Lisboa (24,5)	Steiermark (24,2)	Oberösterreich (25,9)
		Slovenia (27,4)	North East (UK) (24,8)	Salzburg (28,5)
		North West (UK) (29,5)	Wales (29,4)	Tirol (26,2)
		Yorkshire and The Humber (29,1)	Northern Ireland (UK) (26,6)	Vorarlberg (25,8)
		East Midlands (UK) (28,4)		Itä-Suomi (28,7)
		West Midlands (UK) (26,9)		Småland med öarna (29)
				Norra Mellansverige (28,6)

Table 3.1 (continued)

		Students in tertiary education by NUTS2 regions (2008)		
		High (S>111877,67)	*Medium (41513,67<S <111877,67)*	*Low (S<41513*
Human resources in S&T by NUTS2 regions % of economically active population in 2008	Low	Veneto (21)	Canarias (ES) (22,4)	Severozápad (19)
	(A<22,90)	Toscana (22)	Piemonte (21,2)	Strední Morava (22,8)
		Sud (15,8)	Marche (20,8)	Castilla-la Mancha (21,9)
		Campania (15)	Abruzzo (20,5)	Extremadura (20,2)
		Puglia (14,6)	Calabria (17,3)	Illes Balears (21,6)
		Isole (15,6)	Sardegna (16,4)	Champagne-Ardenne (21,4)
		Sicilia (15,3)	Észak-Alföld (16,6)	Corse (18,5)
		Lódzkie (21,7)	Dél-Alföld (18,7)	Valle d'Aosta/ (19,1)
		Malopolskie (22,2)	Podkarpackie (22,1)	Bolzano/Bozen (22)
		Lubelskie (21,6)	Swietokrzyskie (20,8)	Friuli-Venezia Giulia (20,8)
		Wielkopolskie (17,9)	Podlaskie (21,8)	Umbria (21,5)
		Dolnoslaskie (22,5)	Zachodniopomorskie (22)	Molise (18,4)
		Continente (16,5)	Opolskie (18,6)	Basilicata (15,9)
		Norte (13,6)	Kujawsko-Pomorskie (17,7)	Közép-Dunántúl (17,2)
			Warminsko-Mazurskie (20,8)	Nyugat-Dunántúl (19)
			Centro (PT) (12,6)	Dél-Dunántúl (17,5)
			Západné Slovensko (19,8)	Észak-Magyarország (17,2)
			Stredné Slovensko (21,4)	Malta (18,8)
			Východné Slovensko (17,4)	Burgenland (AT) (22,4)
				Lubuskie (19,6)
				Algarve (16,6)
				Alentejo (14)
				Açores (PT) (10,6)
				Madeira (PT) (13,8)

Source: the indicator is indexed by the EU25 average and the respective values are given in parenthesis. Eurostat, Labour Force Survey.

the private business sector and the government sector are not available, some imperfect proxies such as R&D intensities could be applied. This measure, along with the regional endowment of knowledge intensive services, labour mobility, population density and communication infrastructure, enables ideas to be carried from one individual to another (Bessant and Rush 1995; Greunz 2003; Gadray *et al.* 1995).

Globally, the EU15 member states are performing much better in terms of total R&D investments (Table 3.2). Finland, Sweden, Germany, Austria and France also belong to the group of leading countries in terms of R&D intensities. Sweden is clearly Europe's best performing country with an R&D intensity about twice as high as the European average.

For new member states, the EU10 group of countries, R&D intensities are still relatively low despite strong positive tendencies during the considered period. An exception is the Czech Republic and smaller states such as Estonia and Slovenia, which have significantly caught up. Currently, Slovenia and the Czech Republic outperform some EU15 countries, mainly France, Spain, Ireland and Italy in terms of R&D intensity (as in 2007).

Table 3.2 R&D indicators for European regions[a]

EU state/ member country	*Share of total R&D investment*	*Share of R&D investment*			*Knowledge (% of total employment)*
		Higher education	*Business*	*Government*	
Belgium	1.9	0.4	1.32	0.15	38.24
Germany	2.53	0.41	1.77	0.35	34.79
Ireland	1.29	0.35	0.85	0.09	35.48
Greece	0.58	0.29	0.16	0.12	25.07
Spain	1.27	0.33	0.71	0.22	–
France	2.07	0.4	1.31	0.34	36.89
Italy	1.18	0.36	0.61	0.17	30.67
Luxembourg	1.58	0.05	1.32	0.21	43.02
Netherlands	1.81	0.63	0.96	0.22	42.71
Austria	2.52	0.6	1.78	0.14	30.0
Portugal	1.17	0.35	0.6	0.11	23.51
Finland	3.47	0.65	2.51	0.29	40.73
Sweden	3.4	0.75	2.47	0.17	47.83
United Kingdom	1.78	0.46	1.11	0.16	42.85
Czech Republic	1.54	0.26	0.95	0.32	25.66
Estonia	1.1	0.46	0.52	0.1	27.84

EU state/ member country	*Share of total R&D investment*	*Share of R&D investment*			*Knowledge (% of total employment)*
		Higher education	*Business*	*Government*	
Cyprus	0.44	0.2	0.1	0.11	29.2
Latvia	0.59	0.26	0.19	0.14	24.72
Lithuania	0.81	0.41	0.23	0.17	25.97
Hungary	0.97	0.23	0.49	0.23	28.2
Malta	0.58	0.18	0.38	0.02	32.82
Poland	0.57	0.19	0.17	0.20	24.81
Slovenia	1.45	0.23	0.87	0.35	26.27
Slovakia	0.46	0.11	0.18	0.16	24.74

a R&D intensity is defined as the ratio of R&D investment over GDP.

Source: own calculations based on data from Eurostat – REGIO. Knowledge includes the high- and medium-high-tech manufacturing sectors and knowledge-intensive business services (KIBS).

In general, the business sector realizes about two-thirds of total R&D spending on average in Europe. Relatively high R&D expenditure by the business sector occurs in only a few EU member states – Finland, Sweden, Germany, Austria and France. New member states along with some Southern EU15 states have relatively lower performance in business R&D. This fact suggests that for new member states public R&D can still play an important role in the field of knowledge transmission. Government R&D expenditures as a percentage of total R&D is particularly important for Poland.

As far as share of knowledge intensive sectors is concerned, EU15 states have relatively higher shares in this sector for their total employment; however, EU10 states are catching up. These shares are especially high for some smaller EU10 states, such as Malta, Cyprus, Estonia and Slovenia.

Complementary to the positioning of European countries, it is important to answer the question, whether the transition towards a knowledge-based economy accentuates dispersion or concentration tendencies. As far as total R&D intensity is concerned, the figures above show evidence that R&D intensities tend to concentrate geographically, mainly in big metropolitan regions, where they can benefit from the economies of agglomeration and urbanization.

Sweden and Finland are clearly the best performing countries, despite their strong regional differences. Figures 3.1 and 3.2 show that the smallest regional disparities with a relatively high country average are obtained by Belgian regions. German, Austrian and Dutch regions, despite their high country averages, reveal big regional differences. Among the EU10 regions the best country average is achieved by Czech and Hungarian regions, which, however, show huge regional disparities. The smallest difference among R&D intensity was obtained by Slovakia, whose average R&D intensity is relatively weak.

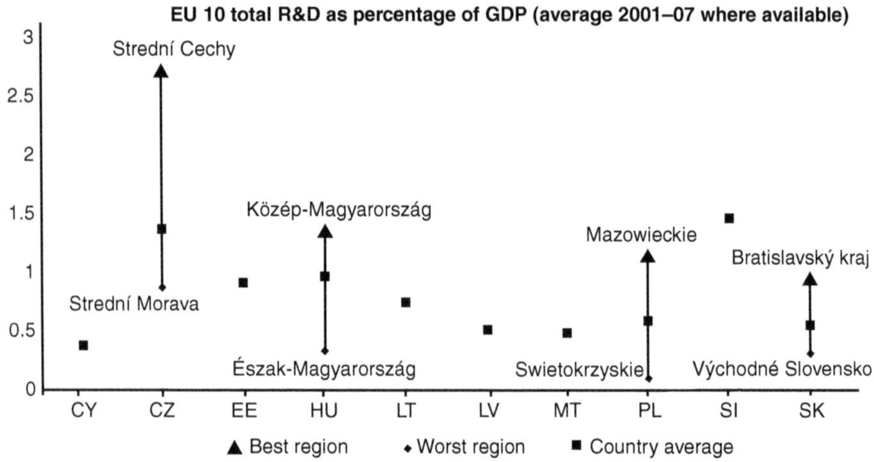

Figure 3.1 EU10 total R&D as percentage of GDP (average 2001–07).

Source: own calculations based on data from Eurostat – REGIO.

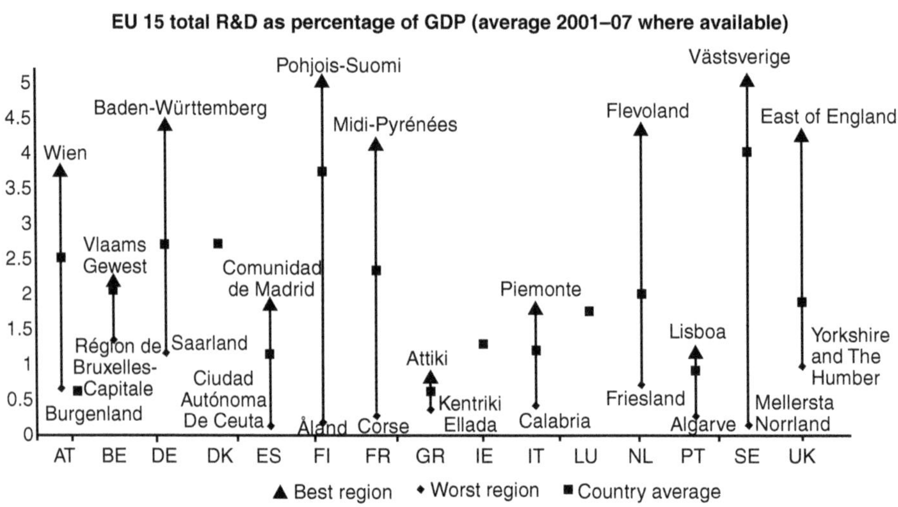

Figure 3.2 EU15 total R&D as percentage of GDP (average 2001–07).

Legend: EU15 = without Danish and Greece regions. Minimum values are indicated by a quadrangle, maximum values by a star and the national average by a dash.

Source: own calculations based on data from Eurostat – REGIO.

5 Framework and model

According to Griliches (1979), innovative input is best reflected by new knowledge, which is primarily embodied in R&D efforts. However, the present model extends beyond them and combines a number of additional variables in the model.

The model estimates a cross-sectional knowledge production model of over 186 geographic regions in the EU in the period 1999–2009.

R&D output = f (R&D input) (1)

The model adopts a general version of the Cobb-Douglas production function, which does not impose any restriction regarding returns to scale, where *a* is a constant while *b* measures the elasticity of knowledge output P with respect to input.

R&D output = a (R&D input)b expε (2)

Taking the natural logarithm of each side leads to equation (3):

ln (R&D output) = ln a + b ln (R&D input) + ε(3)

To compare to what extent business and university R&D explain granted patents and innovations in the EU regions, the following model is estimated:

ln (P) = ln a1 + a2 ln (BR&D) + a3 ln (UR&D) + a4 ln (HQ) + a5 ln (AHQ) + a6 ln (HKIS) + a7 ln (GDP per capita) +ε1 (4)

Following Jaffe (1989), the potential interaction between university and business R&D is captured by extending the base equation (4) to capture the effects of business and university R&D expenditures, the qualification level of the working age population, number of students at the second stage of tertiary education on parent activity. The structure of the knowledge productive system is proxy by the share of the high and medium-high technology industry in total employment. Similarly, it is assumed, based on the regional patent data, that there is a strong relationship between the number of patent applications per capita and GDP per capita, that is richer regions invest more in patenting activity. As far as innovative output is concerned, it is proxy by patent applications to the European Patent Office (EPO), which is the only available innovation measure at the EU25 NUTS2 (Nomenclature of Units for Territorial Statistics) regional level. Even if patent data do not perfectly reflect innovations (Griliches 1979) there is a strong link between patents and inventions (Guellec and van Pottelsberghe 1999; Acs *et al.* 2002).

One of the problems related to the econometric estimation of the knowledge production function is the simultaneity of dependent and independent variables. For example, not all independent variables in the model are independent from other variables, such as previously university–business R&D synergies.

In order to reduce the possible inconsistency of the model the following system of equations is to be estimated:

ln (P) = ln a1 + a2 ln (BR&Dt) + a3 ln (UR&Dt) + a4 ln (HQ) + a5 ln (AHQ) + a6 ln (KIS) + a7 ln (GDP per capita) +ε1 (4)

ln (BR&D) = ln b1 + b2 ln (UR&D t) + b3 ln (GR&D t) +b4 ln (S) + b5 ln (HQ) + b6 ln (AHQ) + ε2 (5)

ln (UR&D) = ln c1 + c2 ln (BR&D t)+ c3 ln (GR&D t) + c4 ln (HQ) + c5 ln (AHQ) + ε3 (6)

Where:

P = patent applications to the EPO by priority year at the regional level per million inhabitants (1999–2009);

BR&D = total intramural business R&D expenditure at the regional level as per cent of GDP (1997–2007);

UR&D = total intramural university R&D expenditure at the regional level as per cent of GDP (1997–2007);

GR&D = total intramural government R&D expenditure at the regional level as per cent of GDP (1997–2007);

HQ = students at ISCED levels 5–6 – as per cent of all pupils and students at regional level (1997–2007);

AHQ per cent of students at the second stage of tertiary education leading to an advanced research qualification – level 6 (ISCED 1997) 1997–2007;

GDP per capita a variable measuring the economic conditions of the regions (1999–2008);

KIS (knowledge intensive services) as a per cent share of total services (1999–2008);

S employment in high and medium high-technology as a per cent of total manufacturing (DG, DK, DL, DM34, DM35.2, DM35.3, DM35.4, DM35.5) in thousands (Source Eurostat – REGIO).

The simultaneous system of equations has been estimated using different econometric methods. Unfortunately, with such a large number of EU regions it is almost impossible to have complete data. Missing data are important to consider, because they may lead to substantial biases in analyses. Before going on to model the study, the following missing value analysis which contains the univariate statistics and the corresponding number of missing cases for all the variables has been done.

In general, it can be assumed that up to 20 per cent missing data shows the good fit of the regression. In the present model for variable lnP there is 12 per cent missing data, which is acceptable. On average, data gaps are estimated at a little more than 22 per cent, which is acceptable although some data would need to be supplemented. That concerns especially lnBRD, lnGRD and lnURD (for Missing Values Analysis results see Annex). For most regions, official R&D data are available for at least the beginning year, the middle year and the end year of the period 1997–2007.

Considering the fact that the innovative process does not happen instantly, the regressions must allow for different time lags applied between the dependent variable (innovation output) and the explantory variable (innovation input).

6. Estimates of the simultaneous system of equations

Globally, the results of the simultaneous system of equations indicate a high similarity between the estimates obtained by the different methods. However, if the estimated coefficients obtained by the two estimation methods are statistically similar, it is relatively difficult to choose a particular one. Klein (1974: 150) and Gujarati (1995: 679) argue that 'single equation methods, in the context of a simultaneous system, may be less sensitive to specification errors in the sense that those parts of the system that are correctly specified may not be affected appreciably by errors in specification in another part', and adopt the 3SLS method.

Table 3.3 Results of the simultaneous system of equations for 186 EU regions: simultaneous equation estimates with 3SLS

Variables	*Without fixed effects*		*With fixed effects*	
	Coeff.	*T-stat.*	*Coeff.*	*T-stat.*
Log patents				
logbusrd_lag3	0.583	(11.14)***	−0.004	(−0.04)
logunird_lag3	0.715	(8.94)***	0.041	(0.27)
Loghq	−0.795	(−5.58)***	1.191	(2.97)***
Loghkis	0.996	(8.68)***	-0.171	(−1.14)
loggdppc	0.649	(6.41)***	-0.576	(−2.53)**
_cons	−2.408	(−2.04)**	0.062	(2.39)**
Log business R&D				
logunird_lag3	0.583	(8.53)***	0.205	(3.41)***
loggovrd_lag3	−0.020	(-0.48)	0.032	(0.83)
htechlshare	23.282	(9.45)***	2.172	(1.22)
Loghq	−0.737	(−5.50)***	0.046	(0.33)
_cons	1.246	(2.68)***	0.070	(6.88)***
Log university R&D				
logbusrd_lag3	0.155	(6.90)***	0.046	(1.27)
loggovrd_lag3	0.082	(3.63)***	0.013	(0.39)
htechlshare	−2.728	(−1.84)*	2.322	(1.48)
Loghq	0.809	(11.59)***	0.189	(1.56)
lpv_growth	−0.060	(−7.22)***	0.008	(1.74)*
_cons	−2.893	(−12.79)***	0.050	(5.43)***
N	488.000	488.000		
r2	0.705	0.020		

Notes: The estimates used the Stata command 'reg3'. For fixed effects estimations, their deviations from the mean values within regions were calculated from original variables (i.e. the fixed effects estimation). Heteroskedasticity-consistent t-statistics in brackets.

* significant at 10% ** significant at 5% *** significant at 1%

The estimations are performed by means of a panel data estimation method that allows for random and fixed effects and assumes a lag structure for business, university and government R&D expenditure at the regional levels. The estimation results for the entire sample of EU25 states and EU10 states' regions are presented in Tables 3.3 and 3.4.

Table 3.4 Results of the simultaneous system of equations with 3SLS: only EU10 regions

Variables	*Without fixed effects*		*With fixed effects*	
	Coeff.	*T-stat.*	*Coeff.*	*T-stat.*
Log patents				
logbusrd_lag3	0.318	(2.04)**	0.176	(0.54)
logunird_lag3	0.439	(2.54)**	0.489	(1.37)
loghq	−0.610	(−1.51)	2.054	(1.86)*
loghkis	1.014	(2.55)**	−0.566	(−1.12)
loggdppc	0.619	(1.37)	−0.766	(−1.15)
_cons	−3.613	(−0.98)	0.200	(2.03)**
Log business R&D				
logunird_lag3	−0.102	(−0.75)	−0.037	(−0.27)
loggovrd_lag3	0.361	(2.98)***	0.124	(1.34)
htechlshare	7.589	(1.58)	2.687	(0.69)
loghq	−0.380	(−1.17)	0.283	(1.01)
_cons	0.114	(0.10)	0.026	(0.85)
Log university R&D				
logbusrd_lag3	−0.125	(−1.72)*	0.072	(0.87)
loggovrd_lag3	0.170	(1.96)*	−0.116	(−1.91)*
htechlshare	3.304	(0.99)	−3.390	(−1.34)
loghq	0.565	(2.56)**	0.286	(1.79)*
lpv_growth	−0.059	(−3.61)***	0.000	(0.05)
_cons	−2.964	(−3.98)***	0.097	(5.02)***
N	93.000	93.000		
r2	0.396	0.145		

Note: The estimates used the Stata command 'reg3'. For fixed effects estimations, their deviations from the mean values within region were calculated from original variables (i.e. the fixed effects estimation). Heteroskedasticity-consistent t-statistics in brackets.

* significant at 10%   ** significant at 5%   *** significant at 1%

As far as the patenting equation is concerned, it is positively and significantly influenced by university and business R&D expenditure (with the best fitting three-year time lag), share of tertiary education students (ISCED levels 5–6), knowledge intensive services (measured as a share of total services) and overall economic conditions of the regions (measured as GDP per capita).

The elasticity of patenting activity for the EU25 regions with respect to university R&D expenditure is relatively significant and important. Roughly, a one percentage increase of university R&D investment per capita generates a 0.72 percentage increase of patent applications per capita with a three-year time lag (with 99 per cent significance), if estimated without fixed effects. In case of the estimates for the EU10 regions, a one percentage increase of university R&D investment per capita generates a 0.44 percentage increase of patent applications per capita with a three-year time lag (with 95 per cent significance).

The elasticity of patenting activity for the EU25 regions with respect to business R&D expenditures is relatively significant and important. Broadly, a one percentage increase of business R&D investment per capita generates a 0.58 percentage increase of patent applications per capita with the best fitting time lag of 3 years (with a 99 per cent of significance). In case of the estimates for the EU10 regions, a one percentage increase of business R&D investment per capita generates a 0.32 percentage increase if introducing the best fitting time lag of three years (with 95 per cent significance).

7 The knowledge production model results for business R&D efforts

The estimates indicate that universities' R&D expenditure has a significant and positive impact on private business R&D efforts (both if the fixed and random effects models are considered). A 1 per cent increase of university R&D expenditure generates a 0.58 per cent increase in business R&D expenditure with a three-year lag. However, the estimates indicate that university R&D efforts are a negative and insignificant function of business R&D efforts for EU10 regions with all time lags considered. This could be explained by the insufficiently developed or ineffective transfer of technology and knowledge from universities to the private business sector in the above-mentioned regions. The causes of this may lie in there being only a few or no industrial incubators and industrial parks, as well as a low level of formal and informal knowledge flows (R&D agreements, faculty consulting in industry, etc.).

As far as the government R&D impact on private business R&D efforts for the entire sample of EU regions it appears to have negative and insignificant influence on private business R&D. However, the role of government R&D efforts seem to be very important for EU10 regions. In fact, a 1 per cent increase of government R&D expenditure generates a 0.36 per cent increase in business R&D expenditures with a three-year lag (with 99 per cent significance). The latter results show that public expenditure for research is a critical factor of innovation activity in the above EU regions, especially if low levels of business and university R&D expenditure are considered.

It is worth noting that the private business R&D activities are positively driven by high-tech and medium-high manufacturing employment (S). Compared to the statistically positive effects of model 3SLS obtained for the entire sample, the increase of high-tech and medium-high manufacturing (S) by 1 per cent generates an increase of business R&D of 23 per cent. The results for the EU10 regions turned out to be positive but insignificant. This could be explained by the relatively smaller employment concentration in high-tech and medium-high tech in the EU10 regions, so the technological and research opportunities of these regions are weaker.

7.1 The knowledge production model results for the university R&D efforts

The estimates of equation clearly indicate a significant and positive interaction between university R&D and business R&D (except for EU10 regions). Generally, a 1 per cent increase of business R&D expenditure generates a 0.15 per cent increase in university R&D in the entire group of EU regions (EU25) (if a three-year time lag is considered). This result is lower than that obtained by Feldman and Florida (1994) for the US case (0.22).

The model estimates for the EU10 regions indicate that an increase in business R&D does not positively influence university R&D. The results could be explained by the fact that the innovation system is not yet in place in these regions, especially in terms of linkages between the industry sector and universities, as well as relatively higher dependence of university R&D in these regions on government R&D funds. And it might take longer before the performance of a transforming regional innovation system of the EU10 regions is comparable to that of the more successful EU15 regions.

Government R&D also contributes to fostering university R&D. In fact for EU25 regions, an increase of government R&D expenditure by 1 per cent causes a 0.08 per cent increase of university R&D with a three-year time lag (with 99 per cent significance). This outcome looks a bit better for EU10 regions, with an increase of 0.17 per cent in five years (with 90 per cent significance).

7.2 The role of education (human qualifications, HQ) in knowledge production

It is generally assumed that education positively influences patent activity in regions. However, the results of the estimations run for the overall sample of the EU regions show that there is a negative relationship between the number of tertiary education students and patent applications. More specifically, a percentage increase in the number of tertiary education students decreases the number of patent applications by 0.79 per cent.[2] Similarly, the same relationship tested on the EU10 regions turned out be negative but insignificant.

This result is somewhat surprising and should be interpreted with caution. It suggests there could be negative or not significant influence of the formation of human capital (which is a major part of the lifelong learning process) on R&D efforts. Furthermore, the results for EU25 regions show a significant but negative relationship between tertiary education (HQ) and private business R&D efforts. A percentage increase in the number of tertiary education students in the entire group of EU regions generates a decrease in business R&D of 0.74 per cent.

The-above mentioned model outcome could be explained by the fact that innovation and R&D activities usually involve people with advanced research skills instead of tertiary qualification level graduates. A similar reasoning applies to the EU10 regions; however, the results turned up to be insignificant for this group of regions.

The number of students with tertiary education (HQ) significantly and positively influences university R&D expenditure. For the entire group of the EU regions a 1 per cent increase in tertiary education students (HQ) generates university R&D growth of 0.083 per cent, whereas the respective number caused by an increase in advanced research capital is 0.81 per cent. The influence of the number of students in tertiary education on the university R&D efforts seemed to be less pronounced for EU10 regions. A 1 per cent increase in high qualifications (advanced human qualifications, AHQ) generates a 0.56 per cent increase of university R&D. The results confirm that student qualification is a prerequisite for basic research mainly realized by universities.

Summing up the main findings of this section, in a nutshell, the patenting activity of European regions depends on both business and university R&D. Since university R&D is essentially concerned with the generation of basic knowledge, its direct impact on patent applications is lower compared to business R&D, which is more market oriented. The relation between university R&D and business R&D is characterized by feedback relations. University R&D and business R&D positively influence each other through various channels of knowledge transmission. University research feeds private business research, while private business research, which is also fuelled by diverse types of expertise coming from customers and supplies, feeds university research in return. While university R&D efforts depend on the region's endowment of tertiary education graduates, the business R&D efforts may require graduates with more specific and high qualifications.

8 Summary and conclusions

The knowledge economy, driven primarily by technological advancement and competitive pressures, increases the demand for knowledge-based services and knowledge intensive sectors. The development of these knowledge-based sectors contributes further to patenting and new knowledge creation activity. Regions with high levels of social capital tend to perform better in terms of creating new knowledge.

The chapter highlighted the low innovation capacities of EU10 regions and clearly indicated their weak performances in terms of university and private business R&D and their impact on patenting activity, despite the fact that these regions are not necessarily disadvantaged in terms of highly qualified human capital compared to the EU15 regions. In the context of EU10 regions, the number of students with tertiary education was shown to be insignificant and have a negative influence on R&D efforts. This result suggests the possibility of competitive activities and lack of balance between research and teaching activities in this group of regional economies, whereas advanced research qualifications seem to have a strong influence on the patenting activity of the EU10 regions. Similarly, the employment in high-tech and medium-high-tech industries seems to be positively and significantly related with the regional innovative capacity of the EU.

While the outcomes of any model should be interpreted with caution and put back into the national or regional context the above-mentioned results suggest that public policies should focus on ensuring a high quality of tertiary education and enhancing opportunities for advanced research programmes (which could also attract foreign talent); enhancing collaborative research between business and universities; creating a friendly business climate for the high-tech companies via attracting foreign investors; and more generally investing in social capital to improve regional absorption capacity and the efficiency of the region's innovation system.

Notes

1 EU25 refers to all the EU current members regions except for the two latest member states: Bulgaria and Romania. The regions are analysed on NUTS2 regional level, with a few exceptions: Ireland, Denmark and Germany, for which data were available only on the national and NUTS1 level, respectively.

2 Unfortunately, the results for EU10 regions show no statistical significance.

References

Abramovitz, M. (1986) 'Catching up, forging ahead and falling behind', *Journal of Economic History*, 46, 385–406.

Acs, Z. J., Anselin, L. and Varga, A. (2002) 'Patents and innovation counts as measures of regional production of new knowledge', *Research Policy*, 31(7), pp. 1069–1085.

Anselin, L., Varga, A. and Acs, Z. (1997) 'Local geographic spillovers between university research and high technology innovations', *Journal of Urban Economics*, 42, 422–448.

Bessant, J. and Rush, H. (1995) 'Building bridges for innovation: The role of consultants in technology transfer', *Research Policy*, 24, 97–114.

Capron, H. (2002) 'Transition towards the Knowledge-based Economy: Growth potential and learning regions', *Belgian Report on Science, Technology and Innovation, Volume II*. Belgian Federal Office of Scientific, Technical and Cultural Affairs.

Dunnewijk, Th., Hollanders, H. and Wintjes, R. (2008) 'Benchmarking regions in the enlarged Europe: Diversity in knowledge potential and policy options', In: *Innovation Policy in Europe; Measurement and strategy*, Nauwelaers, C. and Wintjes, R. (eds.), Edgar Elgar.

Feldman, M. and Florida, R. (1994) 'The geographic sources of innovation: Technological infrastructure and product innovation in the United States', *Annals of the Association of American Geographers*, 84, 210–229.

Gadrey, J, Gallouj, F. and Weinstein, O. (1995) 'New modes of innovation: How services benefit industry', *International Journal of Service Industry Management*, 6(3), 4–16.

Greunz L. (2003) 'The technology gap and European regional growth dynamics', ULB Institutional Repository 2013/9485, Université Libre de Bruxelles.

Griliches, Z. (1979) 'Issues in assessing the contribution of research and development to productivity growth', *Bell Journal of Economics*, 10, 92–116.

Grimpe, Ch. and Patuelli, R. (2011) 'Regional knowledge production in nanomaterials: A spatial filtering approach', *The Annals of Regional Science*, Springer, 46(3), 519–541.

Guellec, D. and van Pottelsberghe De la Potterie, B. (1999) 'Does government support stimulate private R&D?' *OECD Economic Studies*, 29, pp. 95–122.

Guellec, D. and van Pottelsberghe, B. (2001) 'The internationalisation of technology analysed with patent data', *Research Policy*, 30(8), 1253–1266.

Gujarati, D. (1995) *Basic Econometrics*. 3rd edn. New York: McGraw-Hill International Editions.

Jaffe, A. (1989) 'Real effects of academic research', *The American Economic Review*, 79, 957–970.

Klein, L.J. (1974) *A Textbook of Econometrics* (Second Edition), Englewood Cliffs: Prentice-Hall.

Kline, S. and Rosenberg, N. (1987) 'An overview of innovation', in R. Landau and N. Rosenberg (eds), *The Positive Sum Strategy: Harnessing technology for economic growth*. Washington DC: National Academy Press.

Lal, K. and Dunnewijk, Th. (2008) 'Enterpreneurship and Innovation Strategies in ICT SMEs in Enlarged Europe (EU 25)', UNU-MERIT Working Paper Series No. 2008–016.

4 Universities as knowledge nodes in open innovation systems

More than just knowledge providers

Suntje Schmidt

1 Introduction[1]

Unquestionably, universities as public research institutions play an important role in innovation systems. But their role and function in these systems seems not always very clearly defined. They are often at the forefront of or beyond state-of-the-art research and development. Furthermore, they educate young academics who enter labor markets as highly qualified scientists, entrepreneurs and/or employees. Additionally, universities are organizations where highly educated people acquire competencies that enable them to transfer innovative ideas into innovative products and services. Therefore, universities may be considered as condensation points for innovative ideas that might be manifested in patents and publications.

But, if universities are considered to be knowledge nodes in innovation systems, the characteristics of knowledge need to be more highlighted. Knowledge may only partially be externalized. In fact, it is quite difficult to transfer knowledge to an explicit form. Following authors such as Nico Stehr (2001) or Georg Krogh and Marija Köhne (1998), knowledge comprises all experiences, expertise, and skills to solve diverse tasks. Knowledge thus enables people to interpret information and decide for actions according to their interpretation. Knowledge hence has no fixed boundaries, is difficult to clearly describe and is very dynamic as it is constantly altered, updated, revised, and adjusted. One of the central ways to share and transfer knowledge is communication, either oral, written, or with other forms of symbolic encryption. But, taking the above notion on knowledge seriously, knowledge also manifests itself in actions, decisions, products, services, and other forms of processes and outputs. In other words, the way people act also externalizes some forms of their knowledge. In modern economies, knowledge does not only become visible through innovative products or services. In fact, innovations are rather the final outcome of knowledge intensive work. Therefore, when discussing the role of universities in innovation systems, we need to look at knowledge work, knowledge generating processes and knowledge sharing mechanisms too.

Last but not least, the spatial dimension of knowledge work for innovations also needs to be considered. In literature we can find broad debates on the spatial implications of knowledge-intensive works in both the economic as well as academic worlds. Still, one may find proof for both: a very local dimension of knowledge diffusion and no limitations for the spatial diffusion of knowledge. Universities, especially, are characterized by widespread knowledge-generating networks, which is why sometimes universities are considered as lighthouses. They recognize global trends and identify the innovative potential of new knowledge (possibly even generated elsewhere) and support the regional utilization of that knowledge for regional economic development. The spatial complexity of knowledge work can be tracked in innovation system debates too as regional innovation systems are increasingly linked to national and international innovation systems. Furthermore, innovation systems are considered to be open, which indicates the possibilities and necessities for external links for innovations systems to remain dynamic.

This chapter seeks to contribute to the debates on the role of universities for the cities and regions in which they are located. With this objective, the chapter will first briefly review some debates on innovation systems (section 2) and the spatial implications of knowledge work (section 3). Empirically, the chapter draws on semi-structured interviews conducted with entrepreneurs of the biotechnology sector in Brandenburg city regions. The interviews investigate the interviewees' descriptions of different forms of cooperation with universities. Finally, the chapter also argues that the debate on the role of universities benefits from considering knowledge generating mechanisms.

2 Public research institutions in (regional) innovation systems

2.1 Systemic generation of innovations

The innovation system approach underlines innovations as being systemically and systematically generated (Moodysson 2007: 40). Within this context the literature on innovation systems seeks to identify policy frameworks for supporting localized learning processes (Asheim 2004). Caspers (2004: 68) identifies an efficient innovation system as one of four columns of the knowledge economy (the other three being incentives of economic, political, and social institutions for utilizing knowledge, creative and well-educated researchers and entrepreneurs, as well as modern information infrastructures). The systemic approach investigates the interaction between economic agents, companies, research institutions, and the public sector (Lorentzen 2007: 468). Even though different types of innovation systems may be identified (e.g. regional, national, and global innovation systems), what combines them is the attempt to identify important stakeholders as well as the forms and institutional arrangements for cooperating in innovation processes. Such stakeholders may be for instance "buyers, sellers, suppliers, local and national authorities and intermediate organizations, such as universities and R&D institutes" (Andersson and Karlsson 2004: 4). In order for these stakeholders to form and participate in an innovation system, they need to be able to interact with each other. In fact, the interaction is considered the key in innovation, because interacting enables actors to benefit from tacit knowledge (Polanyi 1967) of other entities in the innovation system (Lorentzen 2007: 468). Despite the systemic approach seeming to suggest different forms of formalized interaction (e.g. cooperation agreements, joint projects), Feldman and Massard (2002: 7) point out that the connections in an innovation system largely involve non-market relationships and interactions between agents and include many different means of knowledge transfer. The need for personal interaction, however, and the tacit dimension of innovation-relevant knowledge are often considered to give innovation systems a rather regional dimension.

The key characteristics of innovation systems as roughly sketched out above may be considered as institutional regional innovation systems (in contrast to entrepreneurial regional innovation systems) (Asheim and Gertler 2005: 303). Asheim and Gertler (ibid.: 304) point out that this kind of innovation systems seems to be more typical of German regions and Nordic countries, because these industries draw more on an analytical knowledge base.

The spatial boundaries of innovation systems, however, are also supposed to be permeable, because linkages to national and international innovation systems prevent regional ones from becoming too path-dependent and from running the danger of lock-in effects (Figure 4.1) (Asheim 2004: 404; Fritsch 2009: 11).

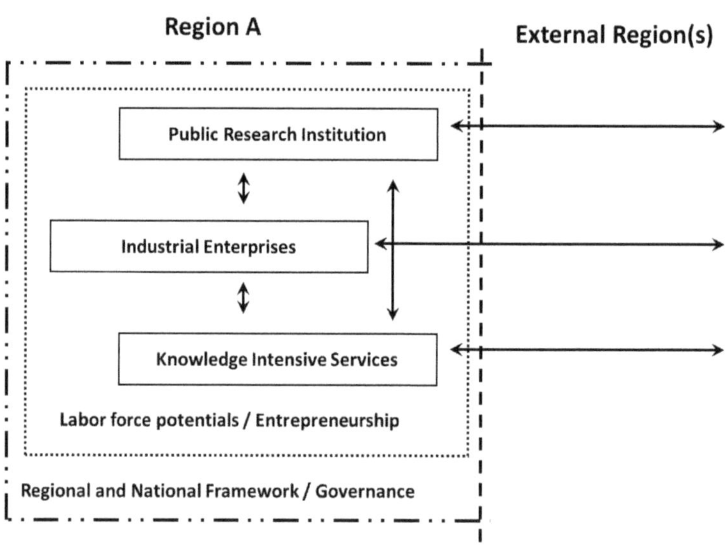

Figure 4.1 Open innovation systems: schematic illustration.

Source: Fritsch (2009: 11), adjusted.

2.2 Universities and other public research institutions in innovation systems

Within innovation systems, universities and other public research institutions (e.g. colleges of applied sciences) might take on different roles. Through the research undertaken at universities, new knowledge is created (Fritsch 2009: 11), especially in knowledge fields not or insufficiently covered by the economic sector. This goes particularly for basic research that is a fundamental precondition for applied research. But, knowledge creation at public research organizations also includes research functions in the form of academic expertise or providing dedicated laboratories for private economic uses.

Public research institutions also accumulate knowledge (ibid.: 12). They accumulate knowledge that was created at the institution, but they also gain access to knowledge generated elsewhere. One might compare public research institutions with an antenna that registers external new knowledge and brings it to the organization. With this function they become a regional node, because public research organizations might offer a bridge toward knowledge generated in other regions.

Furthermore, research organizations such as universities are places of knowledge transfer (ibid.: 12). Students and scientists are educated at universities and gain knowledge, as well as competencies that they integrate in regional, national, or even international labor markets later on. Research projects, staff exchange and conferences further support the knowledge transfer function.

Looking at the "output" of universities that might support regional economic development, one can differentiate between visible and less visible forms. Through research and development, universities register patents that may be the basis for an application in a new product or service. In some cases, this takes place within an existing company, but it is also possible that university spin-offs transfer the basic research to application. Also prototypes might be used for new products or services (e.g. Mowery and Sampat 2005: 212). Universities also externalize generated knowledge with publications and presentations that only have indirect effect for the regions they are located in.

Less visible, but nonetheless with great potential of developing a regional effect are different forms of diffusing implicit knowledge – knowledge that is bound to knowledge carriers and in many cases is difficulty to be transferred to an externalized form. As already mentioned, academically trained graduates enter the labor market. Joint cooperations with other research organizations, but also with enterprises, facilitate joint learning opportunities for all involved actors, either by cooperatively developing new knowledge or by staff exchange among the participating entities.

These latter examples already indicate that universities are not only knowledge providers or knowledge generators. In fact, they function more like a node with the capability of linking diverse knowledge sources. Its boundaries in most cases are not closed, because people interact and relate action to each other. Furthermore, the above discussion also suggests that the impact of universities being a knowledge node does not exclusively unfold a regional dimension, but rather spatially widespread effects. This becomes even clearer when taking a closer look at mechanisms for knowledge transfer and knowledge diffusion.

3 Public research institutions in knowledge generating settings

3.1 Notions on knowledge and space

Knowledge comprises all cognitive and physical skills, competencies and expertise that individuals use to solve tasks and problems, to act in different environments, and to interpret information and data (Krogh and Köhne 1998: 236). Those forms of knowledge that are transferred to a visible form can be considered as explicit knowledge (Polanyi 1958, 1967). Implicit knowledge, in contrast, comprises knowledge bound to individuals (ibid.) and includes cognitive knowledge based on experiences, training, intuition, and emotions and expectations, but also skills and capabilities that can be activated in different contexts (Nonaka 1994; Nonaka and Takeuchi 1995).

Implicit knowledge cannot fully be transferred to explicit forms. It manifests itself in actions and decisions. Therefore it is considered to be quite distance-sensible, while explicit forms of knowledge are supposed to be transferrable over large spatial distances (Schmidt 2012: 38). However, the precondition for explicit knowledge to spread spatially lies in the recipient's capability to recognize and understand that knowledge. Explicit knowledge is a form of codification that can successfully be included in innovative activities only by those who are able to understand this form of codification and who are able to integrate the externalized knowledge in their own line of reasoning, understanding and knowledge. Successfully using patents or licenses, for instance, also implies extensive knowledge within a given field. So, utilizing these forms of knowledge does not have a spatial boundary, but rather a boundary set by language, former knowledge, and expertise within a certain field.

In contrast, transferring implicit knowledge requires personal contact between individuals (Beckert *et al.* 2008: 317). It must also be questioned whether implicit knowledge can be transferred. Following the argumentation laid out here, implicit knowledge shows when actions and decisions are related to each other, to a problem or a task. Therefore it seems to be more appropriate to consider transferring implicit knowledge as knowledge sharing practices or practices that enable participation in other individuals' knowledge (Schmidt 2012: 39). Though bound to personal interaction, implicit knowledge may still diffuse spatially, for example within institutional arrangements and communities of practices.

Transferring, exchanging, as well as sharing both explicit as well as implicit knowledge are thus related to knowledge sharing practices and mechanisms that shall shortly be considered in the next section.

3.2 Knowledge sharing practices

Asheim (2004: 406) argues that the way knowledge is shared and transferred depends on dominant modes for knowledge creation in industries and organizations. He distinguishes two modes in this context: analytic as well as synthetic knowledge generation. Analytical knowledge creation is based on scientific knowledge as well scientific logic and mechanisms for advancing existing knowledge and generating new knowledge. Access to academic expertise (a combination of field specific subject-related knowledge as well as knowledge about utilizing it (e.g. methods, methodologies, practical applications of knowledge)) influences therefore analytical knowledge generation. Synthetic knowledge generation on the other hand is accomplished by applying and combining existing knowledge. Here, starting with existing knowledge fragments, new scopes of application are generated for it. That means individual knowledge such as know-how and competences strongly influence making such connections in this form of knowledge generation.

Analytical knowledge generation is the dominating mode for creating new knowledge in biotechnology research and development services that are the empirical case in this chapter. Based on individual, scientifically based knowledge, new knowledge about cognitive, rational processes is generated. Such new knowledge is based on a deductive way for creating knowledge, based on formal models, methods, and specific, person-related skills (Strambach 2008). For biotechnology services it seems to be characteristic that external knowledge carriers are predominantly of the same epistemic community, which is supposed to be spatially distributed (e.g. Moodysson 2007, 2008).

This brings up the question of institutional arrangements for cooperative knowledge work. Because of the unequal spatial distribution of knowledge, incomplete information and uncertainties for economic activities, institutional economics' approaches assume that organizations establish control and coordination mechanisms that allow reliable access to knowledge (e.g. Hollingsworth *et al.* 1994; Williamson 1994; Richter and Furubotn 1996; Williamson 1996). Markets, hierarchies and networks are institutional arrangements commonly discussed in this context. Between these three forms there are a number of hybrid control mechanisms, such as clans and associations (Mayntz 2006: 14), competitive control (Scott and Meyer 1994: 71), strategic alliances (Sydow 1992), distribution systems, franchising and joint ventures (Wood and Parr 2005: 7). From the viewpoint of knowledge sharing practices, institutional arrangements fulfill two functions. First, they provide a secure framework for knowledge transfer and can try to limit an uncontrolled diffusion of newly generated knowledge. Second, institutions with which these arrangements are set up, formulate rules for generating and utilizing knowledge. Accordingly, projects, alliances, networks, and other institutional arrangements are established for the transfer of knowledge to achieve jointly defined objectives of such institutional arrangements.

In addition to these formal arrangements that are situated mostly in the value chain of knowledge-intensive enterprises, informal institutions also support the spatial diffusion of knowledge. Granovetter (1973), for instance, differentiates between "strong" and "weak" ties. According to Granovetter (ibid.: 1367), the strength of a connection between individuals and social groups derives from a combination of time required to maintain the connection, the intensity of the emotional bonding, mutual trust and mutual benefits. "Strong ties" primarily exist within (knowledge) communities that are constituted by solid relationship and that share a network-specific, objective "codebook" (Jonas 2006: 4). This "codebook" enables members of such knowledge communities to communicate with each other and supports the spatially wide-ranging dissemination of knowledge. As explained above, even if explicit knowledge has the potential to diffuse spatially without problems, the decodification of explicit knowledge depends on the capability of individuals to decipher the code with which the knowledge has been made explicit. According to Granovetter it is the strength of "weak ties" to enable knowledge diffusion across the boundaries of sectors and organizations based on loose, non-binding, sporadic contacts: "Weak ties are more likely to link members of different small groups than are strong ones, which tend to be concentrated within particular groups" (Granovetter 1973: 1376). In Granovetters' argumentation especially weak ties take on a prominent role for sharing and diffusing knowledge.

Granovetter has already laid the ground for identifying knowledge sharing entities that can be clearly identified as neither formal nor informal institutional arrangements. A shared language for externalizing knowledge as well as shared norm, rules, methods, and processes for creating knowledge are preconditions for cooperative knowledge generating forms. These preconditions support the formation of communities within which knowledge is shared and exchanged successfully. Here, one may distinguish, for example, communities of practice (e.g. Lave und Wenger 1991; Wenger *et al.* 2002; Grabher and Maintz 2006; Amin and Roberts 2008b) and epistemic communities (e.g. Knorr-Cetina 2002; Arena *et al.* 2006; Amin and Roberts 2008a). Communities of practices are, for instance, united by a collective meaning system in applying competencies, skills, and knowledge to specific problem solutions. Epistemic communities, in contrast, share cognitive norms and traditions that manifest themselves in a shared scientific language and research rules.

3.3 Public research institutions as knowledge nodes

The argumentation laid out above helps to consider universities as knowledge nodes. As nodes, universities are not mere knowledge providers as is sometimes assumed in innovation models. Rather, universities generate, collect, store, adjust, update, transform, and reformulate knowledge. Instead of being the lead producer of new knowledge, universities are thus also knowledge consumers. Doing so, universities are members in formal and informal institutional arrangements. At the same time, academic staff members are members in knowledge generating communities (Fig. 4.2).

First and foremost, universities are interlinked with other universities and research institutions. Of course there are several ways to do so, the most important one being forms of cooperation in research projects. At the same time, staff exchange supports the spatial diffusion of implicit knowledge as through the exchange the knowledge carrier may plant his knowledge and competencies within another organization. It is important to note linkages among academic partners are supported by a shared academic language—at least as long as partners work within the same or similar academic field. Language in this context is more than mere speech—it is also a knowledge generating routines and practices that are similar among academic partners. Therefore, these partners may easily tune in to partner routines and their line of thinking without too much effort to avoid friction, as is the case for instance in epistemic communities.

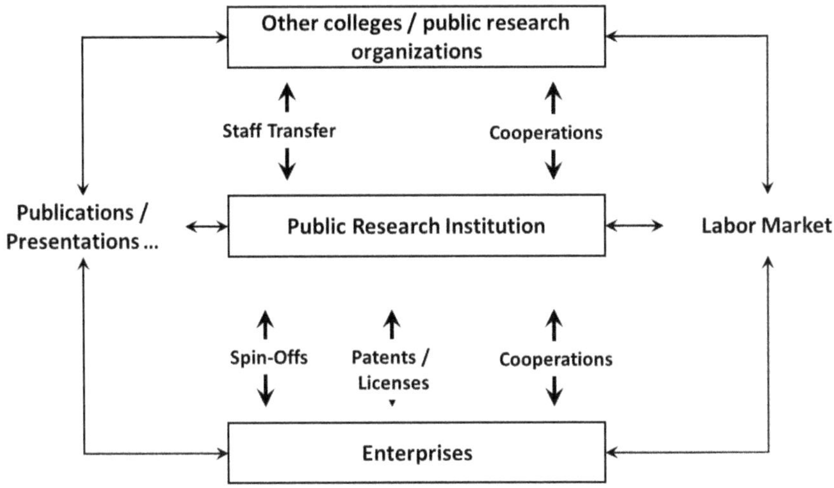

Figure 4.2 Public research institutions in innovation systems.

Source: Fritsch (2009: 15), adjusted.

This is an important notion since forms of collaboration among the academic and economic sectors do not share these similarities. Instead, working and knowledge generating routines between academic and economic organizations differ quite considerably. Forms for knowledge transfer and joint knowledge generation widely discussed not only in academic literature, but also in policy making, are spin-offs, knowledge transfer through licensing and patenting, and again cooperations. Again, such cooperations might take on different forms of institutional arrangements, such as networks and joint research and development projects. In case of spin-offs and knowledge transfer through patenting and licensing, knowledge born at a university is being translated from basic research to applied research or even to new products or services. This means, one stage of knowledge generation is finalized, for instance, at the university and then transferred to the market, as implied by linear innovation models. In contrast, joint cooperation among the academic and economic sectors implicates an ongoing, shared form of knowledge generation. In these constellations, academic and economic actors need to identify and use shared practices to work together, to generate knowledge and to jointly participate in knowledge generating practices. But, the learning effect takes place for all participating entities. Academic partners actively participate in knowledge generating practices by contributing to but also by gaining from the institutional setting.

Universities as organizations for educating highly qualified future employees, entrepreneurs, or academics is also more than a one-way street. Not all graduates leave universities without upholding established ties. Especially in science-based economic sectors such as biotechnologies, contacts to former professors or teachers ensure a knowledge exchange from which both sides may benefit.

This phenomenon so far has rarely been considered in discussions on the role of universities in innovations systems. But the empirical evidence presented below suggests this to be an important mode for knowledge diffusion—with the university taking on a nodal function again.

Finally, publications and presentations coming out of academic research are presented as knowledge providing functions of universities in innovation systems. But again, thinking of universities as knowledge nodes, this function might light up in a new way too. Publications and presentations are of course a prominent way of publishing new academic findings. Collecting, storing and working with publications is not only an academic mode for knowledge generation, but also supports the function of providing access to knowledge in academic archives. However, universities take on a nodal function, because they pointedly provide access to published as well as unpublished works within their networked ties to members of knowledge communities.

4 Cases from the biotechnology sector

The following section will focus on empirical findings made in semi-structured interviews with entrepreneurs in the biotechnology sector. The companies chosen for interviews are categorized regarding the German classification of economic activities as "research and experimental development on natural sciences and Engineering" (DESTATIS 2003: 30). The number of employees subject to social insurance in this category had slightly risen in the federal state of Brandenburg between 1998 and 2008. In 2008, 0.7 percent of all employees subject to social insurances were registered in the category under investigation here. Though this proportion seems quite low, it needs to be pointed out that this percentage was still a little above the German average. One of the spatial cores in the sector is located in Luckenwalde, where most of the interviews were conducted. A technology park with an emphasis in biotechnology has been established in this location. Though there is no university located in the city, it turns out all interviewees describe different forms of linkages with universities and other research and development organizations. The following sections hence will focus on these linkages in order to add some new points to the debate on the role of universities in open innovation systems (Schmidt 2012: 202).

4.1 Universities as socialization places

All interviewees started out their career as academics: all of them carried a PhD title; some even held professorships before entering the private market. For a certain period of time, the interviewees were thus employed at a university or other research or development organization enrolled in fundamental research. With the activities carried out at these organizations, the interviewees established a specialized profile and reputation within the biotechnology community. This also means, all interview partners had been socialized in an academic environment and internalized knowledge generating practices related to this sector.

Furthermore, through their socialization in the biotechnology sector they built up a contact network of colleagues engaged in a very similar knowledge field. These colleagues share knowledge generating practices that are dominated by an analytical modus of knowledge generating processes. Based on the individual, scientifically-based knowledge, new knowledge is generated by deductive cognitive, rational processes. When working cooperatively on generating new knowledge with external entities, these entities primarily belong to the same epistemic community, as pointed out by Moodysson (2007, 2008). Therefore, from the very beginning of their career, the interviewees are embedded in quite a homogeneous epistemic community.

All interview partners have been spatially and organizationally mobile throughout their career prior to becoming self-employed entrepreneurs. The contact network grew with each station in their occupational career. This is an important notion, because as Zucker and Darby (1994: 24) point out, knowledge—especially in biotechnology—cannot be transmitted in its externalized form, but rather requires an active participation in the knowledge generating practices of those who have the knowledge. The interview partners possess a spatially spread network of knowledge carriers that they may activate for accessing external knowledge. Sporadic and informal personal contacts, for example, at conferences and meetings, phone calls, and emails are tools that help the members of the spatially distributed network to stay up to date with others' plans and achievements. Since obviously universities were at least one station in the occupational career of the interviewees, entities still enrolled at universities are potential external knowledge carriers to be activated by the interviewees.

4.2 Universities as customers

The interviewees' companies are highly specialized and very small, most of them are micro sized. These companies contribute with their products and services to just a very small part of a value chain. Main customers for their products and services are usually larger biotechnology companies, pharmaceutical groups, private as well as public research and development organizations, large laboratories, universities, and brokers (Schmidt 2012: 220). Because of the high specialization in biotechnology industries, the value chain for a certain product or service is very fragmented. That means, the value chain is not only divided in many different production units, but is also spatially fragmented. However, that spatial distance between providers and costumers does not seem to be very problematic, because providers and customers share a common scientific language, similar research and working practices, as well as a shared cognitive understanding of methodologies, interpretation of results, and knowledge generating processes. Both, product/service providers and their customers are embedded in the same institutional environment and are familiar with the legal requirements related to products and services in biotechnology industries. Communication between the providing companies and their customers therefore runs smoothly with only minimal danger of misinterpreting information. Therefore, personal meetings between customers and interviewees are not always required for providing products and services, which simplifies market relations across spatial distances.

Universities are one group of customers for small biotechnology companies and compared to large pharmaceutical groups and enterprises not the most important customer with regards to financial benefits. However, in innovation system debates, universities are seldom considered to be the paying receiver of knowledge embedded in services and products. Like other customers of the interviewees, universities are not mere receivers but, through the interaction, are connected with the service provision or production process. Furthermore, as customers, universities are also very much aware of the interviewees' companies' profiles and capacities and may function as multipliers. Being at the forefront of fundamental research, entities at the universities communicate recent trends in the field to the interviewees and identify important stakeholders in the sector. Such information may become important to the interviewees, because that kind of knowledge potentially provides a short knowledge monopole that improves their position in the competitive market. Such information is shared especially among actors that already know each other, either because they used to work together (e.g. as former colleagues) or because they have become acquainted with each other through joint projects. Therefore, the academic career before becoming an entrepreneur still unfolds positive effects even after exiting the academic field. Besides communicating recent trends and acting as important knowledge carriers in the field, entities at universities might also support the interviewees in approaching new customers.

4.3 Universities as active knowledge nodes

Universities are—as explained above—at least one station in the academic and entrepreneurial career of the interviewees. Universities are places of socialization and obviously remain places with which to keep contact. Though, these contacts are not university bound, but rather individual specific. If a former university colleague leaves his or her position at the university, this person takes along all their contacts. In that way, the nodes of sociality networks (Grabher and Maintz 2006: 6) are mobile and change positions along with the occupational career.

University members in informal epistemic communities

Informal relations with former colleagues at universities enable entrepreneurs in the biotechnology sector to access literature and in some cases even to research work not (yet) published. Professional literature is important for the interviewees for staying up to date with recent developments in the sector. However, they seldom subscribe to professional journals or buy recent publications. Rather, they rely on recommended reading by their contacts, most often provided by their customers or through their professional network. Contacts to former colleagues at universities play a key role here, because they do not just recommend publications, in many cases they provide the recommended texts too. Even though universities collect recent publications and provide access to those in libraries, the interviewees seldom use the library. If a certain publication is of interest, they rather contact peers in academic organizations who then mail the requested text to the interview partner. Hence, universities are not just knowledge nodes because they own libraries that can be accessed from the outside. Rather, they have a nodal function, because through the active social relation universities spatially distribute collected externalized knowledge.

Again, spatial distance plays a subordinate role here. In the case presented here, Berlin and Potsdam possess a rich research environment and large professional libraries. But instead of profiting from the spatial proximity to Berlin, the interviewees utilized contacts to former colleagues localized elsewhere for accessing publications rather than using literature sources close by. In fact, the request for literature might be one mode of keeping up contact with a former colleague and keeping ties vital. Furthermore, asking for literature also provides an occasion to exchange one another's recent work, which again might kick-start the exchange of further information. The university contact might then, for instance, provide grey literature and unpublished work, while the entrepreneur might offer some test results for the academic to use.

These examples illustrate that universities as knowledge providers do not just unfold positive effects for the cities they are located in. Instead, individual academics employed at universities are knowledge nodes for former colleagues providing access to publications for peers wherever they are located at that point. At the same time, universities are also knowledge receivers, because these academics also obtain valuable information from entrepreneurs for their own work as they get access to laboratory data, information on certain test arrangements as well as test results. But again, this is detached from spatial proximity.

Universities as participants in temporarily formalized communities

Personal ties and social networks are the foundations for any knowledge sharing arrangement in the biotechnology sector. However, as soon as a promising idea, a joint interest, or a demand is identified, the interview partners describe the tendency of converting these rather informal relations to more formal ones. As soon as a certain degree of formalization is arranged, copy rights, obligations, joint goals, and financial aspects are more clearly defined among the participating entities. Arrangements that include academic organizations very often are frames that create a win–win situation for both, academic partners and entrepreneurs. As one interviewee describes it: research partners receive results that they may use for publications, while entrepreneurs may work on developing diagnostic means subject to patenting which can later be brought to the market (interview FuE_4- 4, translated by the author).

These formal arrangements might take on different forms of institutionalization such as research partnerships or development projects. In most cases, the formalization is only temporal. Once the formalization's objectives are achieved, the ties are brought back to an informal level. In that way, universities are involved in a dynamic circuit of freely sharing knowledge and ideas in personal networks, transferring ideas to projects, implementing projects, terminating project partnerships, and again rather loose and informal contacts until the next idea is converted to funded projects (Schmidt 2012: 257).

4.4 Universities as reputation-references

Since relations to individual knowledge carriers are important, because other knowledge generating arrangements might grow out of these contacts, "know-who" is an important form of knowledge (Lundvall and Johnson 1994). Therefore, entrepreneurs benefit from accessing not only science-related knowledge, but also knowledge about central figures in the biotechnology sector, as well as knowledge about how to approach them and utilize contact with them. Individuals are perceived with their competencies, knowledge, specialization, know-how, methodological expertise, and skills (Schmidt 2012: 255). For the interviewees, knowing who-is-who in their field is of very high relevance. Again, peers at the universities may provide this kind of relational knowledge, because academics are more often at conferences, meetings, congresses etc. than their peers in an economic environment. In some cases, academic peers may even establish contact to central figures in the biotechnology scene which again may provide a competitive advantage for the interviewees.

Personal reputation is very important in the biotechnology field. This even goes beyond individual competencies and knowledge. Instead, in some cases, certain individuals represent specific schools of thought. For instance, one specific professor at the university represents a very specific diagnosis methodology. If a company can create a link to that person, this link may function as a promotion tool, because within the biotechnology sector every other expert will instantly recognize certain aspects and key competencies. Additionally, this link also functions as a quality feature for the company. Entrepreneurs who are able to visibly create a link (e.g. placing respective information on the companies' webpage) to a specific professor at a university may benefit themselves from this professor's reputation. This link may be created by identifying this central knowledge carrier, for example as a PhD advisor, by inviting the professor to become a member of the company's director's board or by having the professor as a scientific advisor for the company.

5 Conclusion

Universities do play an important role in innovation systems, but often they are considered to be knowledge providers within these systems. Such provider functions become visible through publications and presentations, cooperation and staff exchanges, patents and licenses, and university spin-offs. The presented chapter argues since knowledge cannot simply be transferred among different entities, but requires rather knowledge sharing arrangements to do so, the role of universities within innovation systems might appear slightly differently.

Sharing and generating knowledge requires an active participation in knowledge generating activities. Sharing knowledge, especially in its implicit form, is often considered to be space-sensitive and therefore assumed to require spatial proximity. However, if based on personal trust and joint cognitive models (e.g. in epistemic communities), sharing knowledge across spatial distances becomes possible. Academic staff members at universities are actively involved in such arrangements and therefore contribute to the spatial diffusion of knowledge. However, they do so not just as knowledge providers, rather, they are active members in knowledge generating networks and therefore receive knowledge as well.

The above arguments highlighted the nodal function of universities in innovation systems by drawing on empirical findings made in semi-structured interviews with entrepreneurs in the biotechnology sector. The nodal function includes processes of providing, sharing, transforming, receiving, and generating different forms of knowledge. Since entrepreneurs in the biotechnology sector usually started with an academic career, contacts to former colleagues at universities are one important aspect of ensuring access to external knowledge carriers. For both, the academic staff members at universities and the entrepreneurs, personal ties are a precondition for any knowledge sharing arrangement.

Adding to debates on innovation systems, the presented findings underline that universities of course provide academic training and the results of fundamental research. However, they also function as multipliers who open up new market contacts or project partners for entrepreneurs. Knowledge about persons and relations to other individuals thus receives a strategic dimensions not yet fully considered in innovation system debates. However, for the empirical case presented here it is very important to be able to identify important knowledge carriers, gain access to them and find means of further utilizing such contacts. Universities might provide the necessary links, but at the same time professors at universities might be such central entities themselves. Utilizing links to these knowledge characters does not necessarily include joint projects. Instead, benefiting from the individual reputation of these knowledge carriers is also a form of value. Furthermore, libraries so far have been considered to be a traditional and important way of providing access to externalized knowledge, because libraries may be accessed by extra-university individuals too. Again, the presented empirical findings shift this picture a little, because the interviewees seldom use libraries for gaining access to academic literature. Instead, they utilize personal contacts 1) for identifying relevant literature and 2) for receiving published and sometimes unpublished academic works.

In contrast, universities also gain from their contacts with entrepreneurs and therefore receive different forms of knowledge as well. So far, universities have seldom been considered to be customers of biotechnology enterprises. In that way, they are integrated in the enterprises' value chain as receivers of products and services. Additionally, biotechnology enterprises usually possess laboratories and can offer test results sometimes even quicker than universities themselves. Furthermore, through the contacts with entrepreneurs, universities also continue to have knowledge about recent market developments and hence are able to identify future research fields.

Last but not least, the most important foundation for knowledge sharing arrangements between universities and enterprises are personal ties developed along academic and entrepreneurial careers. Joint socialization is more important than spatial proximity for sharing knowledge. Therefore, universities usually unfold positive effects not only for the cities and regions they are located in but also far beyond such territorial boundaries. The interviews analyzed here were conducted in a location that does not host a university. At the same time, however, all interviewees describe contacts to universities in very different locations, depending on, for example, where the interviewees received their academic training or PhD, where they used to work, and with whom they have already implemented joint projects. Therefore, the interaction networks and innovation systems that universities are embedded in are spatially very open.

Summing up, universities contribute to *and* benefit from new knowledge exchanged within the epistemic community, new methodological knowledge, knowledge for new products and services, knowledge on inter-organizational processes, and knowledge for and within relationships (Know-Who).

Note

1 This chapter draws from project results achieved in two INTERREG projects: "Coronas Metropolitanas," financed under INTERREG IIIC and "Know-Man: Knowledge Network Management in Technology Parks," financed under INTERREG IVC. The empirical material presented in this chapter was conducted in 2010 for the author's doctoral thesis.

References

Amin, A. and Roberts, J. (ed.) (2008a) *Community, Economic Creativity, and Organization*. Oxford: Oxford University Press.

Amin, A. and Roberts, J. (2008b) "Knowing in action: Beyond communities of practice," *Research Policy*, 37 (2), 352–369.

Andersson, M. and Karlsson, C. (2004) "The role of accessibility for the performance of regional innovation systems," in C. Karlsson, P. Flensburg and S.-A. Hörte (eds), *Knowledge Spillovers and Knowledge Management*. Cheltenham: Edward Elgar, pp. 283–310.

Arena, R., Lazaric, N. and Lorenz, E. (2006) *Trust, Codification and Epistemic Communities: Implementing an expert system in the French steel industry*. In: DIME WP 12.

Asheim, B. T. (2004) "From clusters to projects: Spatial embeddedness and disembeddedness of learning and knowledge creation in the globalising economy," in C. Karlsson, P. Flensburg and S.-A. Hörte (eds), *Knowledge Spillovers and Knowledge Management*. Cheltenham: Edward Elgar, pp. 396–422.

Asheim, B. T. and Gertler, M. S. (2005) "The geography of innovation: Regional innovation systems," in J. Fagerberg, D. C. Mowery and R. R. Nelson (eds), *The Oxford Handbook of Innovation*. Oxford: Oxford University Press, pp. 291–317.

Beckert, B., Bührer, S. and Lindner, R. (2008) "Verläufe und Motive von 'Seitenwechseln': Intersektorale Mobilität als Form des Wissenstransfers zwischen Forschung und Anwendung," in R. Mayntz, F. Neidhardt, P. Weingart, *et al.* (eds), *Wissensproduktion und Wissenstransfer: Wissen im Spannungsfeld von Wissenschaft, Politik und Öffentlichkeit*. Bielefeld: Transcript, pp. 313–339.

Caspers, R. (2004) "Gesellschaftliche Bedeutung von Wissen und Wissensnetzwerken," in R. Caspers, N. Bickhoff and T. Bieger (eds), *Interorganisatorische Wissensnetzwerke: Mit Kooperation zum Erfolg*. Berlin, Heidelberg: Springer-Verlag, pp. 58–107.

DESTATIS (2003) *German Classification of Economic Activities, Edition 2003 (WZ 2003)*. Wiesbaden: Statistisches Bundesamt.

Feldman, M. P. and Massard, N. (2002) "Location, location, location: Institutions and systems in the geography of innovation," in M. P. Feldman and N. Massard (eds), *Institutions and Systems in the Geography of Innovation*. Boston, Dordrecht, London: Kluwer Academic Publishers, pp. 1–20.

Fritsch, M. (2009) "Wissenstransfer und Innovation im regionalen Kontext," in H. J. Kujath and J. Krupa (eds), *Fachhochschulen als regionales Potenzial: Zur Kooperation von Wissenschaft und Wirtschaft in der Metropolregion Berlin-Brandenburg*. Erkner: IRS, 9–20.

Grabher, G. and Maintz, J. (2006) "Learning in personal networks: Collaborative knowledge production in virtual forums", in Working Papers, Center on Organizational Innovation. Available from www.coi.columbia.edu/pdf/grabher_maintz_lpn.pdf.

Granovetter, M. S. (1973) "The strength of weak ties", in *American Journal of Sociology*, 78 (6), 1360–1380.

Hollingsworth, J. R., Schmitter, P. and Streeck, W. (ed.) (1994) *Governing Capitalist Economies Performance and Control of Economic Sectors*. New York: Oxford University Press.

Jonas, M. (2006) "Brauchen regionale Wirtschaftscluster lebendige 'Kooperation'? Ein Überblick anhand einer Auswahl empirischer Studien zu europäischen Hochtechnologieclustern," in: Reihe Soziologie 79 (Institut für Höhere Studien, Wien).

Knorr-Cetina, K. (2002) *Wissenskulturen: Ein Vergleich naturwissenschaftlicher Wissensformen*. Frankfurt/Main: Suhrkamp.

Krogh, G. and Köhne, M. (1998) "Der Wissenstransfer in Unternehmen: Phasen des Wissenstransfers und wichtige Einflussfaktoren," *Die Unternehmung*, 52(5/6), 235–252.

Lave, J. and Wenger, E. (1991) *Situated Learning: Legitimate peripheral participation*. Cambridge: Cambridge University Press, 19th edition.

Lorentzen, A. (2007) "The geography of knowledge sourcing: A case study of Polish manufacturing enterprises," *European Planning Studies* ,15(4), 467–486.

Lundvall, B.-Å. and Johnson, B. (1994) "The learning economy," *Journal of Industry Studies*, 1(2), 23–42.

Mayntz, R. (2006) "Governance Theory als fortentwickelte Steuerungstheorie," in G. Schuppert (ed.), *Governance-Forschung: Vergewisserung über Stand und Entwicklungslinien*. Baden-Baden: Nomos, 11–20.

Moodysson, J. (2007) *Sites and Modes of Knowledge Creation: On the spatial organization of biotechnology innovation. Meddelanden från Lunds universitets geografiska institution*. Lund: Lund University.

Moodysson, J. (2008) "Principles and practices of knowledge creation: On the organization of 'buzz' and 'pipelines' in life science communities," *Economic Geography*, 84(4), 449–469.

Mowery, D. C. and Sampat, B. N. (2005) "Universities in national innovation systems," in J. Fagerberg, D. C. Mowery and R. R. Nelson (eds), *The Oxford Handbook of Innovation*. Oxford: Oxford University Press, 209–239.

Nonaka, I. (1994) "A dynamic theory of organizational knowledge creation," *Organization Science*, 5(1), 14–37.

Nonaka, I. and Takeuchi, H. (1995) *The Knowledge-Creating Company: How Japanese companies create the dynamics of innovation*. New York: Oxford University Press.

Polanyi, M. (1958) *Personal Knowledge: Towards a post-critical philosophy*. London: Routledge & Kegan Paul.

Polanyi, M. (1967) *The Tacit Dimension*. London: Routledge.

Richter, R. and Furubotn, E. (1996) *Neue Institutionenökonomie: Eine Einführung und kritische Würdigung*. Tübingen: J. C. B. Mohr.

Schmidt, S. (2012) *Wissensspillover in der Wissensökonomie: Kanäle, Effekte und räumliche Ausprägungen*. Münster: LIT-Verlag.

Scott, R. W. and Meyer, J. W. (1994) *Institutional Environments and Organizations: Structural complexity and individualism*. Thousand Oaks, CA: Sage.

Stehr, N. (2001) *Wissen und Wirtschaften: die gesellschaftlichen Grundlagen der modernen Ökonomie*. Frankfurt/Main: Suhrkamp.

Strambach, S. (2008) "Knowledge-intensive business services (KIBS) as drivers of multilevel knowledge dynamics," *International Journal of Services Technology and Management*, 10(2/3/4), 152–174.

Sydow, J. (1992) *Strategische Netzwerk: Evolution und Organisation*. Wiesbaden: Gabler.

Wenger, E., McDermott, R. and Snyder, W. (2002) *Cultivating Communities of Practice: A guide to managing knowledge*. Boston, MA: Harvard Business School Press.

Williamson, O. E. (1994) "Transaction cost economics and organization theory," in N. J. Smelser and R. Swedberg (eds) *The Handbook of Economic Sociology*. Princeton, NJ: Princeton University Press, pp. 77–107.

Williamson, O. E. (1996) *The Mechanisms of Governance*. New York: Oxford University Press.

Wood, G. A. and Parr, J. B. (2005) Transaction costs, agglomeration economies, and industrial location," *Growth and Change*, 36(1), 1–15.

Zucker, L. G. and Darby, M. R. (1994) "Inter-institutional spillover effects in the commercialization of bioscience", in *ISSR Working Papers in the Social Sciences*, 6 (3).

5 Regional cooperation or external links?

Spatial proximity in science–business relations in Poland[1]

Agnieszka Olechnicka

1 Introduction

The main aim of this chapter is to provide a more in-depth understanding of the cooperation phenomenon between the science sphere (mainly universities) and business sphere in Poland and the role of proximity in this respect. We attempt to answer three research questions: 1) How important is cooperation between regionally located scientific partners and innovative enterprises? 2) What is the pattern of knowledge flows between universities and firms in Poland from the regional perspective? 3) What are the factors influencing the role of proximity in building links among universities and firms regarding innovative development?

Science–business cooperation in Poland has been investigated mostly on the basis of company surveys; thus, the aim of the research presented here was to look for some alternative sources of information in this respect. The chapter investigates the significance of close (regionally located) and distant (nationally located and foreign) scientific partners using both quantitative (Web of Science data and a national database regarding projects enhancing science–business links) and quantitative methods (interviews in innovative firms).

The chapter is organized as follows: section 2 presents a literature review regarding the relationship between the presence of a university and innovative development of the given location, with particular attention given to the concept of proximity of partners in such relations. Section 3 gives an overview of the research questions, methodology and data used in the research. Section 4 summarizes the state-of-the-art business–science relations in Poland. Section 5 discusses the results of the research based on bibliographic data, a targeted project database as well as some interviews in innovative firms. Section 6 concludes and discusses the results.

2 University and innovation: literature review

The essential issue in regional economics is how to develop on the basis of innovations and build a knowledge economy. To a considerable degree responsibility for innovative growth is ascribed nowadays to the presence of public and private scientific institutions in the region and their active cooperation with other regional actors, including enterprises. Universities are seen as engines of innovative growth and competitiveness. The rising expectations of the science sector role in innovative regional development have been influenced by several factors, e.g.: 1) growing demand for knowledge – a key driver of economic growth – which may be supplied by scientific institutions, 2) increasing demand for an educated labour-force, particularly in knowledge intensive sectors, 3) rapid changes in innovation processes requiring the cooperation of different agents, openness and networking, 4) increasing awareness of the importance of local contexts and features for the process of individual and collective learning leading to the creation of knowledge (Kukliński 2001; Lawton Smith 2006; Youtie and Shapira 2008). As a consequence, the science sector has been placed in the very centre of the recent academic and political debate about the development and welfare of regional and local economies (OECD 2010; UNESCO 2010).

Cooperation, enhancing links between scientific institutions and business is one of the most acknowledged and apparent alternatives for employing knowledge created at scientific institutions for regional needs. The question is whether the location of scientific partners is important from a business point of view. The answer to the question is closely related to the lively academic discussion on the role of spatial proximity in the cooperation between different actors regarding innovation.

On the one hand, the crucial role of close geographical location of knowledge suppliers and knowledge users for successful innovation cooperation is underlined. For example, the clusters theory or the concepts of regional innovation systems or industrial districts revolve around the conviction that the smaller the distance between firms and scientific institutions the greater precipitation of positive effects (e.g. Boschma 2005; Katz 1994; Marcusen 1996).

Science sector institutions supply knowledge by providing access to research findings as well as by turning out educated graduates. In addition, being situated close to universities, they have easy access to new technology, including global solutions, since these institutions act as intermediaries between the global and local scene (Chang *et al.* 2008). The proximity of the science sector stimulates the establishment of new technology firms (Audretsch *et al.* 2005; Baptista and Mendonça 2009) and influences the innovation activity of that type of enterprise (Costa and Teixeira 2005); it is also a constituent factor of high-tech clusters (Isaksen 2005).

It is stressed that the significance of geographical proximity for innovation stems in particular from the fact that it facilitates the flow of tacit knowledge, that is, knowledge that is inseparably linked to its supplier and recipient – and thus can only be passed on though the interaction that occurs between them. A short distance means that people are near to each other and can share knowledge directly; therefore, the informal links that are so important between universities and businesses are easier to achieve where the partners are closely situated (Lindelöf and Löfsten 2004; Paytas *et al.* 2004). In branches that rely mainly on tacit knowledge, a shorter distance between partners can be of major significance in improving knowledge flows (Bayona Sáez *et al.* 2002; Beugelsdijk and

Cornet 2002; Broström 2010; Levy *et al.* 2009). Equally, the transfer of codified knowledge also requires the close proximity of university–business partners. This is because it is essential that knowledge is correctly interpreted and implemented (Howells 2002), and also, importantly, because it affects the speed of transfer. Thus a university situated in direct proximity can give the knowledge recipient a significant competitive advantage (Beugelsdijk and Cornet 2002; Jaffe *et al.* 1993).

At the same time the critical role of spatial proximity is undermined. It appears that the significance of a university's proximity in the innovative development of a given area depends on various circumstances, and in many analyses appears only to perform a purely supplementary role. Above all, analyses of the situation in different regions of the world have led researchers to the banal conclusion that the idea of proximity is uncommonly relative and does not apply to small economies (Beugelsdijk and Cornet 2002). Moreover, the increasing mobility of people, brought about by advances in the field of transport as well as new forms of communication, mean that the importance of being located at a short distance is no longer of significance. In cases where meetings are necessary, contacts between the cooperating parties can be strengthened by so-called temporary or virtual proximity (Lorentzen 2008; Torre and Rallet 2005).

Scientists have gradually begun to distinguish other forms of proximity than simply geographical, namely, cognitive, organizational, social and institutional proximity, all of which indicate that various forms of proximity are complementary and affect the processes of knowledge spillovers as well as building regional innovation systems (Basile *et al.* 2012; Brinkhoff *et al.* in this volume; Capello 2009; Gaczek 2009). It appears that too close proximity in any dimension can limit interactive learning and innovation creation. With regard to geographical proximity this is linked with a risk of spatial barriers arising. This affects mainly highly specialized fields and can lead to internal closure of a region and gradual deterioration as a result of a weakening local enterprise network (Boschma 2005; Torre and Rallet 2005).

Moreover, there is growing research evidence in favour of distant knowledge sources. In particular, national and international scientific partners are more important when radical innovations are concerned. Thus, innovative firms are looking not for close and easily accessible partners but for valuable new knowledge, adequate research and unique skills. Development in telecommunication technologies and cheap flight connections facilitate cooperation with distant partners. There are several relevant theoretical models that can be applied in this field: *strength of weak ties* (Granovetter 1973), *small worlds concept* (Travers and Milgram 1969) and the notion of *global pipelines* supplementing *local buzz* (Bathelt *et al.* 2004).

A review of the literature leads to the conclusion that the differences between individual regions in terms of the role played by the scientific sector in the innovative development of firms stem on the one hand from the nature of the enterprise, and on the other hand from the strength of the science sector in a given region, and each area in this range possesses a different set of characteristics (ESPON 2011).

Conditions affecting national innovation systems are also varied, particularly in the sphere of financing scientific research from business sources or means of evaluating science-research activities. This significantly affects not only the intensity of cooperation, but also the form it takes and the significance of proximity; this is illustrated by the examples of Germany, Italy and the Czech Republic in comparison with American research (Abramo *et al.* 2010; Beise and Stahl 1999; Žížalová 2010).

Questioning the role of geographical proximity in relations between enterprise and the science sector would seem justified when it comes to peripheral regions. On one hand the limited ability of a region to absorb innovations (absorptive capacity), combined with a weak science sector, together form the conditions responsible for placing national and international ties between science and business over and above regional relations. Researchers studying sources of enterprise innovation located in Norwegian agglomerations do not only confirm the premise that firms should seek sources of innovation outside the region, but equally prove that there is no link between local relations and innovativeness. Only enterprises that make use of international relations are innovative, and the only method of gaining access to new knowledge and stimulating the innovativeness of industry is to make use of so-called global pipelines (Fitjar and Rodriguez-Pose 2011). Equally, analyses of innovative behaviour in Polish industrial enterprises prove that foreign sources of knowledge are the most highly valued by managers. In choosing enterprise cooperation, they look mainly for economic and technological proximity, and are capable of building relations that serve innovation at long distance; the basis for gaining access to knowledge is then virtual or short-term proximity (Lorentzen 2007). Similarly, researchers of Polish clusters reveal a concentration of high-tech firms employing international research (mainly due to a lack of access to this type of knowledge within the country). The presence of national B +R institutions in clusters is largely due to a desire to simplify formal requirements in applying for EU funding (PARP 2010).

3 Research questions, methodology and data

The chapter aims to answer three research questions: 1) How important is cooperation between regionally located scientific partners and innovative enterprises in Poland? Is the regional university more attractive for business than its national or international counterparts and, as a result, more often chosen for innovative cooperation? 2) What is the pattern of knowledge flowing between universities and firms in Poland from the regional perspective? Can we distinguish regions that are suppliers of knowledge and regions that are receivers of knowledge? 3) What are the factors influencing the role of proximity in building links among universities and firms regarding innovative development? Is geographical proximity crucial or an additional factor in this respect?

The chapter investigates the significance of close (regionally located) and distant (nationally located and foreign) scientific partners using both quantitative and qualitative methods. The quantitative analysis is based on two unique purpose-built databases: 1) a database of Polish publications with business affiliations extracted from the Web of Science database (2001–06); 2) a database of special NOT targeted projects (Polish Federation of Engineering Associations) aimed at supporting cooperation between small and medium-size firms with science (2002–05).

The Web of Science database was chosen for verifying the significance of the proximity of knowledge sources in the form of the science sector in the case of Polish enterprises. This choice was dictated by a wish to tap into a source of data that has not been made use of in Poland, and thereby enter into international research streams. Joint publications by scientific workers and enterprises are often used to measure the relations between universities and enterprises, also on the regional scale (Lööf and Broström 2006; Melin and Persson 1996). The purpose of examining co-authored scientific publications is that they reflect a wide range of cooperation, since, in order for the publication to appear, it is essential that there is a flow of people, capital, intellectual copyrights and other resources. The advantage of the bibliometric method is its verifiability and the relatively low cost of obtaining a large collection of data, analysis of which can achieve statistically significant results; it also offers the possibility of international comparison (see Carayannis and Laget 2006; Pechter and Kakinuma 1999).

The starting point for creating a database was to obtain from the Web of Science all articles from the years 2001–06, the bibliographies of which included Poland as the work place (affiliation) of at least one author. Data obtained in this way was then filtered and put through a multiple verification process, in the course of which erroneous entries could be removed or corrected and incomplete addresses filled in with essential place names. In this way a database of 72,817 articles was created, with a total of 201,928 affiliations. A sectoral and spatial identification of all affiliations was conducted. Sectoral analysis was carried out in two stages: first using the most commonly occurring abbreviations of affiliated scientific institutions (e.g. 'univ', 'PAN') and eliminating those affiliations that were obviously not enterprise affiliations; next, bearing in mind that names of institutions entered in the bibliography are not standardized, and thus each institution can appear under an innumerable variety of names, individual verification of remaining affiliations was carried out.

By this means, 488 business affiliations were obtained that were ascribed to 469 published articles in periodicals indexed in WoS in the years 2001–06. Spatial analysis involved ascribing the geographical location of each affiliation appearing in the collection of articles with Polish participation (the city and voivodship for affiliates within Poland, or the country name for foreign affiliates). On the basis of the information contained in the records of these articles a matrix of cooperation relations was drawn up that could be used for further analysis.

The second source of data used to present the cooperation between science and business in Poland was the programme database for the targeted projects of the NOT Centre for Innovation, an initiative realized by the Polish Federation of Engineering Associations – Chief Technical Organization (FSNT-NOT). The programme, which has been underway since 2002, is financed from public funds. The funding programme includes the financing of applied research and development work to activate new products or to introduce new technologies into small and medium-sized enterprises. Applications for targeted project financing must relate to the launching of an innovative new product or technology and include the research-development phase as well as the application; the introduction phase must be wholly financed by the enterprise and is usually realized in partnership with the science sector (Gryzik and Gąsiorkiewicz-Płonka 2007). Thus the programme is a tool for stimulating cooperation between enterprise and the R&D sphere. Up till now the effects of the programme have been very positive (NOT 2010).[2] Analysis of FSNT-NOT projects was conducted as a supplement to the findings of the publications research – it is a good example of a pro-innovation programme stimulating cooperation between business and the science sphere which gives measurable results. After completing the targeted project its social and economic effects are monitored to allow their assessment. One effect of the projects financed by the programme is the fact that awards and commendations are received by the artefacts/technologies that arise as a result. In the second edition of the programme under analysis, 469 targeted projects were entered, of which 88 per cent connected enterprises applying for funding with R&D institutions that carried out research-development services, commissioned and co-financed by the enterprise. On the basis of these applications,[3] a matrix of relations was drawn up for the purposes of our study.

This dual source of quantitative data allowed verification of the results obtained and made it possible to capture different types of cooperation. The cooperation of universities and firms behind the scientific articles contained in WoS is more theoretically oriented while the cooperation within the framework of NOT targeted projects is related more to the application of knowledge.

The results obtained from quantitative data, because of their limitations for interpretation, were supplemented with information of a qualitative nature. This was gained from thirty-five in-length interviews during the latter half of 2009 in enterprises whose affiliations appeared both in the WoS database and the CORDIS database containing information on participation in Sixth European Union Framework Programme projects. The respondents were those responsible for research-development activity, often the authors of publications as well as the managers of research projects that were the basis for the study group. The aim of the interviews was to give some insight into the motivations regarding science–business cooperation and the role of geographical proximity in this respect.

4 State of the art in business–science relations in Poland

The weak cooperation of Polish enterprises with the science sector is widely commented upon in Polish scientific literature (Łapiński 2010; Matusiak and Guliński 2010; Santarek 2008; Starczewska-Krzysztoszek 2008). This phenomenon has commonly been diagnosed on the basis of the results of research into the innovative activity of enterprises, carried out within the framework of the European research project: Community Innovation Survey (CIS).[4] First, Polish innovative enterprises – like their European counterparts – show that the most important sources of knowledge, from an innovation perspective, are of an internal nature (46.1 per cent of firms) with only a small percentage of innovative firms for whom cooperation with the research-development sphere is of major significance: consultants, commercial laboratories and private R&D institutions gained recognition from 6.9 per cent of innovative firms, higher education institutions were indicated by 5.3 per cent of enterprises, and government and public research institutes – 7.2 per cent.[5] The low significance of the scientific sector for the innovativeness of firms is also supported by the low and decreasing percentage of enterprises that declare their cooperation in the field of innovative activity – in the years 2006–08 this amounted to 8.3 per cent of production firms and 6.6 per cent of service providers (compared to 11.1 per cent in the years 2004–06 for both groups). Meanwhile, cooperation agreements entered into by firms in the field of innovative activity rarely involve R&D institutions: just 2.1 per cent of all firms studied held agreements with universities, 1.4 per cent with research-development units and 0.5 per cent held agreements with the Polish Academy of Sciences or with foreign research and development institutions. Assessments of this cooperation are also varied: most firms indicate research-development units (9.8 per cent of firms) and in second place private R&D institutions and universities (about 8 per cent), but the Polish Academy of Sciences and foreign institutions are the least often appreciated (less than 1 per cent of firms) (GUS 2010).

Another characteristic feature of the Polish innovation system is the low level of research funding by the business sector in Poland – merely 0.19 per cent of GDP. Meanwhile, the relations between business funding and budgetary funding are reversed in comparison with global trends – the share of Business Enterprise Research and Development (BERD) in Poland in the years 2005–08 hovered at around 25–26.6 per cent, while in the UE it stood at 55 per cent and in the OECD – 63.8 per cent. The dominance of budgetary funding in R&D in Poland dictates the types of research that are financed by these funds. The public budget generally finances basic research, which in Poland has a share of 38.2 per cent, while the share of applied research has fallen since 1995 to a level of 22.4 per cent (GUS 2010).

Cooperation between the enterprise sector and science can be equally assessed by employing the bibliometric method. Business affiliations can be found in the bibliographic databases indexing the most valued periodicals worldwide, for example the Web of Science or Scopus. A comparison of Poland's performance against global achievements in this regard is possible, thanks to data collected as part of the Innovation Union Scoreboard (IUS). With a total of 2.5 publications written in cooperation with the business sector per million inhabitants, Poland's score is among the worst in the EU27, where the average figure was 36.2 in 2008. Only Malta, Turkey, Latvia and Bulgaria scored less. Yet as regards growth dynamics, Poland was among the leaders in the years 2004–08, when the number of these publications grew by 12 per cent – that is, three times the average of the countries studied. However, it should be observed that this strong dynamic is characteristic almost exclusively of countries with low levels of cooperation (Figure 5.1).

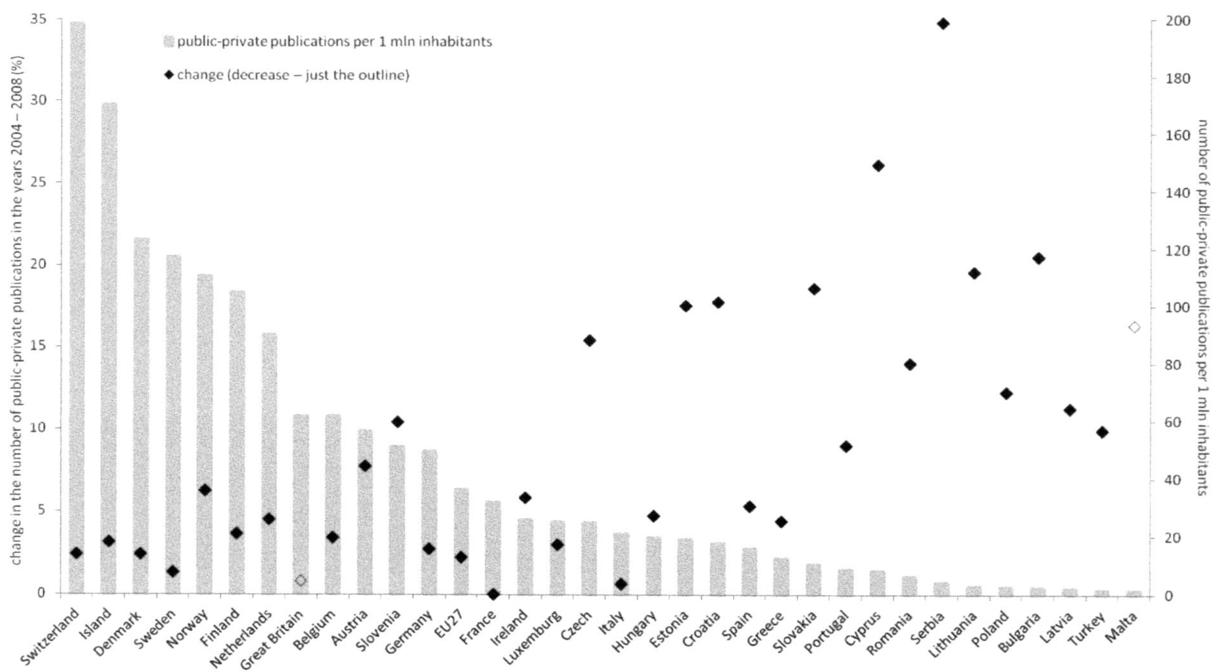

Figure 5.1 Public–private publications in EU27 in 2008.

Source: based on IUS (2011, anex A).

It should be stressed that the research quoted only partly conveys the complexity of the problem, as it presents only a fragment of the cooperation between enterprises and the science sector. To a significant degree, cooperation between firms and researchers takes place informally and thus it is not registered as institutional cooperation, nor is it reflected in scientific products (Olechnicka 2012). Available qualitative analyses conducted in enterprises and in research and development units confirm the dismal picture of cooperation between the science sector and business, while pointing out the reasons for this state of affairs (see e.g. MNiSW 2006). Of particular interest are the conclusions of an analysis conducted in the years 2004–06 as part of work on regional innovation strategies (RIS). Despite the fact that work on RIS was not coordinated methodologically on a national scale, it is possible to trace several common traits (Gorzelak *et al.* 2006):

- Cooperation between science and business is very rare and the level of commercialization of research results is negligible.
- Enterprises very seldom participate in common ventures with science, as they do not expect any benefits from such cooperation.
- Educational programmes in universities are not suited to the needs of the economy, and the offer of science institutions does not meet the technological expectations of enterprises; meanwhile the academic staff does not have the managerial abilities required in order to actively participate in research commercialization.
- Business institutions generally provide services on a basic level, with not much to offer in terms of pro-innovation activity, and their human and economic potential for realizing specialist services is very limited.

These observations have been further developed by recent analyses on the limitations of the effective functioning of the system of knowledge transfer and commercialization in Poland (Matusiak and Guliński 2010). Some of the structural barriers include the excessive formalities and bureaucracy in support mechanisms for cooperation between science and firms as well as irregularities in access to and employment of EU funds. An added problem is the imitational character of innovation in the Polish enterprise sector. The situation is complicated by an academic grey area – which uses the innovation resources for a particular end as well as improperly placing innovation support institutions within university structures, with poor efficiency and lack of coordination. Among the system barriers identified are the inadequate regulations for establishing off-shoot firms and tax liabilities resulting from the transfer of non-material goods from the science sector to the economy. A further barrier is the method of assessing scientific employees as well as the inflexibility of the educational programme which inhibits enterprise initiatives among scientists and students. Cultural-awareness barriers inhibit mutual understanding between the two cooperating spheres: the business world and the science world.

Entrepreneurs regard the research world as closed and only interested in pure science. They accuse universities of being bureaucratic and poorly prepared on an organizational level for cooperation with business. Meanwhile, in some academic circles, cooperation with business is seen as inconsistent with the academic ethos. Overcoming such barriers is difficult, bearing in mind the low level of social trust and lack of social acceptance for innovative attitudes in Poland. This mutual lack of trust between science and business is prolonged by a scant awareness of the possibilities for cooperation and insufficient number of positive role models and examples of good practice in this area. The three types of barrier mentioned are fortified by the lack of competence and effectiveness shown by participants in the innovation system in realizing and managing the process of innovation.

5 Empirical evidence from Poland

5.1 The significance of proximity in business–science relations – business articles

The analysis of Polish articles with business affiliations in the Web of Science database made it possible to verify the significance of the proximity of science-sector knowledge sources for Polish enterprises. It was also possible to recognize the pattern of knowledge flows between universities and firms in Poland from the regional perspective as well as the importance of foreign knowledge sources.

By using the data on publications it was possible to count all instances of firms' cooperation in the field of scientific articles, in other words the co-authoring of individual publications. Out of all firms there were 560 instances of cooperation, of which 53.2 per cent concerned institutions and organizations located in the same viovodship. Relations linking enterprises with research-development institutions constituted the greater part of the research sample – 91 per cent of relations (508). Analysis of the data, after excluding relations between enterprises, indicates that cooperation of firms with the science sector is only slightly more concentrated in the region than cooperation of firms with entities in general, i.e. scientific as well as business entities (54.5 per cent compared to 53.4 percent of all firms' relations).

Analysis of the significance of geographical proximity – understood as location within an administrational unit (voivodship) – in enterprise cooperation on a national level is based on a matrix of enterprise publishing relations revealed in the publications sample. The obtained matrix of relations is non-symmetrical, signifying that the information it contains can be interpreted in two ways – from the perspective of the publishing firms located in a given region (rows) or from the perspective of the publishing R&D institutions which are based in a given region (columns). However, it is difficult to discern a trend in these relations given the material that they relate to. In the case of articles appearing in scientific periodicals, it is hard to define exactly which side was the initiator and what form the cooperation took. Various situations exist. A publication may be the result of a research project, initiated by a firm seeking to introduce a new solution or obtain new equipment. In this case the enterprise will look for cooperation with a R&D institution and commission it to carry out the work. Equally, a university may initiate the contact, when it wishes to make use of enterprise equipment or a machinery park (see Olechnicka 2012).

FIRMS from — REGION (main centre)	R&D INSTITUTIONS from: dolnośląskie (Wroclaw)	kujawsko-pomorskie (Bydgoszcz, Torun)	lubelskie (Lublin)	lubuskie (Zielona Gora)	lodzkie (Lodz)	malopolskie (Krakow)	mazowieckie (Warsaw)	opolskie (Opole)	podkarpackie (Rzeszow)	podlaskie (Bialystok)	pomorskie (Gdansk)	slaskie (Katowice)	swietokrzyskie (Kielce)	warminsko-mazurskie (Olsztyn)	wielkopolskie (Poznan)	zachodniopomorskie (Szczecin)	Share of relation generated by firms sector	Share of relations inside the region
dolnoslaskie (Wroclaw)	27	0	0	1	3	1	4	0	0	0	0	2	0	0	0	0	7%	71,1%
kujawsko-pomorskie (Bydgoszcz, Torun)	1	4	3	0	0	2	2	0	0	0	0	1	0	1	0	0	3%	28,6%
lubelskie (Lublin)	0	0	8	0	1	3	2	0	0	0	4	1	0	0	0	0	4%	42,1%
lubuskie (Zielona Gora)	1	0	0	1	0	0	1	0	0	0	0	0	0	0	1	1	1%	20,0%
lodzkie (Lodz)	2	3	0	0	11	2	4	0	0	0	0	0	0	0	4	1	5%	40,7%
malopolskie (Krakow)	8	1	1	0	1	30	3	0	0	0	0	1	0	0	2	0	9%	63,8%
mazowieckie (Warsaw)	10	8	16	0	1	11	109	0	1	5	2	8	0	3	7	0	36%	60,2%
opolskie (Opole)	1	0	1	0	3	1	0	1	0	0	0	2	0	2	2	0	3%	7,7%
podkarpackie (Rzeszow)	1	0	2	0	0	0	3	0	1	0	0	1	0	0	0	0	2%	12,5%
podlaskie (Bialystok)	0	0	0	0	1	0	1	0	0	4	0	0	0	0	1	0	1%	57,1%
pomorskie (Gdansk)	2	2	1	0	0	1	4	0	1	0	33	0	0	1	2	1	9%	68,8%
slaskie (Katowice)	4	0	0	0	0	7	1	0	0	0	1	29	0	0	6	0	9%	60,4%
swietokrzyskie (Kielce)	0	0	0	0	0	1	1	0	0	0	0	0	0	0	0	0	0%	0,0%
warminsko-mazurskie (Olsztyn)	0	0	1	0	0	0	0	0	0	0	0	0	0	1	1	1	1%	25,0%
wielkopolskie (Poznan)	8	1	1	0	1	1	7	0	0	0	1	2	0	1	16	0	8%	41,0%
zachodniopomorskie (Szczecin)	0	0	3	0	0	0	3	0	0	0	0	0	0	0	0	2	2%	25,0%
Share of relation generated by R&D sector	13%	4%	7%	0%	4%	12%	29%	0%	1%	2%	8%	9%	0%	2%	8%	1%		54,5%

Figure 5.2 Business–science relations from a regional perspective (publications, WoS, 2001–06).

Source: own calculations based on WoS data.

The values given diagonally across in the relations network indicate the number of co-appearances in enterprise and university articles in the same voivodship; for example, the greatest number of university–business relations occurs in the mazowieckie voivodship – with its main scientific centre in Warsaw (109). Meanwhile the remaining values occurring in the matrix describe the connections between regions in the form of firm–university or university–firm.[6] It turns out that the relations between regions in the form of university–firm does not apply to all possible combinations of geographical–institutional profile with regard to publications – amounting to 40 per cent of potential cooperation situations (shaded area, Figure 5.2). The majority of these relations are formed by entities in the mazowieckie voivodship, mainly owing to their number. Even so, scientific institutes in the mazowieckie voivodship have relations with enterprises in all regions apart from opolskie and warminsko-mazurskie, and firms cooperate with universities in 11 voivodships (apart from opolskie, lubuskie, swietokrzyskie and zachodniopomorskie). Moreover, firms based in the mazowieckie voivodship participate in 36 per cent of all business sector relations with science in Poland, and mazowieckie universities take part in 29 per cent of all university relations with Polish firms. Although to a much lesser extent, the universities of dolnoslaskie and małoposkiealso play an integrating role on Polish territory (these areas having 13 per cent and 12 per cent shares in Polish scientific institution relations); so too do enterprises in pomorskie and wielkopolskie (each area with a 9 per cent share in Polish enterprise relations).

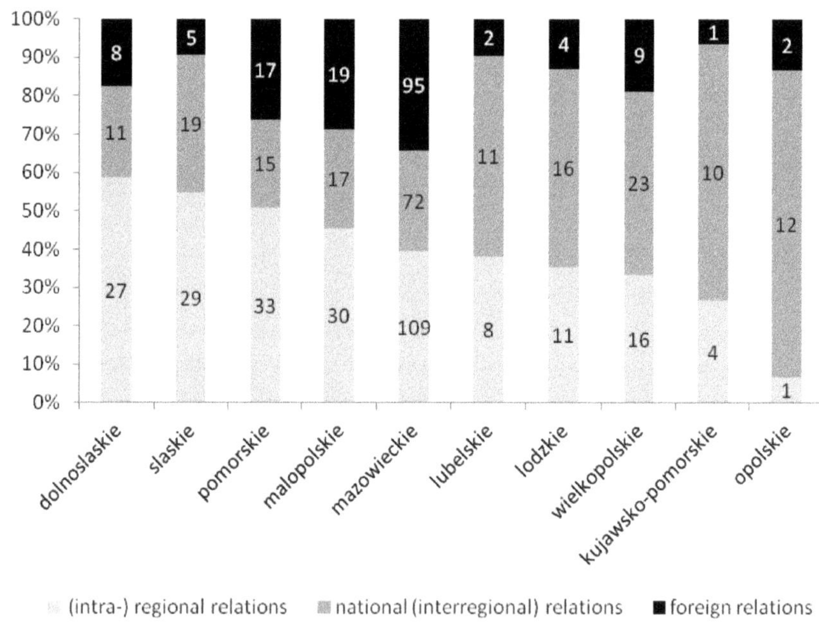

Figure 5.3 Foreign cooperation of Polish firms (number of relations, 2001–06).

Source: own calculations based on WoS data.

In interpreting the above findings it should be taken into account that the presented relations matrix completely ignores contacts with foreign enterprises, which after all form an important part of cooperation related to publications. Foreign connections are shown to be of particular significance in three regions, exceeding one quarter of all relations of firms who publish (national and foreign): these are the voivodships of pomorskie (26.2 per cent), malopolskie (28.8 per cent) and mazowieckie (34.4 per cent) (Figure 5.3).

5.2 The significance of proximity in business–science cooperation – targeted projects

The analysis of NOT targeted projects made it possible to validate the results obtained from the business articles analysis by capturing a different kind of cooperation – related more to the application of knowledge. It also made it possible to distinguish regions that are suppliers of knowledge and regions that are receivers of knowledge.

The analysis of 413 NOT targeted projects, linking up enterprises applying for funding with R&D institutions for carrying out research and development services commissioned and co-financed by them, shows that there is quite a strong link between the number of services commissioned by firms in the region and the number of research services provided in the region. Based on the regional breakdown of research services commissioned and research services provided, it is possible to state that there were only a few regions in which the number of services commissioned paralleled the number of services provided, (dolnoslaskie, podlaskie and lodzkie voivodships). In the majority of regions, the number of services commissioned by firms exceeded the number of research services carried out by scientific institutions (above all in the podkarpackie, pomorskie, swietokrzyskie, kujawsko-pomorskie and lubelskie voivodships). In only three regions – malopolskie, mazowieckie and wielkopolskie – was the situation reversed, that is, the number of scientific research services carried out by commission exceeded the number of commissions received from regional enterprises.

In order to better understand the significance of proximity in realizing this type of project, a non-symmetrical matrix of relations was created – as in the case of publications – taking into account the services commissioned and services realized between individual regions, that is, a region–region perspective (Figure 5.4). The values appearing in the fields placed diagonally across represent relations within regions. The remaining values show the relations between enterprises commissioning research in individual regions and R&D institutions carrying out this research. The relations presented, in contrast to those of publications, show a definite trend. By reading the relation chart by rows (firms), it is possible to identify research services commissioned, while by reading relations by columns (R&D institutions), it is possible to trace the research services realized. This results from the nature of the presented data – in the commissioned projects the enterprise is the party who commissions the work (and simultaneously the applicant and co-financer of R&D work), and the R&D institutions are the contractors.[7]

	REGION (main centre)	dolnośląskie (Wrocław)	kujawsko-pomorskie (Bydgoszcz, Torun)	lubelskie (Lublin)	lubuskie (Zielona Gora)	lodzkie (Lodz)	malopolskie (Krakow)	mazowieckie (Warsaw)	opolskie (Opole)	podkarpackie (Rzeszow)	podlaskie (Bialystok)	pomorskie (Gdansk)	slaskie (Katowice)	swietokrzyskie (Kielce)	warminsko-mazurskie (Olsztyn)	wielkopolskie (Poznan)	zachodniopomorskie (Szczecin)	Share of relation generated by firms sector	Share of relations inside the region
		R&D INSTITUTIONS from																	
	dolnoslaskie (Wroclaw)	6	0	0	0	3	2	1	0	0	0	0	0	0	0	1	0	3%	46,2
	kujawsko-pomorskie (Bydgoszcz, Torun)	0	3	0	0	3	4	7	0	0	0	1	0	0	0	1	0	5%	15,8
	lubelskie (Lublin)	0	1	2	0	2	3	2	0	1	0	0	1	0	0	0	0	3%	16,7
F	lubuskie (Zielona Gora)	0	0	0	0	0	2	3	0	0	0	0	1	0	0	1	0	2%	0,0
I	lodzkie (Lodz)	0	0	0	0	10	8	8	0	0	0	0	4	0	0	4	0	8%	29,4
R	malopolskie (Krakow)	0	0	0	0	1	30	5	0	0	0	0	8	0	0	0	0	11%	68,2
M	mazowieckie (Warsaw)	1	0	0	0	6	6	27	0	0	0	1	2	0	1	7	0	12%	52,9
S	opolskie (Opole)	1	1	0	0	0	6	0	0	0	0	0	1	0	0	1	0	2%	0,0
	podkarpackie (Rzeszow)	0	0	0	0	1	21	10	0	7	0	0	1	0	0	3	0	10%	16,3
f	podlaskie (Bialystok)	0	0	0	0	0	0	0	0	0	2	0	0	0	0	1	0	1%	66,7
r	pomorskie (Gdansk)	1	0	0	0	1	5	2	0	0	0	3	2	0	1	4	1	5%	15,0
o	slaskie (Katowice)	1	4	0	0	0	15	8	0	0	0	1	37	0	0	3	1	17%	52,9
m	swietokrzyskie (Kielce)	0	0	0	0	0	6	2	0	0	0	0	2	2	0	0	0	3%	16,7
	warminsko-mazurskie (Olsztyn)	0	0	0	0	0	0	6	0	0	0	0	1	0	8	0	0	4%	53,3
	wielkopolskie (Poznan)	2	0	0	0	4	6	7	1	1	0	4	2	0	0	20	0	11%	42,6
	zachodniopomorskie (Szczecin)	0	0	0	0	0	1	1	0	0	0	0	0	0	0	7	3	3%	25,0
	Share of relation generated by R&D sector	3%	2%	0%	0%	8%	28%	22%	0%	2%	0%	2%	15%	0%	2%	13%	1%		38,8

Figure 5.4 Business–science relations from a regional perspective (SME projects, NOT, 2002–05).

Source: own calculations based on NOT (Polish Federation of Engineering Associations) data.

Analysis of the matrix confirms our earlier observations – that inter-regional cooperation between individual regions occurs only among a few voivodships. Cooperation occurs in only 35 per cent of all possible cooperation links between voivodships – that is, possible region–region connections (shaded area, Figure 5.4). It also appears that the degree of regional 'self-sufficiency' in Poland in terms of meeting the demand for knowledge – measuring the number of research commissioned in a given region and the number realized in the given region – is low, amounting to 38.8 per cent. This means that three out of four relations in terms of innovation between firms and R&D institutions within the programme framework take place outside regions, not within them.

The participation of enterprises located in individual voivodships in the programme, their measured share in the general pool of submitted applications, is very diverse, the greatest number coming in from firms located in the voivodships of slaskie (17 per cent), mazowieckie (12.4 per cent) and wielkopolska (11.4 per cent). Meanwhile, research services are undertaken mainly by R&D institutions in the voivodships of malopolskie (27.9 per cent), mazowieckie (21.6 per cent) and slaskie (15 per cent) (figure 5.4). This graphic depiction allows us to discern the key significance of the malopolskie and mazowieckie voivodships in the integration of science and business world links. Both voivodships have ties with the majority of other regions regarding applications (with the exception of two in each case). This means de facto that the cooperation network between the enterprise sphere and science sphere in realizing innovative solutions is centralized from the scientific research system perspective, and more dispersed from the perspective of research demand.

5.3 The significance of proximity in business–science cooperation – interviews with innovation enterprises

The information obtained from interviews supplemented the results of the quantitative analysis. The interviews gave some insight into the motivations regarding science–business cooperation and the factors influencing the role of geographical proximity in this respect.

Only a quarter of respondents among the innovation firms studied state decisively that proximity to scientific institutions (and also other enterprises) is significant for cooperation regarding innovation, at least in certain instances. In this respect there was nothing to distinguish enterprises – neither size (firms numbering ten employees as well as those employing several thousand), nor branch (footwear manufacture, construction, fibre-optics) nor location country wide. However, a decisive majority (twenty-four firms, including all those studied in the fields of IT, electronics and pharmaceuticals) claim that distance from partners in innovation activity, including research-development institutions, was of no significance whatsoever. The primary reason why the significance of proximity for the innovation activity of enterprises is lower than anticipated is that partners are sought on the basis of their expertise. Enterprises form ties not for reasons of spatial proximity but based on the adequacy of the partner's offer in meeting their needs; they look for contacts with 'interesting' entities with suitable levels of scientific–technological expertise ('we select partners on a purely thematic basis – we know where to find a specialist or team dealing with a problem that interests us, and that's all that guides us'). This is often unavoidable, when suitable partners are located far away, as an enterprise in Rzeszow explains: 'As far as cooperation in the experimental field is concerned, there is only one pharmacy faculty in Poland – Gdansk – which is on a suitable level, and we can cooperate with them regarding the form of medicines.'

Enterprises seem to look for a reasonable balance between that which is near and readily available without incurring additional costs, and that which is the best option in terms of expertise. This is supported by the findings of other authors indicating that firms try to strike a happy medium between geographical proximity to a university and its research standing (Mansfield 1995; Schartinger *et al.* 2002). This is well put by one of the respondents:

> If we think of innovation, of the firm's development, it's hard to think in the category of the region or even the nation. Unfortunately, we need to think globally. Today, when looking for partners to cooperate with, we can't afford to be led by the sentiment that someone is from the same region. There's no point. We need someone with ideas, who knows what they are doing; someone in cooperation with whom we can profit the most. That's the main criteria in looking for partners.

The interviews conducted with enterprises confirm the significance of foreign cooperation partners. Cooperation with Polish scientific institutions strictly in the area of innovation is of less significance than contacts with foreign institutions. In the opinions of one quarter of the enterprises involved in publishing and participating in projects, cooperation with Polish universities is in many ways unsatisfactory. They blame poor levels of organization and expertise for this state of affairs, in particular: lack of research or inadequate levels of expertise in certain scientific disciplines, failure of R&D offers to meet the needs of enterprise, imprecise terms of cooperation and bureaucracy as well as competition on the part of the university on the product and services market.

A firm's choices for cooperation are shaped by several factors (Figure 5.5). First, means of communication are helpful in cooperating efficiently at long distance. Among those mentioned in particular were telephones (with video function), Internet and Skype, use of video camera and digital photography. Second, temporary proximity in the form of meetings and mutual visits can compensate the enterprise for a lack of geographical proximity with innovation partners. As one respondent jokingly remarked, quoting the chairman of Sun Micro Electronics: 'the best information technology back-up to the Internet is the Boeing 747'. Third, the importance of proximity to meritorious partners depends on the specific phase of the innovation project. For instance, in the initial stages meetings are essential, allowing on the one hand to establish concepts and definitions that will be held to as the project evolves. Moreover, 'live contact is clearly needed in scientific and

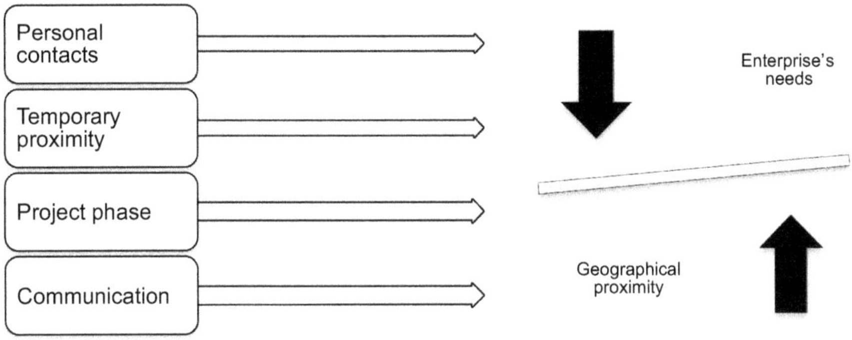

Figure 5.5 Determinants of the geographical proximity role in science–business relations.
Source: own elaboration based on interview.

technological debates, in negotiating and settling contracts, as well as inspection and overseeing of the investment at different stages'. Communication via telecommunications works better if direct contacts have already been made. Face-to-face meetings should be arranged 'in order to get to know each other better, outline each other's expectations, and agree the nature of the cooperation', and also to gain 'confidence in the reliability and professionalism of the partners'. Personal contacts are decisive in choosing a regional partner. Several respondents indicated that their cooperation with a regional R&D sector was influenced by earlier personal relations with specific representatives from the science world. This might be the result of earlier cooperation, or equally the result of ties with a specific regional scientific partner owing to a worker or, more commonly, the owner of a firm having previously been employed in the science sector.

6 Conclusions and discussion

Thanks to the matrix of relations created on the basis of publications data and NOT targeted projects, it can be stated that relations between firms and the science sector in Poland do not remain within the voivodship borders; quite the opposite – a significant proportion of these relations take place between regions and also on an international scale. However, with regard to the question as to which regions are more internally integrated and which internal relations predominate, there is no clear-cut answer: in some regions only a few relations were identified, and the results obtained from the two data sources were conflicting, as in the case of the pomorskie voivodship (the share of inter-regional relations revealed in publications was 68.8 per cent, and in NOT projects – 15 per cent, see Figures 5.2 and 5.4). A comparison of the analysis findings allows us to determine the situation of selected regions in this respect. The group of regions where internal relations are rather more important include the malopolskie and mazowieckie voivodships – the indicators in each situation are higher than average. Meanwhile the kujawsko-pomorskie voivodship is characterized by a relatively greater emphasis on external ties with the science sector (along with a greater number of relations).

Second, inter-regional relations from a business–scientific institution perspective involve only a portion of the possible regional pairs. Thus university–firm knowledge transfer takes place between certain regions, where some act as suppliers of knowledge, namely universities located there conduct more research than is commissioned by local firms (malopolskie, mazowieckie and wielkopolskie) while others, quite the reverse, are receivers of knowledge (podkarpackie, pomorskie, swietokrzyskie, kujawsko-pomorskie, lubelskie).[8] Third, relations between regions in terms of flows allow us to discern the key significance of the malopolskie voivodship (with its scientific centre in Krakow) and mazowieckie voivodship (with its scientific centre in Warsaw) in the integration of the worlds of science and business. Both regions have relations with most other regions in the area of publications and targeted projects. The integrating role of mazowieckie and malopolskie is more apparent for R&D institutions (more simply – contractors), than for firms (more simply – those commissioning research). This means, de facto, that the network of cooperation between the enterprise sphere and the science sphere in developing and realizing innovative solutions is centralized in terms of the scientific research system, and more dispersed in terms of demand for research. Consequently enterprises seeking cooperation in innovation try to strike a balance between that which is near at hand and that which provides the adequate expertise for the needs and interests of the firm. Important factors in making this choice include modern communication methods of the realization stage of the project, temporary proximity as well as personal contacts.

By comparing the findings of the empirical research with available knowledge (Gorzelak *et al.* 2006), including a review of selected updated regional innovation strategies for certain regions (Olechnicka 2012), it is possible to state that the significance of geographical proximity between cooperating institutions is overestimated in strategic documents relating to innovation-based development on a regional level in Poland. The supra-regional environment as well as the national and international dimension for cooperation between entities is barely mentioned (if not totally overlooked) in regional innovation politics. Discussion regarding the building of regional innovation in Poland should not be restricted to individual regional frameworks for several reasons: first, considering the nature of the Polish scientific sector, which includes: low emphasis on the R&D sector (particularly from the enterprise sector), a concentration of scientific potential in the eight most prominent centres (Warsaw, Krakow, Wroclaw, Poznan, the Slaskie agglomeration, Lublin, Lodz and Gdansk), a relatively weak profile as regards scientific products compared to other European countries; second, considering the nature of innovation processes in Poland, such as: a constant lack of importance attached to development based on enterprise innovation and little readiness of universities to cooperate with the business sector; third, considering the necessity of taking into account emerging tendencies in the global economy (the necessity of internationalizing research and innovation cooperation). In the light of the research conducted on the cooperation of Polish firms with the scientific sector, the answer to the question: 'In what sense is the contribution of universities to innovation and economic development a regional problem?' (Power and Malmberg 2008) would seem clear. In Polish conditions, it is more important to build Polish economic potential through strengthening those relations that are most effective from the point of view of an innovation-based economy, even if such relations do not emerge within the limits of a single region. However, if this is to occur, a closer analysis of the situation in individual regions is required and, equally, of existing science–business relations, as well as of the science sector potential and the absorptive capacity of firms. This chapter should be treated as a point of departure for further analysis in this area.

Notes

1 This chapter is based on the book published in Polish in 2012 entitled: *Potencjał nauki a innowacyjność regionów (Science potential and regional innovativeness)* (Olechnicka 2012).
2 http://centruminnowacji.org/projekty/nagrodzone/.
3 Taking into account all applications made (not just those accepted for funding) it appears justified that, in the case of this programme, partnership between enterprises and R&D institutions must already be entered into and documented at the stage of applying for funding (NOT 2010). As it also turns out, considering these aspects on the level of agreements, without taking account of applications which for various reasons were not given funding, does not significantly alter the situation.
4 CIS research conducted in Poland by the General Statistical Office encompasses innovation enterprises employing more than nine people.
5 Calculations based on: http://epp.eurostat.ec.europa.eu.
6 For example, firms from the lubelskie voivodship form nineteen relations with Polish scientific sector institutions, of which two with those located in the mazowieckie voivodship. Whereas lubelskie universities together formed thirty-seven relations with the Polish business sector, of which sixteen in mazowieckie (Figure 5.2).
7 For example, malopolskie voivodship has one relation going into the lodzkie voivodship, i.e. one commission for research from a firm in malopolskie for a university in lodzkie, and at the same time malopolskie has eight relations coming in from lodzkie – eight commissions from lodzkie firms for malopolskie universities (Figure 5.4).
8 The regional examples given in the two categories relate to targeted projects, since only in these cases is it known that relations are formed between the client and the executer of the research, whereas in the case of publications it is less obvious, as co-authored articles may be the effect of cooperation activities other than realizing research commissioned by firms.

References

Abramo, G., D'Angelo, C. A., Di Costa, F. (2010) 'University–industry research collaboration: a model to assess university capability', *Higher Education, opublikowany on-line*, 1–19.

Audretsch, D. B., Lehmann, E. E. and Warning, S. (2005) 'University spillovers and new firm location', *Research Policy*, 34(7), 1113–1122.

Baptista, R., Mendonça, J. (2009) 'Proximity to knowledge sources and the location of knowledge-based start-ups', *The Annals of Regional Science*, 45, 5–29.

MNiSW (2006) *Bariery współpracy przedsiębiorców i ośrodków naukowych. Raport* Warszawa: Ministerstwo Nauki i Szkolnictwa Wyższego.

Basile, R., Capello, R. and Caragliu, R. (2012) 'Technological interdependence and regional growth in Europe', forthcoming in papers in *Regional Science*, doi:10.1111/ j.1435-5957.2012.00438.x.

Bathelt, H., Malmberg, A., Maskell, P. (2004) 'Clusters and knowledge: Local buzz, global pipelines and the process of knowledge creation', *Progress in Human Geography* (28), 31–56.

Bayona, Sáez C., Garcia, Marco T., Huerta, Arribas E. (2002) 'Collaboration in R&D with universities and research centres: An empirical study of Spanish firms', *R&D Management*, 32(4), 321–341.

Beise, M. and Stahl, H. (1999) 'Public research and industrial innovations in Germany', *Research Policy*, 28(4), 397–422.

Beugelsdijk, S. and Cornet, M. (2002) '"A far friend is worth more than a good neighbour": Proximity and innovation in a small country', *Journal of Management and Governance*, 6(2), 169–188.

Boschma, R. (2005) 'Proximity and innovation: A critical assessment,' *Regional Studies*, 39(1), 61–74.

Broström, A. (2010) 'Working with distant researchers: Distance and content in university–industry interaction', *Research Policy*, 39(10), 1311–1320.

Capello, R. (2009) 'Spatial spillovers and regional growth', *European Planning Studies*, 17(5), 639–658.

Carayannis, E. and Laget, P. (2006) 'The emergence of regional technological capabilities and transatlantic innovation networks: A bibliometric study of public–private, EU–U.S. R&D Partnerships', in E. Carayannis and D. F. J. Campbell (eds), *Knowledge Creation, Diffusion, and Use in Innovation Networks and Knowledge Clusters*. Westport CT, London: Praeger, pp. 123–143.

Chang, Y.-C., Lin, B.-W., Liu, M.-C., Hung, S.-C., Ou, Y.-P. (2008) 'Innovation symbiosis among geographical knowledge network: The Case of the Hsinchu science-based industrial park, Taiwan (1991–2002), in E. Carayannis, D. Assimakopoulos and M. Kondo (eds), *Innovation Networks and Knowledge Clusters: Findings and insights from the US, EU and Japan*. New York: Palgrave McMillan, pp. 343–360.

Costa, J. and Teixeira, A. (2005) 'Do universities influence innovative efforts and location choices of technology based firms? The case of Portugal', *DRUID Academy*.

ESPON (2011) *KIT. Knowledge, Innovation, Territory. Interim Report.* Version 24/02/2011.

Fitjar, R. D. and Rodriguez-Pose, A. (2011) 'When local interaction does not suffice: Sources of firm innovation in urban Norway', *Environment and Planning A*, 43(6), 1248–1267.

Gaczek, W. M. (2009) *Gospodarka oparta na wiedzy w regionach europejskich* (Vol. TOM CXVIII). Warszawa: Studia KPZK PAN.

Gorzelak, G., Bąkowski, A., Kozak, M., Olechnicka, A., przy współpracy. Płoszaj, A. (2006). *Polskie Regionalne Strategie Innowacji: Ocena i wnioski do dalszych działań.* Warszawa: Ministerstwo Rozwoju Regionalnego.

Granovetter, M. (1973) 'The strength of weak ties', *American Journal of Sociology*, 78(6).

Gryzik, A. and Gąsiorkiewicz-Płonka, P. (2007) 'Innowacyjne Projekty Celowe', in T. Baczko (ed.), *Raport o innowacyjności gospodarki Polski w 2007 roku*. Warszawa: Instytut Nauk Ekonomicznych Polskiej Akademii Nauk.

GUS (2010) *Nauka i technika w Polsce w 2008 r.* Warszawa: Główny Urząd Statystyczny.

Howells, J. R. L. (2002) 'Tacit knowledge, innovation and economic geography', *Urban Study*, 39(5–6), 871–884.

Isaksen, A. (2005) Regional clusters building on local and non local relationships: A European comparison, in A. Lagendijk and P. Oinas (eds), *Proximity, Distance and Diversity: Issues on economic interacion and local development*. Aldershot: Ashgate, pp. 129–152.

Jaffe, A. B., Trajtenberg, M. and Henderson, R. (1993) 'Geographic localization of knowledge spillovers as evidenced by patent citations', *The Quarterly Journal of Economics*, 108(3), 577–598.

Katz, J. S. (1994) 'Geographical proximity and scientific collaboration', *Scientometrics*, 31(1), 31–43.

Kukliński, A. (2001) The role of universities in stimulating regional development and educating global elites', *Higher Education in Europe*, 26(3), 437–445.

Lawton Smith, H. (2006) *Universities, Innovation and the Economy*. London and New York: Routledge.

Levy, R., Roux, P. and Wolff, S. (2009) 'An analysis of science–industry collaborative patterns in a large European University', *The Journal of Technology Transfer*, 34(1), 1–23.

Lindelöf, P. and Löfsten, H. (2004) 'Proximity as a resource base for competitive advantage: University–industry links for technology transfer', *Journal of Technology Transfer*, 29(3/4), 311–326.

Lööf, H. and Broström, A. (2006) *Does Knowledge Diffusion between University and Industry Increase Innovativeness?* : Royal Institute of Technology, CESIS – Centre of Excellence for Science and Innovation Studies.

Lorentzen, A. (2007) 'The geography of knowledge sourcing: A case study of Polish manufacturing enterprises', *European Planning Studies*, 15(4), 467–486.

Lorentzen, A. (2008) 'Knowledge networks in local and global space', *Entrepreneurship and Regional Development*, 20(6), 533–545.

Łapiński, J. (2010) 'Źródła pochodzenia innowacji', in *Innowacyjność*. Warszawa: Polska Agencja Rozwoju Przedsiębiorczości, pp. 43–62.

Mansfield, E. (1995) 'Academic research underlying industrial innovations: Sources, characteristics, and financing', *The Review of Economics and Statistics*, 77(1), 55–65.

Marcusen, A. (1996) 'Sticky places in slippery space: A typology of industrial districts', *Economic Geography*, 72(3), 293–313.

Matusiak, K. B. and Guliński, J. (eds) (2010) *System transferu technologii i komercjalizacji wiedzy w Polsce – Siły motoryczne i bariery*. Poznań, Łódź, Wrocław, Warszawa: Polska Agencja Rozwoju Przedsiębiorczości.

Melin, G., and Persson, O. (1996) 'Studying research collaboration using co-authorships', *Scientometrics*, 36(3), 363–377.

NOT (2010) *Centrum Innowacji NOT. Broszura.*

OECD (2010) *Higher Education and Regions. Globally competitive, locally engaged*: OECD.

Olechnicka, A. (2012) *Potencjał nauki a innowacyjność regionów*. Warszawa: Wydawnictwo Naukowe Scholar.

PARP (2010) *Benchmarking klastrów w Polsce – 2010. Raport z badania*. Warszawa: Polska Agencja Rozwoju Przedsiębiorczości.

Paytas, J., Gradeck, R., Andrew,s L. (2004) *Universities and the Development of Industry Clusters*. Pittsburgh and Washington, DC: US Economic Development Administration.

Pechter, K. and Kakinuma, S. (1999) 'Coauthorship linkages between university research and Japanese industry', in L. M. Branscomb, F. Kodama and R. Florida (eds), *Industrializing Knowledge. University–industry linkages in Japan and the United States*. Cambridge, MA, London: MIT Press, pp. 102–127.

Power, D. and Malmberg, A. (2008) 'The contribution of universities to innovation and economic development: In what sense a regional problem?' *Cambridge Journal of Regions, Economy and Society*, 1(2), 233–245.

Santarek, K. (ed.) (2008) *Transfer technologii z uczelni do biznesu: Tworzenie mechanizmów transferu technologii.* Warszawa: PARP.

Schartinger, D., Rammer, C., Fischer, M. M. and Frohlich, J. (2002) 'Knowledge interactions between universities and industry in Austria: Sectoral patterns and determinants', *Research Policy*, 31(3), 303–328.

Starczewska-Krzysztoszek, M. (2008) *Ranking najbardziej innowacyjnych firm w Polsce.* Warszawa: Lewiatan.

Torre, A. and Rallet, A. (2005) 'Proximity and localization', *Regional Studies*, 39(1), 47–59.

Travers, J. and Milgram, S. (1969) 'An experimental study of the small world problem', *Sociometry*, 32(4), 425–443.

UNESCO (2010) *UNESCO Science Report: The current status of science around the world.* UNESCO Publishing.

Youtie J. and Shapira P. (2008) 'Building an innovation hub: A case study of the transformation of university roles in regional technological and economic development', *Research Policy*, 37(8), 1188–1204.

Žížalová, P. (2010) Geography of knowledge-based collaboration in a post-communist country: Specific experience or generalized pattern?' *European Planning Studies*, 18(5), 791–814.

6 'What do you offer?'

Interlinkages of universities and high-technology companies in science and technology parks in Berlin and Seville

Sascha Brinkhoff, Lech Suwala and Elmar Kulke

1 Introduction[1]

The innovation process is characterized by interactive learning involving multiple entities (Lundvall 1988). The flow and exchange of information and knowledge spurs R&D activities and innovation enhancing regions' economic growth and competitiveness. Being aware of this process, government policies aim to stimulate and even plan inter-organizational interaction (Ibert 2010; Hewitt-Dundas 2011). In this respect, science parks have become prominent instruments of government infrastructure support to promote business-to-science relationships based on geographical proximity. Universities as sources of knowledge play a major role in science parks and in regions in general. They primarily contribute to the regional economy's competitiveness and innovativeness by providing access to new scientitifc knowledge, developing solutions for specific problems and by offering access to skilled talent and equipment, as well as research techniques (Hewitt-Dundas 2011).

The purpose of this chapter is to discuss the role of universities in two science parks in Berlin and Seville and their contribution to localized interactive innovation processes. Both case studies provide similarities in their development process, although, as we shall see later, distinct conceptual framework conditions regarding business-to-university linkages differ. In the case of science parks, the literature predominantly stresses the great significance of geographical proximity. In this chapter, however, we focus on the discussion of the multidimensional character of proximity – integrating 'relational proximity' to the strict geographical interpretation of proximity. Thus, we aim to add further aspects to the 'soft architecture of learning' (Thune 2009: 9) within the regional and science park-related network of university–industry–government relations ('Triple Helix') (Etzkowitz and Leydesdorff 2000: 109).

Finally, we make some proposals for policy implications regarding how to more successfully stimulate business-to-university linkages and associated localized innovation processes in science parks in particular, as well as in regions.

2 Theoretical background of business-to-university linkages

Based on a revival of initial ideas by Marshall (1919, 1920) and his notion of the industrial district, new ideas in the search for endogenous regional growth models began to flourish among regional economists, geographers and planners some thirty-five years ago. This tendency was propelled both by the economic crisis in the production logic of large-scale enterprises and criticism of traditional national state-led regional economic policies in the late 1970s.

Alternative local and regional economic development strategies arose mainly from successfully operating SME-based regions in the so-called 'Third Italy' (Bagnasco 1977) and were adopted by different academic circles. New concepts such as innovative milieus (Aydalot 1986), localized production systems (Bouchrara 1987), new industrial spaces (Storper and Scott 1989), clusters of innovation (Porter 1990), regional innovation systems (Lundvall 1988, 1992; Braczyk *et al.* 1998) and learning regions (Florida 1995; Asheim 1997; Morgan 1997) were created. The mutual ideas of these concepts emphasized the importance of spatial proximity in generating innovation and knowledge (Kulke 2008). Although further crucial growth and developmental factors such as qualified human capital, local business culture, the educational system and infrastructure were identified, the firm still was considered to be the nucleus for economic progress in the capitalist logic (Moulaert and Sekia 2003).

This perspective was questioned and relativized by the Triple-Helix approach (e.g. Leydesdorff and Etzkowitz 1996; Etzkowitz and Leydesdorff 1997). This concept suggests a closer look at university–industry–government relations and the different constellations between these three stakeholders in territorial innovation systems. Accordingly, dynamics of innovation evolve out of close linkages between the private sector, government laboratories and academic research groups (Etzkowitz and Leydesdorff 2000). In spite of the importance of the different linkages, it is observed that universities in particular 'can play an enhanced role in innovation in increasingly knowledge-based societies' (Etzkowitz and Leydesdorff 2000: 109) and inhibit a 'knowledge-hub-function' (Youtie and Shapira 2008: 1189). Two policy measures are held responsible for the increase of knowledge transfer from universities to companies: on the one hand, purposive legislation designed to stimulate research joint ventures between universities and firms (e.g. the European Union Framework Programmes) and on the other hand, a major shift in the intellectual property regulation in favour of universities (e.g. enactment of the Bayh-Dole Act in 1980) (Mowery *et al.* 2001; Rafferty 2008; Grimm 2011).

Since then, manifold studies have concentrated on the nexus between universities and companies as a constituent property in realizing innovative environments (e.g. Adams *et al.* 2001; Poyago-Theodoky *et al.* 2002; Bercovitz and Feldman 2006). Most of them were carried out in the context of science parks (e.g. Siegel *et al.* 2003a,b,c; Kulke 2008), which are interfaces between industry, universities and other research institutions. Such territorially bounded entities provide an excellent research setting when examining the spatial dimension of these linkages.

With regard to the research question, science parks refer to property-based initiatives that consist of (1) a technology park, (2) an incubator and (3) at least one higher education institution (HEI). All three elements need governance that can be carried out by an informal team, a single on-site manager or an on-site management company (Grayson 1993).[2] Independent of the mode of governance, science parks pursue several objectives in terms of their impact on localized business–university relationships and the region in general. Objectives associated with inter-organizational linkages comprise the promotion of knowledge and technology transfer, facilitating the formation of new technology-based firms (NTBF), propelling the growth of existing high-technology firms, attracting new firms involved in cutting-edge technologies, and fostering strategic alliances. Regarding their regional impact, science parks aim to contribute to the regional economic development and the growth of employment, as well as to the regional identity and brand formation (Siegel *et al.* 2003c).

Academic discussion regarding whether spatial proximity is sufficient for university–industry knowledge transfer and technology offers ambiguous results. On the one hand, numerous empirical studies provide evidence that industrial innovations result from significant knowledge spillovers, which originated from the spatial proximity of universities as knowledge pools (Jaffe 1989; Acs *et al.* 1992; Jaffe *et al.* 1993; Fritsch and Franke 2004, Fritsch *et al.* 2008). These findings are coherent with the widely accepted explanation that knowledge spillovers tend to be localized owing to their implicit and non-codified nature, of which the exchange requires face-to-face contacts between universities' and companies' representatives (Fugukawa 2006). On the other hand, several studies rather contradict this perception by underlining that geographical proximity is neither a necessary nor a sufficient condition for innovation (Boschma 2005; Gallaud and Torre 2005). Furthermore, some researchers show that geographic proximity does not seem to be an important factor in science parks for strenghening business–science realationships (Vedovello 1997; Huber 2012).[3]

Thus, recent research has focused on detailed analysis of the complex and diverse relationship between learning, innovation and the multidimensional nature of proximity (Boschma 2005; Torre and Rallet 2005; Knoben and Oerlemans 2006). It is argued that a better understanding of the interplay of the different dimensions of proximity may shed some more light on the complexity of collective learning on various spatial scales (Thune 2009). Different dimensions of proximity are considered in the literature (in particular geographical, social, cognitive, organizational and institutional proximity).

Maskell *et al.* (2006) distinguish three dimensions of geographical proximity – vertical, horizontal and social. The vertical dimension emphasizes learning by interacting among companies in vertically integrated relationships, while the horizontal dimension underlines the process of learning by comparing and observing co-located competitors. Finally, learning processes created by unintentional exchange of information and frequent face-to-face interaction, also referred to as 'local buzz' (e.g. Bathelt *et al.* 2004), define the social dimension of geographical proximity. The idea of temporary geographical proximity adds another aspect to this debate (Gallaud and Torre 2004; Torre and Rallet 2005; Torre 2008; Maskell *et al.* 2006). It takes the changes in production and the advancements in transportation as well as communication technologies, which have modified the perception of distance, into account. As a consequence, the need for geographical proximity has become more temporary, which is realized through increased mobility (Torre 2008). Temporary geographical proximity in terms of industry gatherings such as conferences and trade shows is critical in developing and maintaining inter-organizational relationships. Therefore, these platforms are labelled 'temporary clusters' (Maskell *et al.* 2006: 2).

Social proximity is strongly linked to the concept of embeddedness (e.g. Granovetter 1985; Boschma 2005). It argues that economic relations and outcomes are heavily affected by social ties and social capital, respectively. Socially embedded relations between actors are defined by trust based on friendship, kinship and experience. The embeddedness literature emphasizes that the level of social embeddedness of a company's relationships influences the firm's interactive learning and innovativeness. Lundvall and Johnson (1994: 28) also refer to the great importance of 'know-who' in terms of social settings as a specific category within the concept of knowledge,[4] while, at the other end of the spectrum, too much social proximity may reduce learning capability.

Cognitive proximity is highlighted as another fundamental dimension of proximity that facilitates communication and learning between different entities. The effective transfer of knowledge requires the absorptive capacity to identify, understand and exploit new knowledge. Thus, firms and research institutions, for instance, search for partners with similar references and knowledge bases in order to process the exchanged knowledge successfully (Nooteboom 2000). It is, however, also argued that a certain cognitive distance should be maintained to enable interactive learning (Boschma 2005; Thune 2009).

Organizational proximity is defined as the extent to which relations, namely, the rate of autonomy and control, are shared in an organizational arrangement – within or between organizations. It facilitates interactive learning, since new knowledge creation is characterized by uncertainty and opportunism. For example, strong control mechanisms are needed to ensure intellectual property rights. Tightly managed networks with high organizational proximity not only coordinate transactions, but also facilitate the exchange of information and knowledge (Knoben and Oerlemans 2006). Too much organizational proximity, however, results in a lack of flexibility (Boschma 2005).

In contrast to the previously illustrated social proximity stressing social embeddedness on the micro level, the concept of institutional proximity is related to the institutional framework such as norms and values of conduct, which reduce uncertainty and transaction costs at the macro level. Institutions are distinguished between formal institutions (e.g. laws and regulations) and informal institutions (e.g. common language, cultural norms and habits). Furthermore, social, institutional and organizational proximity are strongly interrelated, as inter-organizational relations are deeply embedded in institutional settings (Boschma 2005; Thune 2009). Institutional set-ups influence the creation of new knowledge by defining what kind of knowledge is perceived as useful, finding knowledge, assessing its usefulness and enabling its exploitation to solve distinct problems. Thus, cognitive and normative rules, values, norms and traditions affect the perception of what is considered valuable and how to find and implement solutions (Thune 2009).

This chapter aims to examine the significance of proximity in its multiple dimensions in business-to-university relations in the science parks in Berlin-Adlershof and Seville-Cartuja.

3 The Berlin-Adlershof and Seville-Cartuja science parks

3.1 The case study Berlin-Adlershof

The Adlershof Science and Technology Park was founded in 1991. Today about 866 companies, eleven non-university research institutions[5] and six natural science departments of the Humboldt-Universität zu Berlin (HUB) are located at the site. Approximately 425 high-technology companies primarily operate in the six key clusters: optics and photonics, material and micro system technologies, ICT, environmental technologies, energy and biotechnology. In total, about 14,100 employees work in the science park – about 11,500 of them in the private sector and 2,600 at the university and non-university research institutions. Furthermore, roughly 7,900 students study and work in the science park. In 2010, all companies generated a total of €2.1 billion in turnover (see Table 6.1) (WISTA-Management GmbH 2011).

Since its establishment in 1991, high-technology companies in particular have shown impressive growth. In the time period between 1996 and 2010, the number of these companies has more than doubled from around 200 to 425. Turnover has increased threefold from approximately €190 million to about €580 million, while employment has also more than tripled from approximately 1,600 employees to about 5,000 employees (Suwala and Dannenberg 2009; WISTA-Management GmbH 2011). The large majority of the businesses, however, are relatively small. Nevertheless, most of them seem very successful on the national and international market. Insolvency rates have been very low (Kulke 2008). The OECD (2010) ranked the science park in Berlin-Adlershof among the fifteen biggest science and technology parks worldwide and named it as one of the most successful high-tech locations in Germany.

The history of research and sciences in Adlershof dates back to the beginning of the twentieth century. In 1909, the opening of the first airport for engine-powered aircrafts, Johannisthal, was the starting point for the location of the aeronautical industry and related research institutions at the site. Several research institutions – for example, institutions of the German Laboratory of Aviation ('Deutsche Versuchsanstalt für Luftfahrtforschung') – and further aerospace-related production companies (e.g. Erich Rumpler Aircraft Construction Company and the Fokker Airplane Factory) followed in the 1920s and 1930s (Mieg and Mackrodt in press).

Table 6.1 Overview of the science parks' characteristics and figures in 2010

	Berlin-Adlershof	*Seville-Cartuja*
	Science and Technology Park	*Science and Technology Park*
Date of establishment	March 1991	October 1993
Area (in km²)	4.2 km²	6.7 km²
Used space (m²) in 2011	390,000 m²	349,000 m²
Total employment	14,148	15,065
No. of companies (total)	866	336
No. of high-technology companies	425	177
No. of service companies	302	159
No. of media companies	139	n.a.
Employment in high-technology companies	5,007	7,857
Employment in services etc.	7,839 (incl. media city)	3,996
Turnover of companies	€2.03 billion	€1.91 billion
No. of higher education institutions (HEIs)	1 (six departments of Humboldt-Universität zu Berlin)	5 (e.g., University of Seville, EOI Business School)
No. of students	7,900	8,000
No. of non-university R&D institutions	11	36
Employment in HEIs and R&D institutions	2,626	3,212
Budget of HEIs and R&D institutions	€230.3m	€202.5m
Investments	€2 billion (1991-2010)	€2 billion (1992-2010)

Sources: WISTA-Management GmbH (2011); Cartuja 93, S.A. (2011).

After 1945, a great shift in the scientific and economic orientation of the site took place, as research and production in aerospace and aeronautics was strictly prohibited in post-war Germany. From the 1950s onwards, the Academy of Sciences, which was the central research institution in the GDR, and several associated R&D centres, predominantly in natural sciences (physics, chemistry etc.), were located in Adlershof. A great share of research was conducted in close cooperation with industry. By 1989, more than 5,400 people worked in fifteen research institutions of the Academy of Sciences in Adlershof. Adlershof also became the centre of TV production and broadcasting of the national television of the GDR, starting from 1956. In addition, the site was home to the Security Regiment Felix Dzerzhinsky of the armed forces of the GDR (Kulke 2008).

After 1990, Adlershof experienced major changes. The facilities of the army were closed immediatedly. Facilities related to TV activities were dissolved and privatized. The Academy of Sciences of the GDR was also liquidated, and the restructuring of its research resources went in different directions. Eight of the former Academy of Science institutions that predominantly focused on basic research in natural sciences and technology were integrated into the public, non-university research institutions of the Max-Planck Society, Fraunhofer-Gesellschaft and Leibniz Society. More than 30 per cent of the former roughly 5,400 researchers of the Academy of Sciences were absorbed by these R&D institutions (Mieg and Mackrodt in press). In contrast, R&D institutes that carried out industry-oriented research were dissolved. In some cases, they re-emerged as newly founded companies. In fact, more than 100 new companies were founded by former Academy of Sciences' employees and also located at the science park (Mieg and Mackrodt in press; Suwala and Dannenberg 2009).

Since German reunification, the state of Berlin has primarily pursued an innovation-oriented and science-led regional economic development strategy to compensate for the severe loss of employment in manufacturing. As a consequence, several technology parks and incubators have been established. The Adlershof science park is the largest and most prominent project. In 1990, an urban planning concept for the restructuring of Adlershof was developed. Based on its tradition in natural sciences and media, the objective was to create an international centre of science, business and media, designed as an integrated city (Kulke 2008; Mieg and Mackrodt in press) (see Figure 6.1).[6]

It was aimed at generating a dynamic and innovative growth process based on the geographical proximity of complementary units – universities, R&D institutions and the private sector (Kulke 2008). As such, the decision to integrate a university campus of the Humboldt-Universität zu Berlin was of fundamental significance. In 1991, the state of Berlin and the university agreed to relocate the spatially dispersed natural sciences' departments to the newly developed science park in Adlershof (Kulke 2008). As a result, since 1998, six major natural sciences' departments of the Humboldt-Universität zu Berlin have moved sucessively to the new university campus in the southeast of the city.[7] Research institutions and university departments predominantly complement the science park's technology clustering strategy. As an illustration, BESSY II and the Leibniz Institute for Crystal Growth (IKZ) conduct research in material technologies and photonics, while the Fraunhofer Institute FIRST focuses on research in ICT. At the beginning of the science park's development, however, the proximity-oriented approach did not include any network formation strategy (Kulke 2008).

3.2 The case study Seville-Cartuja

The Cartuja Science and Technology Park was established in October 1993 at the former site of the Expo 1992 in Seville. Today, about 340 companies are located there. Approximately 177 of them are high-technology companies in the science

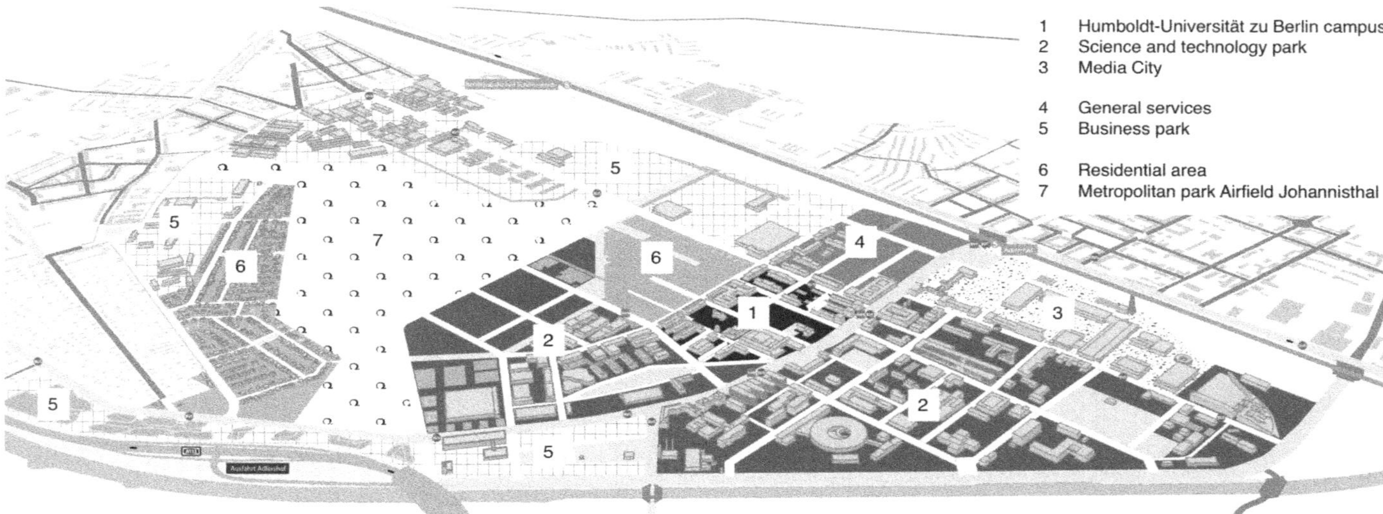

Figure 6.1 Conceptual plan of the Berlin-Adlershof science park.

Source: WISTA-Management GmbH (2012) (modified by authors).

park's major clusters: ICT, engineering, environmental technologies, biotechnology/agro-food, energy and health. Furthermore, thirty-six R&D institutions and five universities (e.g. the University of Seville and EOI Business School) with approximately 8,000 students reside in the science park as well.[8] In total, roughly 15,060 employees work in the STP Cartuja – about 7,860 of them in high-technology companies. In 2010, the companies generated a total turnover of approximately €1.9 billion (see Table 6.1) (Cartuja 93, S.A. 2011).

In particular since 2003, the science park has experienced successful growth. In the time period between 2003 and 2010, the number of companies increased from 245 to 377. The turnover grew by more than 50 per cent to around €1.91 billion, while the total employment also increased by about 60 per cent to 15,065 employees. About 342,000 m² of space is currently used, compared to only about 77,200 m² at the very beginning in 1999 (Cartuja 93, S.A. 2005, 2011).

The Cartuja science park derived its name from the monastery that is situated on the island. Already in 1989, before the Expo took place in 1992, the regional government of Andalusia contracted a team of scientists, which was led by Manuel Castells and Peter Hall, to design a redevelopment plan for the Expo site. According to PINTA,[9] a new model of technology parks was to be pursued. The model combined public and private applied research centres, institutions dedicated to the transfer of technology to the Andalusian economy and training centres in new technologies (Cartuja 93, S.A. 2010a). Two primary goals were defined: first, to use the high-quality infrastructure constructed for the Expo to promote applied research in the strategic technology fields in Andalusia (e.g. ICT, new materials, biotechnology applied to agriculture, food technology and environmental technologies), and second, to transfer and apply the new knowledge to the regional economy and on the international scale (Vázquez-Barquero and Carrillo 2004). The concept included three major elements (see Figure 6.2): 'Tecnopolis' – the science and technology park; a service hub (including public administration, business services and general services) and cultural as well as recreational activities (including a theme park, museums, sport facilities such as the Olympic Stadium and a park) (Cartuja 93, S.A. 1994).

After the end of the World Fair, the restructuring process began. In October 1993, the science park's management body, Cartuja 93, was established, and the science park's operation started.[10] The unique characteristics of the Cartuja science park are threefold. First, the science park is based on the conversion of the Expo'92 site equipped with high-quality facilities and infrastructure. Second, it is situated in the vicinity of Seville's historic city centre, which also transfers a certain urbanity to the science park. Third, the shareholders of the managing company Cartuja 93 were positioned at various adminstrative levels, namely, the regional government of Andalusia (51 per cent), the Spanish government (34 per cent), the City of Seville (10 per cent) and the provincial administration of Seville (5 per cent) (González Romero 2002).

Similar to Berlin-Adlershof, the concept of Seville-Tecnópolis is based on the combination of a technology park with high-technology companies alongside complementary research intitutions and university faculties. Geographical proximity is aimed at lowering regional deficits in business-technology-science interaction (González Romero 2002). Consequently, the co-location of regional public

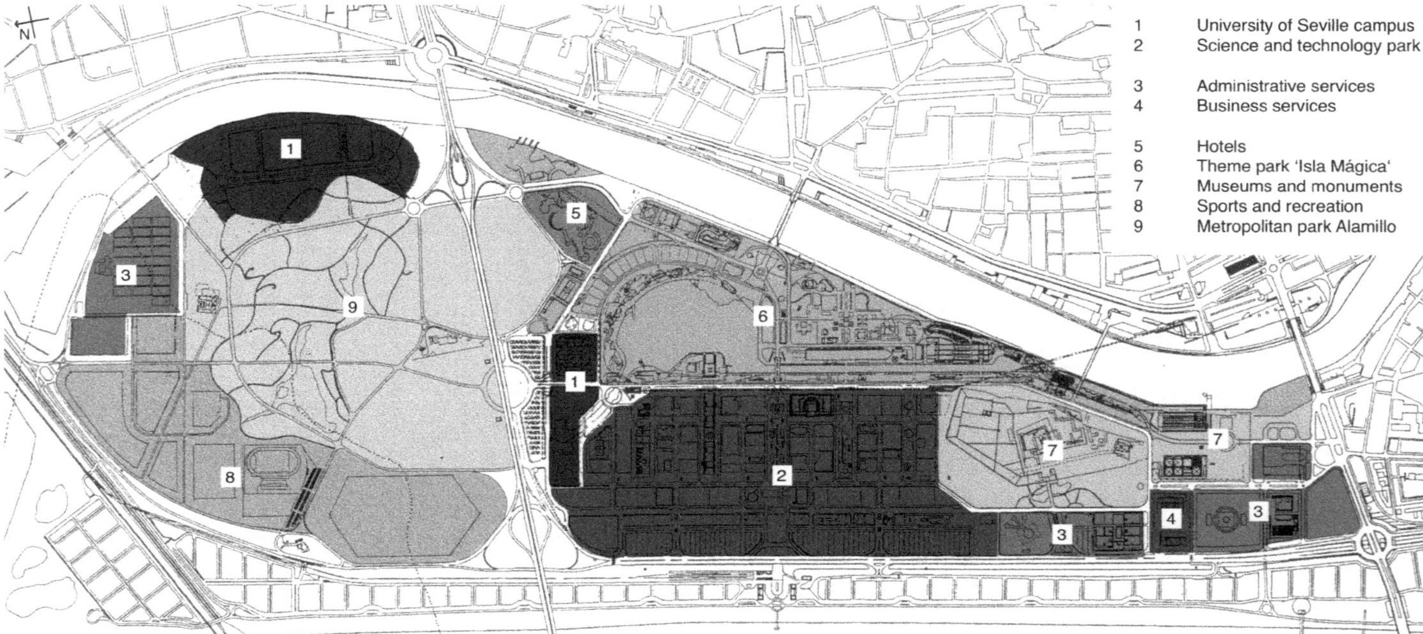

Figure 6.2 Conceptual plan of the Seville-Cartuja science park.

Source: Cartuja 93, S.A. (1994) (modified by authors).

administration institutions, multiple R&D centres and several faculties, as well as research units of the University of Seville (School of Engineering and Faculty of Communication)[11] are all critical prerequisites to the stimulation of innovation at the Cartuja science park (Cartuja 93, S.A. 2010b).

In summary, both science parks represent examples of large-scale redevelopment landmark projects, which are aimed at spurring interactive innovation processes and to promote endogenous regional economic growth in specific high-technology sectors. Universities and further R&D institutions are fundamental elements of the science parks. In the recent past, Berlin-Adlershof and Seville-Cartuja have evolved successfully, which can also be illustrated in quantitative terms, for example, by the number of companies, growth of employment and the companies' turnover (see Table 6.1). The Adlershof science park is characterized by a large number of micro and small companies. In contrast, Cartuja 93 also heavily focused on the attraction of large national and international firms at the project's outset (González Romero 2002; Vázquez-Barquero and Carrillo 2004). Consequently, its company portfolio today includes several large national companies and branches of prominent international firms.

4 Inter-organizational networks at the Adlershof and Cartuja science parks

The data were collected within the EU Interreg project 'Knowledge network management in technology parks (Know-Man)'. In total, fifty-four NTBFs of different sizes and ages were surveyed at the two innovation sites from October 2010 to April 2011. The two case studies are characterized by a methodological mix of standardized questionnaires and expert interviews that aimed to analyse business-to-science interaction in the science parks with a special focus on co-located universities. In the questionnaire, the companies assessed the intensity of interorganizational interaction quantitatively. Here, relations were distinguished as informal, formal and human resources-related, in which various activities were analysed (see Vedovello 1997, among others).

The expert interviews examined the pecularities of the businesses' relationships to universities and other scientific institutions in more detail. Important apects of analysis were the origin, the evolution and the geographical scope of these linkages. In this regard, it was specifically focused on the importance of social and geographical proximity. Furthermore, additional pre-requisites and framework conditions associated to other dimensions of proximity that define successful business–university interaction were assessed.

The majority of the companies surveyed at the two science parks are characterized as R&D-intensive high-technology companies that maintain strong or even multifaceted long-term relationships to scientific institutions in general. Typically, firms take great advantage of the broad scientific and research infrastructure in the respective science park and region.[12]

At the Adlershof science park, however, the intensity of business–university linkages appears to be relatively weak. Only in rare cases do companies strongly collaborate with specific university departments (e.g. the Department of Physics,

Department of Computer Science and Department of Chemistry) and related research groups, respectively. Interaction mostly takes place in terms of informal relationships, such as personal contacts and student placements. Rarely have joint research projects or even general cooperation agreements been implemented. In contrast, many of the interviewed companies have instead stressed their informal and formal relationships to non-university research institutions co-located in the science park. Other studies have also detected a low significance of business-to-university interlinkages amidst strong relations between the private sector and non-university R&D institutions. Personal networks of former employees of the Academy of Sciences, who now work at R&D institutions and companies have remained well intact. Furthermore, strong (horizontal and vertical) links among co-located businesses have developed rapidly at the science park (Kulke 2008; Kulke and Wessel 2008).

The survey at the Cartuja science park reveals similar findings, although important details diverge. In contrast to González Romero (2002), who assessed the science park as having a weak role as a platform for interactive innovation activities caused by weak business-to-science interaction, we found instead that profound business–university ties have been established. The School of Engineering of the University of Seville is the central scientific partner in the science park overall, which was also underlined by González Romero (2002). Similar to Adlershof, rather informal forms of collaboration dominate (i.e. personal contacts, student placements and the use of equipment and laboratories). Nevertheless, mutual formal efforts such as general cooperation agreements, joint laboratories, contract research, joint R&D projects, and commercialization activities of research results in terms of university spin-offs, have all been realized on a larger scale. In contrast, cooperation with other co-located R&D institutions is rather weak and comprises mainly personal linkages.

In addition, the vast majority of the companies in both case studies also maintain strong relationships with scientific institutions outside the science parks – both regionally and even on a national and international scale. In fact, external relationships even appear to be more important to the companies than science park-based links. This corresponds with the discussion among scholars about the importance of permanent geographical co-location and the abolition of distance in collective learning (Boschma 2005; Gallaud and Torre 2005). The case studies reveal, however, that the dimension and purposes of business–university interaction diverge as a result of distinct geographical scales. Informal interaction, in terms of personal contacts and access to talent as well as equipment, is more important at the science parks. This supports other findings that highlight the access to skilled labour as a key driver of business links to universities (Hewitt-Dundas 2011). In contrast, formal cooperation, for instance, in publicly funded R&D projects is realized instead through scientific partners located outside the science park. Hewitt-Dundas (2011) finds that non-local financial funding sources, such as national and EU research programmes (e.g. 7th EU Framework Programme), strongly affect non-local cooperation of business–university partnerships. Additionally, highly specialized high-technology companies, in particular those that cater to very specialized niche markets, must operate on a national or international scale. This considered, research partners also are most likely to spread beyond the local scale.

Figures 6.3 and 6.4 illustrate the geographical scope and interaction modes of the surveyed companies' business-to-science networks at the Berlin-Adlershof and Seville-Cartuja science parks.[13]

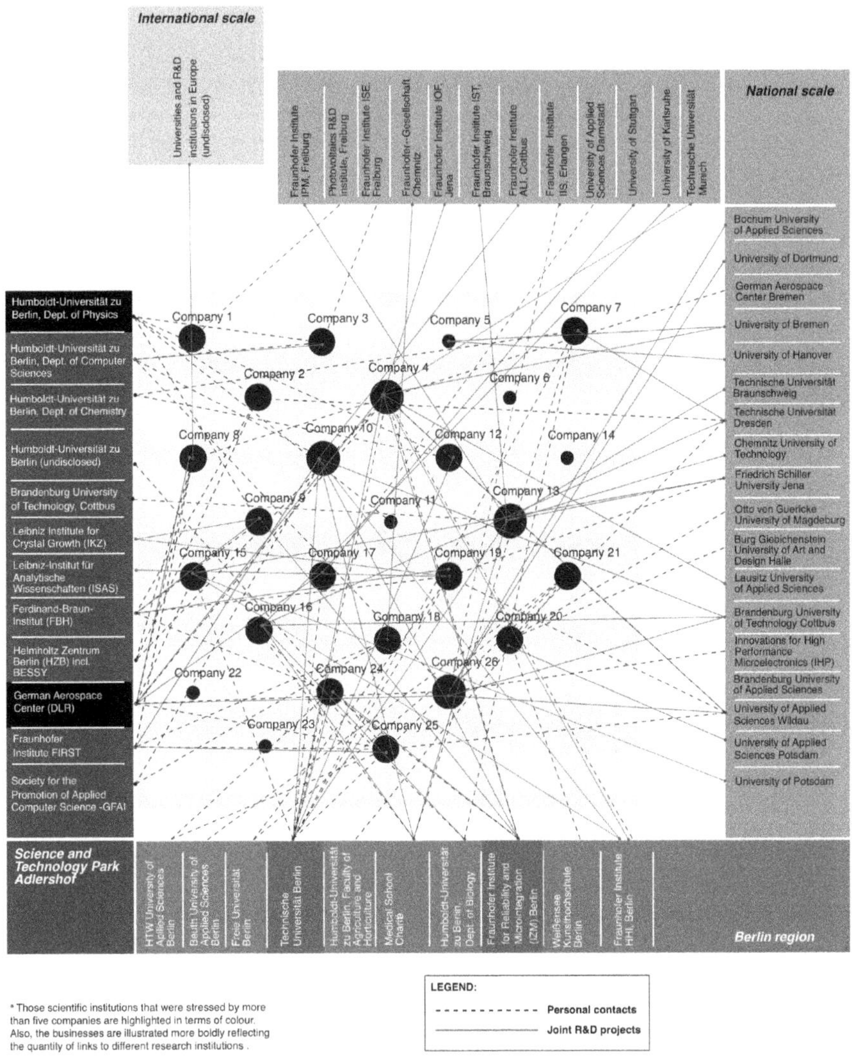

Figure 6.3 Business-to-science network at the Berlin-Adlershof science park.

Source: Authors' illustration.

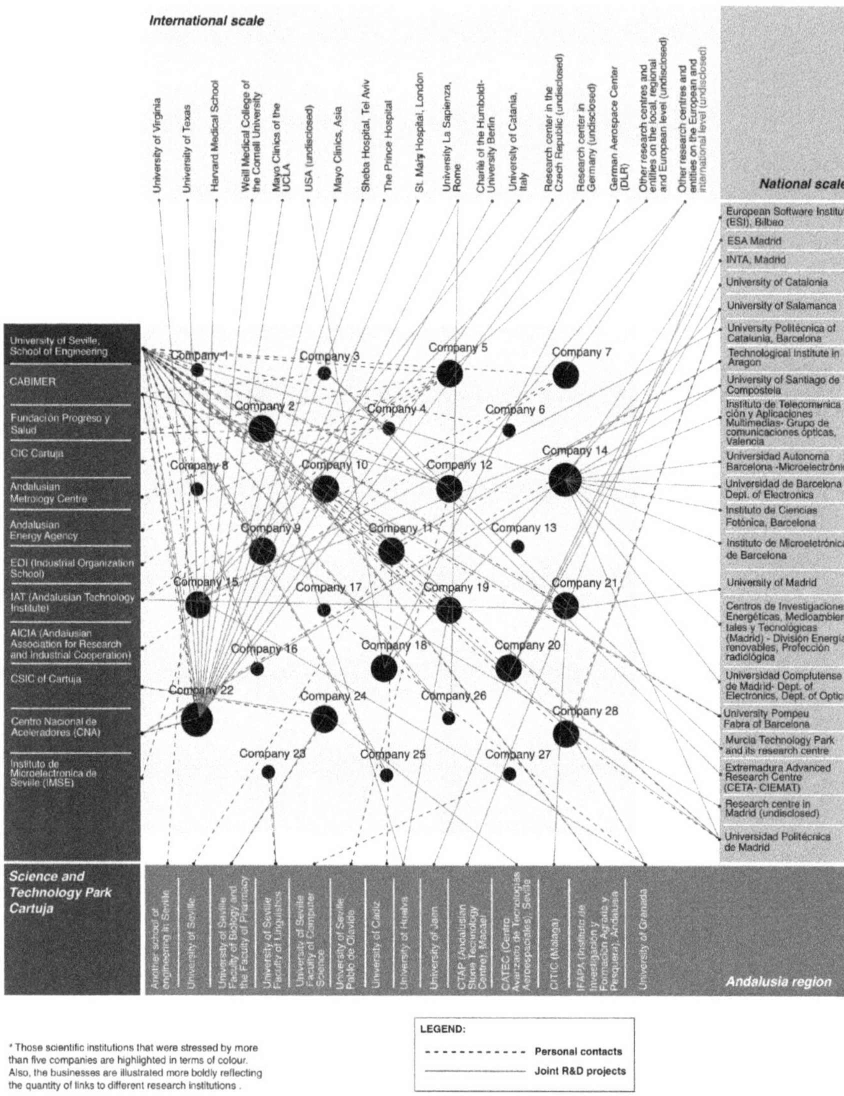

Figure 6.4 Business-to-science network at the Seville-Cartuja science park.

Source: Authors' illustration.

In contrast to the initial concepts at the science parks' development, the geographical co-location of complementary scientific and business units alone does not suffice in establishing strong inter-organizational relationships (González Romero 2002; Kulke 2008). Although the university has been located for more than eight years entirely at the Adlershof science park,[14] linkages between businesses and university chairs remain relatively weak. The findings at the Cartuja science park, however, rather support the concept stressing the significance of geographical co-location.

The literature that discusses the relevance of the different dimensions of proximity and their specific interrelations, perceives geographical proximity instead as a complementary framework condition. It is rather argued that additional criteria (and dimensions of proximity) are critical for successful business-to-university interaction in science parks. Cognitive and social proximity are considered fundamental prerequesites for inter-organizational relations (Oinas 1997; Boschma 2005). Hewitt-Dundas (2011) even argues that the need for geographical proximity, in order to exchange knowledge, decreases as cognitive proximity among partners increases.

The two case studies show that social proximity is central to most of the innovation-related activities realized with universities and other scientific institutions in the science park and beyond. The overwhelming majority of interviewees consider previously established contacts to be fundamental for collaborative activities. Personal relationships are utilized frequently and intensely to establish inter-organizational relationships. These linkages are primarily based on personal ties to former work colleagues at universities or non-university research institutions or fellow students that currently hold a research position. Even referrals by other staff members are additional sources. Two examples from both sites illustrate the origin of university ties as a result of mutual professional and educational biographies:

> Many of these relationships are based on personal contacts, in particular to former colleagues at the Academy of Sciences [of the former GDR] that work at Humboldt-Universität zu Berlin or at research institutions now.
>
> (Company 2, STP Adlershof)

> Well, we have a lot of employees that have studied at the School of Engineering [of the University of Seville]. And their fellow students now work for some research groups at the university.
>
> (Company 11, STP Cartuja)

In particular, university and R&D spin-offs take advantage of their scientific background. Typically, interaction with mother scientific institutions at the science parks on informal, formal and human resource-based levels (e.g. personal relationships, joint research projects and staff exchange) remain very strong.

> We have very good relationships with the Departments of Physics and Computer Science based on our previous university contacts. This comprises

> talent, but we also realize joint research projects and staff exchanges, which is facilitated by the TransferBONUS programme.[15]
>
> (Company 7, STP Adlershof)

> The basis of the company is basically the School of Engineering of Seville, because the knowledge comes from there. [...] For us, the university is the key in everything, regarding the installations, possibilities of knowledge, possibilities to do projects, cooperation etc. [...] Pretty much priceless, also for the philosophy, as two of the partners are professors there.
>
> (Company 20, STP Cartuja)

At the same time, several spin-offs that were interviewed in Berlin-Adlershof originate instead from research and university institutions located outside the science park. Their linkages to these external institutions also are very strong. Thus, our findings strongly correlate with previous studies that underline the strong interrelation of social embeddedness and economic relations in localized innovation networks. Accordingly, these networks are less likely to evolve because of a need for face-to-face relations to exchange knowledge. Their formation is rather owing to mutual professional, academic biographies or the simple fact of belonging to the same social and family network (Torre 2008). In addition, we have found that companies not possessing these kind of personal ties generally experience great difficulties in gaining access to the universities at the science parks.

In some cases, collaborative relationships evolved without prior personal contacts. Companies were looking for additional competences in order to gain access to new expertise, to find specific solutions and to join larger research projects. Here, organizational proximity embodied by specific intermediary organizations played an important role in facilitating the formation of business-to-science and business-to-university linkages, respectively. At the Berlin-Adlershof science park, for example, the science park's management organization WISTA-Management GmbH, as well as locally rooted networks such as the business organization Technologiekreis Adlershof and the optical cluster OpTecBB, generally spur the development of personal relationships, which may result in interactive learning. At the science park in Seville, mostly young micro-companies, which have recently developed at the site, stress the broker role of the science park management body Cartuja 93 and other public intermediary institutions such as CTA and FIDETIA.[16] Generally, intermediary organizations facilitating the interaction between industry and sciences play a slightly stronger role at Seville-Cartuja than at Berlin-Adlershof.

> It's always easier when you know each other beforehand, when you know who you have to talk to than when you have to ask through the university in order to find the right person. [...] That's the reason why we are in various boards and networks such as OpTecBB. Not because we're expecting big business, but to get to know beforehand who works in what field and who could contribute to solutions in certain problems at some point.
>
> (Company 9, STP Adlershof)

> About 50 per cent of the connections came from the contacts I had from my previous research work experiences. The other 50 per cent of the relationships came through contacts to CSIC or are referred by the CTA, for example. Also, through a networking event initiated by the research group of CABIMER, they facilitated the meeting of companies and researchers.
>
> (Company 12, STP Cartuja)

In addition, financial support schemes of the respective regional and national governments as well as the European Union (e.g. the 7th Research Framework Programme) strongly contribute to enhanced business-to-science interaction (e.g. transfer of talent, spin-off creation and joint R&D projects). They provide some certainty to the involved organizations owing to the high degree of control (e.g. related to intellectual property). Joint activities, however, are not limited to the science park level. The geographical scope rather correlates with the institutional scope of the support programmes. Industry gatherings such as conferences and trade shows are important 'temporary clusters' (Maskell *et al.* 2006) to create and maintain non-local inter-organizational relationships ('global pipelines') (Bathelt *et al.* 2004: 41).

Another central criteria that defines the norms of the institutional framework of localized business-to-university relations is a mutual understanding of work culture and the capability to pursue a systematic business approach. In the case studies, the institutional distance between the companies and the universities is percieved as one of the major obstacles inhibiting stronger interaction. Two examples showcase the difficulties in combining the different work ethics of the universities and the NTBFs:

> The institution is the problem. They are structured that you can't work with them – to the point of very different ideas about the costs for certain activities.
>
> (Company 20, STP Adlershof)

> Sometimes it's like the university and the company don't speak the same language. [...] What happens is that the University [of Seville] is a very rigid element, which takes a lot to understand the concepts of a private company.
>
> (Company 16, STP Cartuja)

The lack of institutional proximity is generally a universal obstacle in both case studies. Interaction usually takes place on the personal level between university researchers and company staff. As a consequence, too much institutional distance can be supplemented by social proximity, which is indicated by individual interviewees.[17]

The relativeness of the universities' research and training foci to the companies' activities is critical for interaction as well. The companies' demand primarily focuses on scientific partners in practical-oriented and applied research. These are basic prerequisites in order to create market-oriented solutions. In fact, one of the reasons for the relatively weak business–university interaction at the Adlershof science park appears to be the great cognitive distance between companies and the university's natural sciences' departments, which heavily concentrate on basic research. At Berlin-Adlershof, specific non-university R&D institutions instead match the firms' need for more market-oriented research, which has resulted in stronger interaction. One entrepreneur in Berlin illustrates the greater cognitive proximity to co-located non-university R&D institutions in comparision to the university:

> Products that we develop have to be marketable. [...] We want the scientific background being proved. We also want the people's expertise, if it is useful and first applications that we can bring on the market, of course. There we want to be able to say we have tested this device there and there, and that's why it's good. [...] It works very well with such institutions that work in the same field like Ferdinand-Braun-Institut. But you can't do it with the Humboldt-Universität that, for example, focuses on nuclear physics.
>
> (Company 19, STP Adlershof)

Generally, it is evident that the entrepreneurial role has not been fully incorporated by most universities owing to the primary focus on teaching, research and third-party funds-related criteria in university benchmarking (Back and Fürst 2011). Already, the university's deficits and passive behaviour have resulted in a search for co-located companies and for other opportunity partners in the Berlin region. In particular, universities of applied sciences, which are characterized by a more demand-led approach, appear to be more suitable partners, for instance, in talent-related cooperation[18]:

> We have tried to support master theses, but the professors didn't want to, because it is too practical-oriented. [...] Now, we only do it with universities of applied sciences. [...] First, I thought, it's just my impression regarding the willingness for cooperation by HUB. Though, in the meantime there's an increasing indication that professors from HUB do not want to cooperate in general.
>
> (Company 22, STP Adlershof)

In terms of the competence and knowledge base, the university at Cartuja appears to match the firms' needs better than in the case of Berlin-Adlershof. The School of Engineering's strong role is based on its alignment towards practical-oriented research and training in engineering, which caters to the large number of engineering companies (e.g. civil and mechnical engineering) and other firms that operate in related fields (e.g. environmental technologies, ICT and aerospace) at the science park Cartuja.[19] An example for the demand-led approach at Cartuja is the creation of specific study programmes for engineers. The university's School of Engineering created specific training programmes in aeronautics and aerospace in response to the sector's strong position in the region and the firms' demand for skilled labour and industry-related research.[20]

The cognitive dimension of proximity is also strongly interrelated to social proximity, as it has been the case for many of the personal relationships established in universities and R&D institutions. Thus, a shared understanding based on a similar educational and professional background facilitates the formation of collaborative ties (Grossetti and Bes 2001). In particular, the strong university ties of the university spin-offs in both science parks illustrate the interrelation of social, cognitive and institutional proximity.

Nevertheless, the generally weak business-to-university ties at the science park Adlershof are largely caused by the lack of 'non-spatial proximity' (Knoben and Oerlemans 2006: 73). First, the university's natural science departments primarily focus on basic research, and consequently, rarely complement the companies' need for demand-led solutions. Second, prejudices about the rigid work culture and long bureaucratic decision-making processes remain strong. Third, many companies simply lack the personal relationships with university researchers. Finally, we argue that, in particular, the lack of transparency and openness regarding potential opportunities for collaborative activities is responsible for the university's weak local integration at the science park. One entrepreneur underlines the necessity for a more active approach to show the university's commitment at the science park:

> The last thing that I saw was a clean room, where two people work [...]. It was completely empty. Now the question is, what are the opportunities for others? Who will organize it, and does the Humboldt-Universität zu Berlin actually wants it? [...] It's not clear to me, how the HUB wants to integrate into this SME landscape here.
>
> (Company 17, STP Adlershof)

At the science park Cartuja, the School of Engineering appears more successfully integrated. Not only do the university spin-offs maintain strong informal and formal relationships, but multifaceted interlinkages with other co-located firms have also evolved. Research and training foci better match the companies' needs for market-oriented knowledge and solutions. Despite the important role of the School of Engineering, the University of Seville's Faculty of Communication and other non-university R&D institutions at the science park have not yet shown a great impact on localized interactive innovation processes.

5 Discussion and conclusion

The goal of this chapter was to contribute to the discussion of universities' role in interactive learning and innovation processes in science parks by analysing the significance of proximity in its multidimensional character. In the case studies of Berlin and Seville, geographical proximity between universities and business was an integral element of the science park concept.

In the comparative analysis, we find that geographical co-location alone is not enough to develop strong interdependencies between universities and businesses. Four particular dimensions of proximity have been more crucial for business-to-university linkages in the two science parks. First, personal relationships between entrepreneurs and scientists are fundamental (social proximity). Second, research and training at the universities must correspond to the businesses' activities (cognitive proximity). Third, the universities have to embody their role as 'entrepreneurial universities' (Bercovitz and Feldman 2006: 175; Etzkowitz and Leydesdorff 1997: 141) by adapting to the private sector's work culture (institutional proximity). Finally, support schemes and intermediaries also spur the inter-organizational knowledge transfer (organizational proximity).

The impact of the University of Seville on localized inter-organizational innovation activities at the science park Cartuja stands out in particular owing to the multifaceted interconnectedness of the University's School of Engineering. In particular, the strong personal contacts between entrepreneurs and researchers as well as a demand-led approach in research and training are identified as responsible ingredients for strong integration in the science park. Intermediaries that facilitate business-to-university cooperation and specific support schemes enhancing the development of a 'spin-off culture' (e.g. the CAMPUS programme of the Andalusian regional government) in the region are additional success factors.

At Berlin-Adlershof, a particular lack of convergence of university research and companies' needs is observed ('unrelated variety', OECD 2010: 77). The OECD (2010) concludes that the science-led economic development strategy in Berlin has only been sucessful in distinct industries such as the life science sector, in which innovations are more science-based (for example, in the case of the Berlin-Buch science park). More practical-oriented industries represented by the key cluster in the Adlershof science park, which are dominated by SMEs, show weaker results.

For a stronger role of universities in existing science parks and future science park development projects, we want to outline several policy recommendations. First, a convergence of university research and industry clusters has to be ensured. Second, universities have to adapt their work culture as well as research and training to their designated entrepreneurial role. Third, they have to communicate how they plan to contribute to the science park's objectives and the co-located entities of the private and research sector. Here, proactive marketing of the accessible university competencies and assets (e.g. knowledge, infrastructure and talent) is required. The university staff's activities and motivation also have to be aligned to the overall strategy and vision that define the university's third role (Etzkowitz and Leydesdorff 1997; Goddard *et al.* 2011). Finally, informal and low-barrier forms of business-to-science cooperation, such as student placements and use of university equipment, may serve as 'ice-breakers' for the development of continuous, multifaceted inter-organizational relationships. Additionally, mixed teams of researchers and company staff as well as shared co-working spaces can contribute to enhanced inter-personal work experiences and, consequently, in the building of personal trust.

In addition, we emphasize that superior factors, particularly time, are critical for developing social capital and for fostering interactive relationships in territorial innovation systems (Hewitt-Dundas 2011). In comparison to the step-by-step relocation of the various departments of the Humboldt-Universität zu Berlin between 1998 and 2003, the School of Engineering became part of the Cartuja science park in July 1997 (see Figure 6.5). Other case studies of science parks, for example, Sophia Antipolis in France, have also shown that the passage of time is a key factor for the formation of localized interactive innovation processes (Longhi 1999).

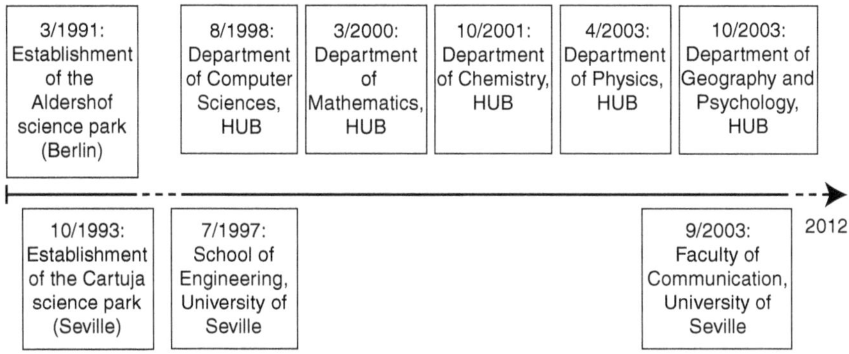

Figure 6.5 Time frame of university relocation at the two science parks.

Source: Authors' illustration.

Notes

1 Special thanks to David Brinkhoff for his support in graphic design.
2 A comprehensive overview of the definitions and properties of science parks is provided by Chan *et al.* (2009).
3 For a further discussion and a literature review about the role of the university as a driver for regional development see Back and Fürst (2011).
4 Lundvall and Johnson (1994) suggest a differentiation of economically relevant knowledge in four categories: 'know-what' referring to knowledge about facts, 'know-why' referring to scientific knowledge, 'know-who' referring to specific social relations and networking, as well as 'know-how' referring to skills.
5 For example, Helmholtz Zentrum Berlin (HZB) incl. BESSY II, German Aerospace Centre (DLR), Ferdinand-Braun-Institut – Leibniz-Institut für Höchstfrequenztechnik (FBH), Fraunhofer Institute FIRST, Leibniz Institute for Crystal Growth (IKZ), Max Born Institute for Nonlinear Optics and Short Pulse Spectroscopy (MBI), Leibniz-Institut für Analytische Wissenschaften ISAS e.V., Brandenburg University of Technology, Cottbus (branch institution), Physikalisch-Technische Bundesanstalt (PTB) and Federal Institute for Materials Research and Testing (BAM).
6 In 1991, the science park management company WISTA-Management GmbH was established. Mixed uses such as commercial activities, social infrastructure, residential areas and a park also were integrated into the development concept (Kulke 2008).
7 The new campus of the Humboldt-Universität zu Berlin in Adlershof successively grew in the following chronology: Department of Computer Sciences (8/1998), Department of Mathematics (3/2000), Department of Chemistry (10/2001), Department of Physics (4/2003), Erwin-Schrödinger Centre (5/2003), which contains the library, media centre, lecture and conference rooms as well as a cafeteria and a bookstore, and the Departments of Geography and Psychology (10/2003).
8 Prominent R&D institutions include the Andalusian Molecular Biology and Regenerative Medicine Centre (CABIMER), Andalusian Technology Institute (IAT), Centro Nacionales de Aceleradores (CAN), Scientific Research Centre 'Isla de la Cartuja' (cicCartuja) and the Institute of Microelectronics of Seville (IMSE).
9 Under the leadership of Manuel Castells and Peter Hall, researchers from various universities (e.g. Autonoma de Madrid, California-Berkeley, University of Seville and University of Malaga) and technicians of the Andalusion government participated in the project 'Proyecto de Investigacion sobre Nuevas Tecnologias en Andalucia

(PINTA)' (González Romero 2002). For further information see Castells, M. and Hall, P. (1992).

10 Cartuja93 is a public company. It is responsible for the development and management of the science park's infrastructure and facilities as well as for the network management activities at the science park (Interview Cartuja93, 23.11.2010).

11 The university's School of Engineering relocated to the Cartuja science park in July 1997, while the Faculty of Communication followed in September 2003 (Interview with Cartuja93 management on 15 February 2012).

12 In the states of Berlin and Brandenburg, in total seven universities, twelve universities of applied sciences, one medical school, three art schools and more than eighty-five R&D institutions (including nine institutes of the Fraunhofer Gesellschaft, nine institutes of the Max Planck Society, seven institutes of the Helmholtz Association of German Research Centres, and twenty-one institutes of the Gottfried-Wilhelm-Leibniz Scientific Community) are located (Sources: Websites of Berlin Senate Department for Education, Sciences and Research (accessed 5 March 2012); Brandenburg Ministry for Sciences, Research and Culture (accessed 5 March 2012)). In Andalusia, in total ten public universities and nine business schools with around 241,000 students and 34,200 graduates as well as more than 100 R&D institutions are located. Altogether, approximately 2,000 research groups conduct research at the scientific and higher education institutions in the region (Sources: Websites of Junta de Andalucía, Consejería de Economía, Innovación y Ciencia; Agencia IDEA (accessed 9 November 2011)).

13 The data and information material associated with the interviews were analysed to the best of our abilities. However, the illustration of the companies' networks only provides a snapshot and does not claim to provide a complete picture.

14 Depending on the specific university department (see paragraph 3.1).

15 The TransferBONUS programme is a financial support scheme provided by the Investitionsbank Berlin (IBB). The innovation voucher is primarily dedicated to initiate small joint projects (max. €15,000) between SMEs and scientific institutions in the State of Berlin (Source: Website of IBB, accessed 2 April 2012).

16 Corporación Tecnológica de Andalucía (CTA) is a private foundation promoted by the regional government of Andalusia. It focuses on the promotion of inter-organizational collaboration between science and the private sector in the region. Fundación para la Investigación y el Desarollo de las Tecnologías de la Información en Andalucía (FIDETIA) is a teaching and research foundation. It is dedicated to knowledge transfer and innovation between science and businesses in particular in ICT. It is part of the Faculty of Computer Science of the School of Engineering at the University of Seville (Sources: Websites of CTA and FIDETIA, accessed 5 March 2012).

17 'Usually there aren't any obstacles, when the personal level works. It's important that you know each other. [...] Openness for dialogue is important, especially in case of different objectives.' (Company 19, STP Adlershof).

18 OECD (2010) finds that universities of applied sciences, which are more application-oriented, are more suitable partners to match the needs of companies in the Berlin region. Typically, research-intensive universities in metropolitan regions such as Berlin focus on their national and international roles and show low interest in cooperation with regional SMEs. Thus, geographical co-location among innovation entities (universities, R&D institutions, companies etc.) in metropolitan regional innovation systems sometimes is not matched with social and institutional proximity (Asheim 2010; Boschma 2005). In addition, it is argued that a high share of SMEs work in more practical-oriented industries, where innovation is more incremental and demand-led than science-based (OECD 2010).

19 Additionally, nine spin-off companies of the University of Seville participated in the survey.

20 Helice Foundation – Andalusian Aerospace Cluster (Meeting within Know-Man project, 8 February 2012).

References

Acs, Z., Audretsch, D. and Feldman, M. (1992) 'Real effects of academic research: Comment', *American Economic Review*, 82, 36–367.

Adams, J. D., Chiang, E. P. and Starkey K. (2001) 'Industry–University Cooperative Research Centers', The Journal of Technology Transfer, 26(1–2), 73–86.

Asheim, B. T. (1997) 'Learning regions' in a globalised world economy: Towards a new competitive advantage of industrial districts?, in S. Conti and M. Taylor (eds) *Interdependent and Uneven Development: Global–local perspectives*. Aldershot, Ashgate, 143–176.

Asheim, B. T. (2010) 'La política regional de innovación de la próxima generación', *Ekonomiaz: Revista Vasca de Economía*, 70, Gobierno Vasco – Departamento de Hacienda y Administración Pública, Vitoria-Gasteiz, 106–131.

Aydalot, P. (1986) *Milieux Innovateurs en Europe*. Paris: GREMI.

Back, H.-J. and Fürst, D. (2011) Der Beitrag von Hochschulen zur Entwicklung einer Region als 'Wissensregion', No. 11, Hanover: ARL.

Bagnasco, A. (1977) *Tre Italia: La problematica territoriale dello sviluppo economico Italiano*. Bologna: Il Mulino.

Bathelt, H., Malmberg, A. and Maskell, P. (2004) 'Clusters and knowledge: Local buzz, global pipelines and the process of knowledge creation', *Progress in Human Geography*, 28, 31–56.

Bercovitz, J. and Feldman, M. P. (2006) 'Entrepreneurial universities and technology transfer: A conceptual framework for understanding knowledge-based economic development', *Journal of Technology Transfer*, 31(1), 175–188.

Boschma, R. A. (2005) 'Proximity and innovation: A critical assessment', *Regional Studies*, 39(1), 61–74.

Bouchrara, M. (1987) 'L'industrialisation rampante: ampleur, mécanismes et portée', *Economie et Humanisme*, 297, 37–49.

Braczyk, H.-J., Cooke, P. and Heidenreich, M. (eds) (1998) *Regional Innovation Systems*, London: UCL Press.

Cartuja 93, S. A. (1994) *El Proyecto Cartuja93*. Sevilla: Cartuja 93, S. A.

Cartuja 93, S. A. (2005) *2004 Inventario y Evaluación Tecnológica de las Empresas y Organizaciones Instaladas en el Parque Científico y Tecnológico Cartuja93*. Sevilla: Cartuja 93, S. A.

Cartuja 93, S. A. (2010a) *Evolución del Parque Científico y Tecnológico Cartuja 93 1993–2009*. Sevilla: Cartuja 93, S. A.

Cartuja 93, S. A. (2011) *2010 Inventory and Technological Review of the Companies and Organizations Located at the Cartuja Scientific and Technological Park*. Sevilla: Cartuja 93, S. A.

Castells, M. and Hall, P. (1992) *Andalucía: Innovación tecnológica y desarrollo económico*. Madrid: Espasa Calpe.

Chan, K. Y., Oerlemans, L. A. G. and Pretorius, M. W. (2009) 'Explaining mixed results of science parks performance: bright and dark sides of the effects of inter-organizational knowledge transfer relationships', *South African Journal of Industrial Engineering*, 20(2), 53–67.

Etzkowitz, H. and Leydesdorff, L. (eds) (1997) *Universities in the Global Economy: A Triple Helix of university–industry–government relations*. London: Cassell Academic.

Etzkowitz, H. and Leydesdorff, L. (2000) 'The dynamics of innovation: From national systems and "Mode 2" to a Triple Helix of university–industry–government relations', *Research Policy*, 29(2), 109–123.

Florida, R. (1995) 'Toward the learning region', *Futures*, 27(5), 527–536.

Fritsch, M. and Franke, G. (2004) 'Innovation, regional knowledge spillovers and R&D cooperation', *Research Policy*, 33, 245–255.

Fritsch, M., Henning, T., Slavtchev, V. and Steigenberger, N. (2008) 'Hochschulen als regionaler Innovationsmotor?' *Arbeitspapier 158*, Düsseldorf: Hans-Böckler Stiftung.

Fukugawa, N. (2006) 'Science parks in Japan and their value-added contributions to new technology-based firms', *International Journal of Industrial Organization*, 24(2), 381–400.

Gallaud, D. and Torre, A. (2004) 'Geographical proximity and circulation of knowledge through inter-firm cooperation', in R. Wink (ed.) *Academia–Business Links: European policy strategies and lessons learnt.* Basingstoke: Palgrave Macmillan, 137–157.

Gallaud, D. and Torre, A. (2005) 'Geographical proximity and the diffusion of knowledge: The case of SMEs in biotechnology', in G. Fuchs, P. Shapira and A. Koch (eds), *Rethinking Regional Innovation*. Dordrecht: Kluwer, 127–146.

Goddard, J., Vallance, P. and Kempton, L. (2011) 'Research impacts – the perspective of the academic community: Summary of the main findings from an online survey – June 2011', Part of the Civic University Study Programme (CUSP), unpublished report, Newcastle University.

González Romero, G. (2002) 'Actividad innovadora y creación de sinergias en el complejo innovador de Sevilla-Technópolis', *Archivo Hispalense*, 259–260.

Granovetter, M. (1985) 'Economic action and social structure: The problem of embeddedness', *American Journal of Sociology*, 91, 481–510.

Grayson, L. (1993) *Science Parks: An experiment in high technology transfer.* London: The British Library.

Grimm, H. (2011) 'The diffusion of Bayh-Dole to Germany: Did new public policy facilitate university patenting and commercialization?', *International Journal of Entrepreneurship and Small Business*, 12(4), 459–478.

Grossetti, M. and Bes, M. P. (2001) 'Encastrements et découplages dans les relations science–industrie', *Revue Française de Sociologie*, 42(2), 327–355.

Hewitt-Dundas, N. (2011) 'The role of proximity in university–business cooperation for innovation', *Journal of Technology Transfer*, July, 1–23, doi:10.1007/s10961-011-9229-4.

Huber, F. (2012) 'Do clusters really matter for innovation practices in information technology? Questioning the significance of technological knowledge spillovers', *Journal of Economic Geography*, 12, 107–126.

Ibert, O. (2010) *Dynamische Geographien der Wissensproduktion. Die Bedeutung physischer wie relationaler Distanzen in interaktiven Lernprozessen*, Working Paper. Erkner: Leibniz-Institut für Regionalentwicklung und Strukturplanung.

Jaffe, A.B. (1989) 'Real effects of academic research', *American Economic Review*, 79 (5), 957–971.

Jaffe, A. B., Trajtenberg, M. and Henderson, R. (1993) 'Geographic localization of knowledge spillovers as evidenced by patent citations', *Quarterly Journal of Economics*, 108(3), 577–598.

Knoben, J. and Oerlemans, L. A. G. (2006) 'Proximity and inter-organizational collaboration: A literature review', *International Journal of Management Reviews*, 8(2), 71–89.

Kulke, E. (2008) 'The technology park Berlin-Adlershof as an example of spatial proximity in regional economic policy', *Zeitschrift für Wirtschaftsgeographie*, 52(4), 193–208.

Kulke, E. and Wessel, K. (eds) (2008) 'Innovatives Milieu Adlershof', *Geographische Arbeitsmaterialien*, 142, 3–5.

Leydesdorff, L. and Etzkowitz, H. (1996) 'Emergence of a Triple Helix of university–industry–government relations', *Science and Public Policy*, 23, 279–86.

Longhi, C. (1999) 'Networks, collective learning and technology development in innovative high technology regions: The case of Sophia-Antipolis', *Regional Studies*, 33(4), 333–342.

Lundvall, B. A. (1988) 'Innovation as an interactive process: From user–producer interaction to the national system of innovation', in G. Dosi, C. Freeman, R. Nelson, G. Silverberg and L. Soete (eds), *Technical Change and Economic Theory*. London: Pinter, 349–369.

Lundvall, B. A. (ed.) (1992) *National Systems of Innovation: Towards a theory of innovation and interactive learning*. London and New York: Frances Pinter.

Lundvall, B. A. and Johnson, B. (1994) 'The learning economy', in *Journal of Industry Studies*, 1, 23–42.

Maskell, A., Bathelt, H. and Malmberg, P. (2006) *Building Global Knowledge Pipelines: The Role of Temporary Clusters*, DRUID Working Paper No. 05–20, Danish Research Unit for Industrial Dynamics.

Marshall, A. (1919) *Industry and Trade*, London: Macmillan.

Marshall, A. (1920) *Principles of Economics*, 8th edition. London: Macmillan.

Mieg, H. A. and Mackrodt, U. (in press) 'Science and technology park management: From business integration to technology and urbanity', in R. Holländer, C. Wu and N. Duan (eds) *Sustainable Development of Industrial Parks*, Berlin: Logos.

Morgan, K. (1997) 'The Learning Region: Institutions, innovation and regional renewal', *Regional Studies*, 31(5), 491–503.

Moulaert, F. and Sekia, F. (2003) 'Territorial innovation models: A critical survey', *Regional Studies*, 37(3), 289–302.

Mowery, D., Nelson, R., Sampat, B. and Ziedonis, A. (2001) 'The growth of patenting and licensing by U.S. universities: An assessment of the effects of the Bayh-Dole Act of 1980', *Research Policy*, 30, 99–119.

Nooteboom, B. (2000) *Learning and Innovation in Organizations and Economies*, Oxford: Oxford University Press.

OECD (2010) *Higher Education in Regional and City Development. Berlin, Germany 2010*. Paris: OECD Publishing.

Oinas, P. (1997) 'On the socio-spatial embeddedness of business firms', *Erdkunde*, 51(1), 23–32.

Porter, M. (1990) *The Competitive Advantage of Nations*, London: Macmillan.

Poyago-Theotoky, J., Beath, J. and Siegel, D. (2002) 'Universities and fundamental research: Reflections on the growth of university–industry partnerships', *Oxford Review of Economic Policy*, 18(1), 10–21.

Rafferty, M. (2008) 'The Bayh-Dole Act and university research and development', *Research Policy*, 37(1), 29–40.

Siegel, D. S., Waldman, D. and Link, A. N. (2003a) 'Assessing the impact of organizational practices on the productivity of university technology transfer offices: An exploratory study', *Research Policy*, 32(1), 27–48.

Siegel, D. S., Westhead, P. and Wright, M. (2003b) 'Assessing the impact of university science parks on research productivity: Exploratory firm-level evidence from the United Kingdom', *International Journal of Industrial Organization*, 21 (9), 1357–1369.

Siegel, D. S., Waldman, D., Atwater, L. and Link, A. N. (2003c) 'Commercial knowledge transfers from universities to firms: Improving the effectiveness of university–industry collaboration', *Journal of High Technology Management Research*, 14, 111–133.

Storper, M. and Scott, A. J. (1989) 'The geographical foundations and social regulation of flexible production complexes', in J. Wolch, and M. Dear (eds), *The Power of Geography*. London: Allen & Unwin, 19–40.

Suwala, L. and Dannenberg, P. (2009) 'Cluster-und Innovationspolitik maßgeschneidert – Das Beispiel Adlershof in Berlin', *Standort*, 33(4), 104–112.

Thune, T. (2009) 'Proximity and interactive learning in university–firm relationships', *Industry and Higher Education*, 23(1), 7–16.

Torre, A. (2008) 'On the role played by temporary geographical proximity in knowledge transmission', *Regional Studies*, 42(6), 869–889.

Torre, A. and Rallet, A. (2005) 'Proximity and localization', *Regional Studies*, 39(1), 47–59.

Vázquez-Barquero, A. and Carrillo, E. (2004) 'Cartuja 93, a Technological Park located at the Site of Sevilla's World's Fair', in paper presented at conference of European Regional Science Association. Available PDF: www-sre.wu-wien.ac.at/ersa/ersaconfs/ersa04/PDF/486.pdf.

Vedovello, C. (1997) 'Science parks and university–industry interaction: Geographical proximity between the agents as a driving force', *Technovation*, 17(9), 491–502.

WISTA-Management GmbH (2011), Annual report 2010, Berlin: WISTA-Management GmbH.

Youtie, J. and Shapira, P. (2008) 'Building an innovation hub: A case study of the transformation of university roles in regional technological and economic development', *Research Policy*, 37, 1188–1204.

Websites

Agencia IDEA (accessed 9 November 2011).

Berlin Senate Department for Education, Sciences and Research (accessed 5 March 2012).

Brandenburg Ministry for Sciences, Research and Culture (accessed 5 March 2012).

Corporación Tecnológica de Andalucía (CTA) (accessed 5 March 2012).

Fundación para la Investigación y el Desarollo de las Tecnologías de la Información en Andalucía (FIDETIA) (accessed 5 March 2012).

Investitionsbank Berlin (IBB) (accessed 2 April 2012).

Junta de Andalucía, Consejería de Economía, Innovación y Ciencia (accessed 9 November 2011).

WISTA-Management GmbH (accessed 1 March 2012).

Further sources

Cartuja 93, S. A. (2010b): Interview with STP management (23 November 2010).

Cartuja 93, S. A.: Interview with STP management (15 February 2012).

Helice Foundation – Andalusian Aerospace Cluster (Meeting within Know-Man project, 8 February 2012).

7 Relations between international research collaboration, scientific production and the impact of research at Polish universities

Krzysztof Klincewicz

1 Universities and international research collaboration

The present chapter focuses on the public universities in Poland and the relations between their international research collaborations, measurable outcomes of scientific production (published articles) and impact of the research on the worldwide scientific community (citations).

In the triple helix model (Etzkowitz and Leydesdorff 1995), universities assume a particularly responsible role, especially when developing new research areas and emerging technologies. The higher education sector is only one part of the national innovation system, but in many countries universities are among the most active players on the innovation scene. Polish observations confirm the special role of academics in creating and diffusing innovations, as universities in Poland dominate the national innovation system both in terms of scientific publications and filed patent applications. The research will thus focus on this selected part of the triple helix or the 'science pole' in the techno-economic network model (Bell and Callon 1994).

Research cooperation with other organizations in the innovation system can stimulate creation and transfer of knowledge. International research collaboration seems particularly valuable, as it is based on weak ties (Granovetter 1974), thus 'building bridges' between diverse communities of researchers, extending beyond the closely intertwined, national network of academics and bringing new insights, relevant for local research projects. The cooperation often results in a joint authorship of scientific papers and books. Co-authorship of articles is one of many forms of scientific collaboration (Royle *et al.* 2007: 63–64), is relatively easy to measure and focused on quantifiable collaboration results, not merely inputs into the process. In modern science, openness and physical mobility of scientists are implicit requirements for academic success as they enhance a researcher's credibility within her discipline (Mahroum 2000: 514, 516). Collaboration with other researchers became 'a scientific value' (Duque *et al.* 2005: 756), and in many cases institutions sponsoring scientific projects display preference for funding international initiatives as opposed to individual projects (Raymond 1997: 9). Empirical research demonstrates strong correlations between collaboration and research productivity (Lee and Bozeman 2005: 687). Cross-country cooperation in science was contributing to the reduction of political tensions even during the Cold War (Skolnikoff 2001: 462), and it paralleled the processes of European integration by exchange of knowledge among scientists and proliferation of European-wide research organizations (Lebeau 2001).

International collaboration can be explained by the presence of relative strengths and weaknesses of local research teams (Nagpaul 2003: 414). Scientists form *invisible colleges* – geographically dispersed networks of specialists, who analyse similar problems by means of related techniques (De Solla Price and Beaver 1966). Surveys among scientists indicate that they resort to research collaboration and article co-authorship when their research partners have special competences, data or equipment (Melin 2000: 34). The observed growth in collaboration can also be attributed to the continuous increase in the complexity of research problems, prohibiting costs of necessary equipment, as well as the relative availability of the enabling infrastructure (such as modern information and communication technologies or cost-efficient transportation) (Duque *et al.* 2005: 756). New technologies facilitate dispersed scientific collaborations (Anderson 1997) and form the basis for 'distributed intelligence' (Cummings and Kiesler 2005: 704). For detailed lists of factors, contributing to the increase in scientific collaboration, compare Katz and Martin (1997), Raymond (1997), van Raan (1998), Duque *et al.* (2005), or Lee and Bozeman (2005). International research collaboration entails also coordination problems and the related increase in transaction costs linked to research projects, so merely economic arguments are not always sufficient to justify the efforts (Landry and Amara 1998; Cummings and Kiesler 2005: 714).

International co-authorship with representatives of a prestigious institution enhances the reputation and credibility of researchers and their employing university (Mahroum 2000: 517). It leads to 'a considerable "broadening" of the audiences around the authors' (van Raan 1998: 427). Empirical research confirms that the total credit, given by the scientific community to co-authors of papers is on average greater than the citations of single-author papers, and internationally co-authored papers are cited up to two times as frequently as papers authored by researchers from the same country (Katz and Martin 1997: 6). Therefore, it should come as no surprise that scientists from countries with less developed academic infrastructure demonstrate a particular inclination to engage in co-authorship relations with scientists from the leading countries (National Science Board 2010: 5–35). The tendency might differ among higher education institutions, as prestigious research-oriented universities prefer working with equal partners, expecting that the cooperation would generate valuable scientific contributions (Hewitt-Dundas 2012: 265), even though their reputation attracts also less prestigious research partners from other geographical regions (Hewitt-Dundas 2012: 273). Surprisingly, the international collaboration does not always increase the count of an article's citations, as the empirical analysis of Chinese publications demonstrated (Royle *et al.* 2007: 83), and the reputation-enhancing effects of having foreign co-authors clearly differ by countries and employing institutions.

Bibliometric studies demonstrate the continuous increase in the shares of internationally co-authored articles in the US, EU and Japan, as well as a surprising, slight decline in China (National Science Board 2010: 5–34 – 5–35). In 1998, 20 per cent of scientific articles written by researchers from the US had international co-authors (Zitt and Bassecoulard 2004: 410), while in 2008 the rate increased to 30 per cent (National Science Board 2010: 5–37). Corresponding rates for European countries were: France – 38 per cent (1998) and 52 per cent (2008), Germany – 36 per cent (1998) and 51 per cent (2008), United Kingdom – 32 per cent (1998) and 49 per cent (2008). Chinese international co-authorship rations were: 26 per cent in 1998 and 25 per cent in 2008, while Japan increased its cross-country collaboration rates from 17 per cent in 1998 to 26 per cent in 2008 (National Science Board 2010: 5–37).

2 Research problem

The present chapter outlines the most active participants in the system of innovation among universities, presenting individual differences in the scientific intensity and impact of research. It documents previously unpublished research, based on large sets of aggregated data, corresponding to all scientific articles published by researchers from selected, most active universities, published in the years 2000–09, collected from a representative bibliometric database. It offers a comparison of twenty-four Polish universities and focuses on three aspects of their research activities: scientific production, impact of research (interpreted as forward citations to articles, authored by researchers from a given university) and international research cooperation (evidenced by co-authorship of articles with representatives of foreign institutions). Internationalization of research activities is hypothesized as an important driving force, improving the status of individual academic institutions, as well as the significance of national scientific research and competitiveness of the national system of innovation. Basu and Aggarwal (2001) conducted a similar study concerning Indian universities, while Inzelt *et al.* (2009) focused on Hungarian higher education institutions.

Polish researchers are increasingly engaging in international co-authorship. In 2006, co-authored articles of Polish scientists, indexed in the Web of Science database, were prepared jointly with researchers from the EU (66.3 per cent of co-authored articles, including Germany – 23.3 per cent, France – 15.3 per cent, United Kingdom – 11.5 per cent, Italy – 9.2 per cent, Spain – 6.2 per cent, the Netherlands – 5.0 per cent), the US (25.6 per cent) and Japan (6.4 per cent) (OST 2008: 273–275). Polish shares of publications with co-authors from other European Union countries are higher than the corresponding shares for France, Germany, United Kingdom, Spain, Italy and Sweden (OST 2008: 273), indicating the inclination of Polish researchers to cooperate with their European colleagues. Similarly, Poland's co-publication shares with Japan belong to the highest among the EU countries (OST 2008: 273).

The study investigates the phenomenon of international co-authorship and presents results related to the articles of the selected Polish universities, with a subset of the data including articles with foreign co-authors, as well as citations to the articles, with a subset encompassing citations to articles with foreign co-authors. A longitudinal analysis, covering the most recent decade for which data are available, eliminates possible annual variation, linked to temporary changes in the levels of funding or other environmental factors.

There have been a limited number of previous studies, related to the scientific production and impact of Polish research. Olechnicka and Płoszaj (2008) present publication counts and cross-institutional collaboration data for years 2001–06, divided by geographical subregions rather than individual research institutions; their subsequent research focuses on institutions performing scientific research from Warsaw only (Olechnicka *et al.* 2010). Aggregate measures of productivity for thirty-one Polish universities are presented by Parteka and Wolszczak-Derlacz (2011), and corresponding bibliometric indicators for the entire Polish academic system by Klincewicz (2008). Wolszczak-Derlacz and Parteka (2010) conduct an analysis of thirty-four universities and their publications from only one year, 2008. Bibliometric research techniques are still relatively unpopular in Poland, and analyses of empirical data related to scientific publications could strengthen the management practices in the Polish higher education sector as well as help improve the government's science and innovation policies.

3 Research methods

The research uses bibliometrics – the quantitative analysis of scientific publications – and is based on a dataset derived from the Scopus database in June 2011. The data represent the scientific publication activities of the twenty-four most active academic institutions in Poland, and the analysed institutions include universities, technical colleges and medical schools.

Elsevier Scopus is one of the most comprehensive collections of the abstracts of scientific articles, covering all disciplines of research. Earlier research suggests that Scopus indexes more sources and offers better coverage of European publications than its main competitor, Thomson Reuters Web of Science/Web of Knowledge database (Leydesdorff *et al.* 2010). The analyses were conducted for years 2000–09, representing the most recent decade for which all empirical data were available (collections from year 2010 were still incomplete in Scopus in June 2011).

Scopus lists altogether 566 affiliations in Poland, but many of the less productive affiliations turn out to be misassigned as they form part of larger entities already included in the list. According to Scopus, only twenty-five of these institutions published more than 5,000 articles, books and conference papers each, and the remaining affiliations displayed significantly lower levels of output. One publicly funded research institution, the Polish Academy of Sciences, was excluded from the further analysis owing to its distinctive character and the difficulty of cross-institutional comparison. The Academy enjoys a special status in the Polish research system, employing more researchers than any individual universities, with no teaching responsibilities, and facilitated access to research funds.

The twenty-four selected universities are the most prolific Polish institutions, identified by means of relevant queries in Scopus, looking for affiliations in Poland and compiling a ranking of affiliations with the highest numbers of publications. Scopus offers unique IDs for each affiliation, thus facilitating the identification of relevant assets. Below is an example of a query used to identify publications authored by researchers from the Warsaw University of Technology (‘*Politechnika Warszawska*’ in Polish):

> (AF-ID(‘Politechnika Warszawska’ 60003675)) AND (LIMIT-TO(PUBYEAR, 2000) OR (LIMIT-TO(PUBYEAR, 2001) OR (LIMIT-TO(PUBYEAR, 2002) OR (LIMIT-TO(PUBYEAR, 2003) OR (LIMIT-TO(PUBYEAR, 2004) OR (LIMIT-TO(PUBYEAR, 2005) OR (LIMIT-TO(PUBYEAR, 2006) OR (LIMIT-TO(PUBYEAR, 2007) OR (LIMIT-TO(PUBYEAR, 2008) OR (LIMIT-TO(PUBYEAR, 2009)

In the case of some affiliations, more complex queries were required, as relevant publications were mistakenly assigned to several unique affiliation identifiers in Scopus.

The downloaded data required additional, laborious cleaning procedures since the dataset contained multiple abstracts assigned to wrong affiliations (apart from Scopus’ affiliation ID, the dataset also contains the full affiliation names as listed in the original publication). In order to guarantee the cross-institutional comparability of publications and citations, only published articles were used for the subsequent analysis, and other sources included in Scopus were excluded (such as conference papers and books).

4 Publication and citation counts

The list of twenty-four universities in Poland is presented in Table 7.1, and their geographical locations are represented by Figure 7.1.

The total numbers of articles are based on the so-called *whole counting*, that is, each article written in co-authorship with researchers from other institutions is counted as one publication towards the total number of the institution’s publications (Okubo 1997: 21).

The use of publication counts for cross-institutional comparison has important limitations. The mere fact of publishing an academic article is no evidence of the scientific ‘value added’ of the research, and a substantial percentage of publications in each discipline are of limited scientific merit. This merit varies depending, among other factors, on the scientific rank of the journal that publishes the specific contribution. This negative phenomenon might be limited in the present study, as Scopus tends to index only the most prestigious journals. Excessive reliance on publication counts might also be risky in research systems, where higher numbers of publications in recognized outlets translate into better individual and institutional evaluation. Polish research institutions used to be evaluated in exactly this way, as the Ministry of Science and Higher Education allocated research funds based on publication counts in highly ranked journals, rather than on citations or other indicators of scientific impact. This evaluation scheme can lead to duplication of publications (auto-plagiarism) and encourage the publication of many lower-quality contributions.

Table 7.1 Articles and citation counts for twenty-four Polish universities, 2000–09

Abbreviation	*Name*	*No. of articles*	*No. of citations*	*No. of citations/ no. of articles*
UW	University of Warsaw	9,735	104,706	10.76
UJ	Jagiellonian University, Kraków	12,722	122,374	9.62
PW	Warsaw University of Technology	6,078	41,882	6.89
PWr	Wrocław University of Technology	5,791	35,069	6.06
UAM	Adam Mickiewicz University, Poznań	6,465	42,625	6.59
UWr	University of Wrocław	4,873	46,295	9.50
WUM	Medical University of Warsaw	4,995	39,192	7.85
UMLo	Medical University of Łódź	5,048	33,999	6.74
PL	Technical University of Łódź	4,117	21,549	5.23
UMSl	Medical University of Silesia	4,429	28,354	6.40
USl	Silesian University	4,350	28,436	6.54
UMP	Poznan University of Medical Sciences	3,938	22,935	5.82
UMCS	Maria Curie-Skłodowska University, Lublin	3,417	21,870	6.40

Abbreviation	*Name*	*No. of articles*	*No. of citations*	*No. of citations/ no. of articles*
AGH	AGH University of Science and Technology, Kraków	5,845	30,126	5.15
UMG	Medical University of Gdańsk	3,785	34,889	9.22
UML	Medical University of Lublin	3,844	21,438	5.58
PG	Technical University of Gdańsk	3,182	22,961	7.22
UL	University of Łódź	3,232	24,433	7.56
UMWr	Wrocław Medical University	4,873	21,824	4.48
UMK	Nicolaus Copernicus University, Toruń	5,582	40,086	7.18
UMB	Medical University of Białystok	3,671	21,742	5.92
PSl	Silesian University of Technology	2,891	15,248	5.27
UG	University of Gdańsk	3,217	27,150	8.44
PP	Poznań University of Technology	2,348	19,601	8.35
	Total for 24 universities	118,428	868,784	7.34

Source of data: Scopus database.

Table 7.1 also lists the total number of citations received by individual universities (forward citations to the university's publications). The figures were calculated as sums of citations to each publication included in the dataset (articles in academic journals from years 2000–09), authored by researchers from a specific institution.

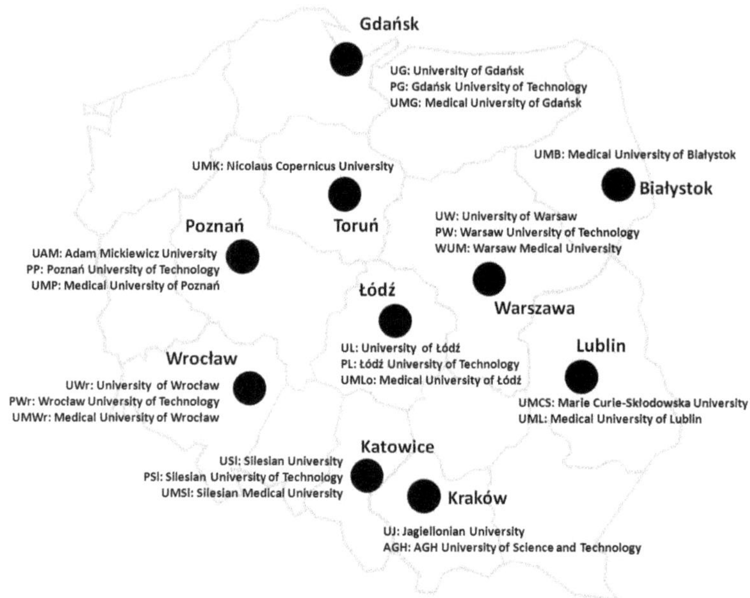

Figure 7.1 Geographical location of twenty-four Polish universities included in the research.

The adopted analytical technique was based on direct citations (Small 1973: 265; Boyack and Klavans 2010: 2390). There are several important restrictions of this approach. Citation-based indicators cannot be regarded as objective measures of quality or scientific impact. The results might be influenced by co-ordinated reciprocal citations, intended to increase the status of certain research teams in international rankings (Zhou and Leydesdorff 2006: 89–91). A forward citation to an article by a low-impact scientific article will yield the same result as a citation by an internationally recognized specialist, as long as all three publications are included in the same bibliometric database (Ding *et al.* 2009: 2229). Some researchers go so far as to question the underlying principles of the citation analysis (MacRoberts and MacRoberts 1989; MacRoberts and MacRoberts 2010). Scientific citations tend to be 'ritualized' in many academic cultures (Cockburn and Henderson 1998: 165) and some junior researchers might, for example, be required to cite professors from their own institutions regardless of the actual academic merits of their works. The citation counts listed by a specific bibliometric database are also limited by the coverage of this database (Bar-Ilan *et al.* 2007: 26). Scopus citation data is extensive but obviously restricted to citing sources, which are indexed in the Scopus database, and for obvious reasons cannot include all existing citations.

The data aggregation technique used in the research does not differentiate between older and newer publications, even though one could assume that the most recent publications cannot be frequently cited owing to long publication cycles. For a sizeable set of bibliometric data from multiple institutions, it would also be difficult to normalize the citation data, for example, by dividing the citation counts by the number of years that elapsed from the publication date (Guo *et al.* 2010: 201). The adopted approach was a compromise between the need to assess the impact of scientific publications and the limitations of available analytical techniques.

The ratio of total citations received to the total number of articles shows the average number of forward citations to each article, authored by researchers from a specific university and published in years 2000–09. It is important to remember the substantial variations in the numbers of citations to individual articles, as well as specific citation patterns related to publications from various scientific disciplines. Experimental scientific disciplines tend to have substantially higher citation counts than social sciences, humanities and arts (Moed 2005: 138).

The data reveal substantial differences among the universities. Two institutions – the University of Warsaw and Jagiellonian University, Kraków – lead in the number of publications and received citations, leaving the remaining twenty-two institutions far behind. Jagiellonian University accounts for 10.74 per cent of all publications from the twenty-four universities and 14.08 per cent of all citations in the dataset, while the University of Warsaw generates 8.22 per cent of all publications and 12.05 per cent of all citations. Each of the remaining universities yields significantly lower results, corresponding to 2–5.1 per cent of publications and 1.8–5.3 per cent of citations.

Comparison of the two top universities yields potential analytical problems. Jagiellonian University includes Collegium Medicum (in Scopus listed as a separate institution '*Collegium Medicum Uniwersytetu Jagiellońskiego*' with affiliation ID 60021323), whereas the University of Warsaw does not have its own faculty of medicine. The Medical University of Warsaw is historically linked to the University of Warsaw, but has operated as a separate entity since 1950. As of June 2011, over 21 per cent of all publications from the Jagiellonian University were authored by researchers representing the Collegium Medicum.

Even though the total counts of publications and citations for Jagiellonian University exceed the comparable counts for the University of Warsaw, the average number of citations to each publication is higher for publications from the University of Warsaw, thus demonstrating that its academic community has exerted the highest global scientific impact among all Polish universities.

Universities taking subsequent positions in the ranking of publication counts are: Adam Mickiewicz University in Poznań, Warsaw University of Technology, AGH University of Science and Technology in Kraków, Wrocław University of Technology and Nicolaus Copernicus University in Toruń. Interestingly, these universities lead in the total numbers of publications, but do not receive proportional numbers of citations – that is, their researchers write a lot, but their articles do not in general have a corresponding impact in relevant disciplines (even though some individual publications might be deemed very influential and scientifically important). There are universities enjoying better international scientific recognition, which have more limited publication counts. These institutions can be regarded as particularly well focused and conducting high-impact research. The high citation to publication ratios characterize: the University of Wrocław, the Medical University of Gdańsk, the University of Gdańsk and Poznań University of Technology. The Medical University of Warsaw and the University of Łódź are above the country-wide average of 7.34 citations per published article from years 2000–09.

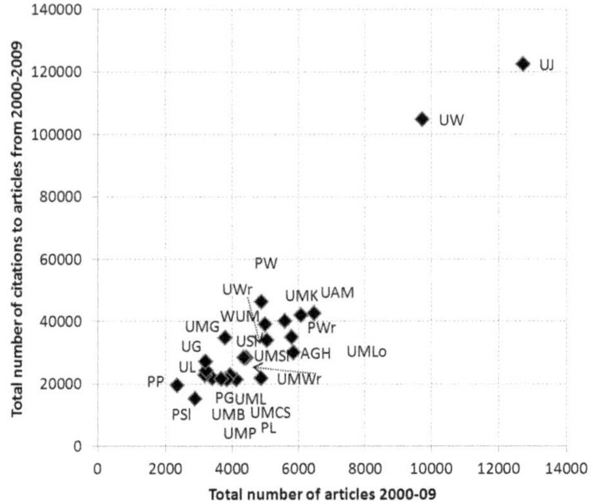

Figure 7.2 Total numbers of articles and citations to articles from twenty-four Polish universities, 2000–09.

Source of data: Scopus database.

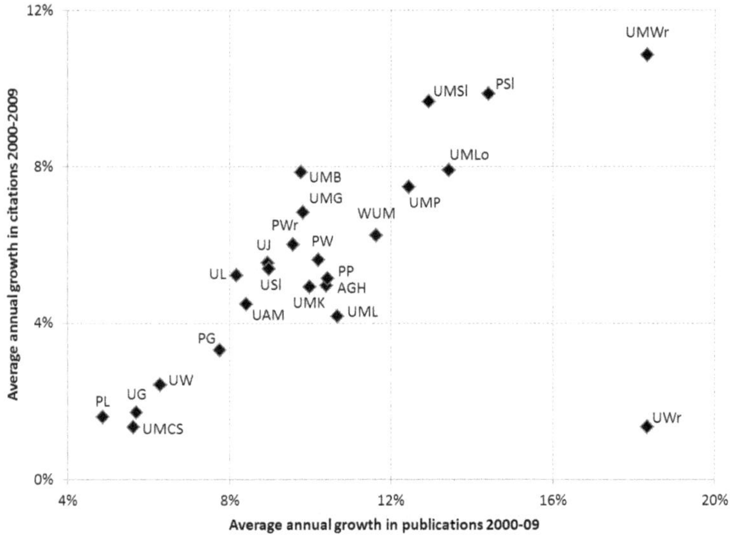

Figure 7.3 Average annual growth in articles and citation counts for twenty-four Polish universities, 2000–09.

Source of data: Scopus database.

Figure 7.2 shows the outstanding positions of two leading Polish universities as well as a cluster of the remaining twenty-two universities. An interesting source of variation among universities is the annual growth in publication and citation counts. For the decade 2000–09, each of the analysed universities increased its annual counts. These positive developments are linked to increased financing for scientific research, including the availability of public funds from the state budget, the European Union, as well as from corporate sponsors (the latter source of funding is still relatively seldom used in Poland).

The changes had varying impact on the scientific activities of various universities. Figure 7.3 compares the average annual growth in publications with the average growth in citations for the period of 2000–09. The Medical University of Wrocław turns out to be the growth leader, followed by the Silesian University of Technology and Silesian University of Medicine. Leaders of the publication and citation rankings do not experience substantial growth rates, but this can be attributed to the fact that they might already have gained a 'critical mass'. Nevertheless, it's worth noting that the University of Warsaw belongs to the 'laggards' in the comparison, experiencing particularly low annual increases in the numbers of publications and citations. The University of Wrocław is another interesting case, as the institution has experienced a dramatic increase in publication counts (on average, 18.31 per cent per annum), while the corresponding growth in citations is the lowest among the analysed Polish universities (annual value of 1.35 per cent). The University of Wrocław belongs to leaders of the citation rankings, being the third most frequently cited university in Poland (an average of 9.50 forward citations to each published article) and over the decade it managed to maintain its high citation level, while not being able to further increase the score.

5 International research collaboration

Scientific research is becoming increasingly internationalized, and universities in Poland are active participants in international cooperation projects. The subsequent analysis concerned the degree of internationalization of a university's research, measured by the share of articles co-authored with representatives of foreign institutions.

The co-authorship of scientific articles is a widely observed trend, since most experimental research initiatives are nowadays team-based. Globalization, the use of information and communication technologies and the increased availability of funds stimulate international cooperation patterns and the mobility of researchers. Some articles are co-authored as a result of jointly conducted research projects, while others are created as the outcome of ad hoc cooperation, such as specialist consultation or making specific empirical data available for purposes of another study. International scholarships, fellowships and research visits can also lead to the cross-national authorship of publications.

Co-authorship of scientific articles with international partners can be regarded as the best measure for the internationalization of research activities, representing quantifiable outcomes rather than inputs and thus superior to other indicators such as the number of cooperation projects, foreign visits or funding allotted to international projects.

Table 7.2 Articles for twenty-four Polish universities co-authored with researchers from foreign institutions, 2000–09

Abbreviation	*University*	*No. of articles co-authored with researchers from foreign institutions (A)*	*Share of total articles co-authored with researchers from foreign institutions (B)*	*No. of citations to articles with foreigners/no. of articles with foreigners*
UW	University of Warsaw	4,332	44.50%	15.59
UJ	Jagiellonian University, Kraków	4,964	39.02%	15.12
PW	Warsaw University of Technology	1,939	31.90%	10.90
PWr	Wrocław University of Technology	1,739	30.03%	8.76
UAM	Adam Mickiewicz University, Poznań	2,320	35.89%	8.81
UWr	University of Wrocław	516	10.59%	45.11
WUM	Medical University of Warsaw	789	15.80%	27.33
UMLo	Medical University of Łódź	574	11.37%	20.38
PL	Technical University of Łódź	1,049	25.48%	8.42
UMSl	Medical University of Silesia	564	12.73%	24.76
USl	Silesian University	1,435	32.99%	11.02
UMP	Poznań University of Medical Sciences	710	18.03%	13.35
UMCS	Maria Curie-Skłodowska University, Lublin	962	28.15%	9.00

Abbreviation	*University*	*No. of articles co-authored with researchers from foreign institutions (A)*	*Share of total articles co-authored with researchers from foreign institutions (B)*	*No. of citations to articles with foreigners/no. of articles with foreigners*
AGH	AGH University of Science and Technology, Kraków	2,152	36.82%	7.98
UMG	Medical University of Gdańsk	902	23.83%	24.23
UML	Medical University of Lublin	426	11.08%	17.47
PG	Technical University of Gdańsk	889	27.94%	11.44
UL	University of Łódź	1,002	31.00%	11.28
UMWr	Wrocław Medical University	516	10.59%	21.10
UMK	Nicolaus Copernicus University, Toruń	1,750	31.35%	13.30
UMB	Medical University of Białystok	369	10.05%	17.05
PSl	Silesian University of Technology	623	21.55%	10.92
UG	University of Gdańsk	1,194	37.12%	13.20
PP	Poznań University of Technology	586	24.96%	16.34
	Total for 24 universities	32,302	27.28%	14.03

Source of data: Scopus database.

Table 7.2 presents the degree of internationalization measured by the share of published articles written jointly with authors representing foreign institutions. The University of Warsaw leads the ranking with 44.50 per cent of its publications from years 2000–09 being co-authored with researchers affiliated abroad. The second place is taken by Jagiellonian University (39.02 per cent). Other universities with highly internationalized research are: the University of Gdańsk (37.12 per cent), AGH University of Science and Technology in Kraków (36.82 per cent), Adam Mickiewicz University in Poznań (35.89 per cent), Silesian University (32.99 per cent), Warsaw University of Technology (31.90 per cent), Nicolas Copernicus University in Toruń (31.35 per cent) and the University of Łódź (31.00 per cent).

The least internationally oriented universities among the analysed group of institutions are: the Medical University of Białystok (10.05 per cent), the University of Wrocław (10.59 per cent), Wrocław Medical University (10.59 per cent), the Medical University of Lublin (11.08 per cent) and the Medical University of Łódź (11.37 per cent). The results are puzzling, as worldwide research in medicine belongs to the most internationalized and conducted by the largest research teams. It is also interesting that the University of Wrocław, one of the most productive and high-impact institutions, belongs to the less internationalized universities. Its share of internationally co-authored articles is only 10.59 per cent, but these articles are particularly highly cited, with the record average of 45.11 forward citations.

The above-mentioned ratio of citations to counts of articles co-authored with foreigners varies for individual universities. Figure 7.4 presents the relations between citations of all articles and citations of the articles written jointly with researchers representing foreign institutions.

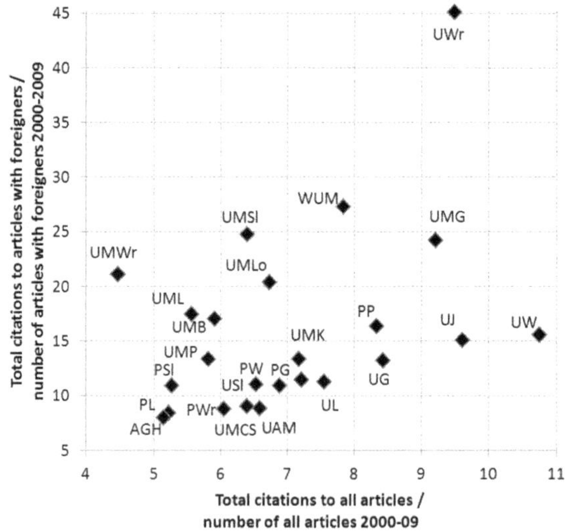

Figure 7.4 Citation ratios for all articles of twenty-four Polish universities and articles co-authored with researchers from foreign research institutions, 2000–09.

Source of data: Scopus database.

International cooperation turns out to be particularly beneficial for several universities, where cooperation with foreign authors increases the impact of research. These universities are: the University of Wrocław (a ratio of 45.11 citations per article co-authored with foreigners as compared to the ratio of 9.50 citations for all articles, a difference of 35.61), the Medical University of Warsaw (27.33 with foreign co-authors, 7.85 all, a difference of 19.48), the Silesian Medical University (24.76 and 6.40, a difference of 18.36), the Medical University of Gdańsk (24.23 and 9.22, difference of 15.01), the Medical University of Wrocław (21.10 and 4.48, difference of 16.62) and the Medical University of Łódź (20.38 and 6.74, difference of 13.64). Even the universities with a relatively low level of internationalization enjoy the positive effect of cross-national cooperation, for example, the Medical University of Lublin had on average 11.90 more forward citations to articles written with international co-authors, and the Medical University of Białystok received 11.13 more citations.

Large and established universities from the top of the ranking don't experience dramatic growth in citations thanks to international co-authorship, but each of the analysed twenty-four universities showed average increases in citation counts for internationally co-authored articles. Table 7.3 also gives the results of a two-sample z-test, which was used to verify the existence of statistically significant differences between the citations of all publications from a given university and the citations of publications with foreign co-authors. The value of the two-sample z-test is calculated based on the equation:

$$z = \frac{x^1 - x_2}{\sqrt{\frac{\sigma 1^2}{n1} + \frac{\sigma 2^2}{n2}}}$$

where: n^1, n_2 are counts; x^1, x_2 are mean values and σ^1, σ_2 are standard deviations of variables no. 1 (citations to all articles from a given university, 2000–09) and no. 2 (citations to articles from a given university written with co-authors from foreign institutions, 2000–09).

The results were compared with the normal distribution for statistical significance. In the case of each university, the test results indicate that distinctive differences could indeed be observed and that co-authorship increases the scientific impact of Polish articles.

Figure 7.5 depicts the relations between the average annual growth in the number of articles written with foreign co-authors and the impact on the citation counts of universities. Several universities enjoying a particularly high positive effect of international collaboration had also been constantly increasing the scope of such collaborations. These were: the University of Wrocław and the Medical Universities of Silesia, Wrocław, Łódź, Gdańsk and Lublin. For every analysed university, an average annual growth in the number of internationally co-authored articles could be observed.

Table 7.3 Two sample Z-test values for citation data of twenty-four Polish universities

	No. of citations to articles with foreign co-authors/no. of articles with foreign co-authors (A)	*No. of citations/no. of articles (B)*	*Difference (A–B)*	*Two sample z-test*	*Statistically significant?*
UW	15.59	10.76	4.83	26.23	*
UJ	15.12	9.62	5.50	29.97	*
PW	10.90	6.89	4.01	34.64	*
PWr	8.76	6.06	2.70	35.49	*
UAM	8.81	6.59	2.22	30.53	*
UWr	45.11	9.50	35.61	19.60	*
WUM	27.33	7.85	19.48	17.20	*
UMLo	20.38	6.74	13.64	28.65	*
PL	8.42	5.23	3.19	32.15	*
UMSl	24.76	6.40	18.36	16.59	*
USl	11.02	6.54	4.48	25.72	*
UMP	13.35	5.82	7.53	30.06	*
UMCS	9.00	6.40	2.60	32.97	*
AGH	7.98	5.15	2.83	32.04	*
UMG	24.23	9.22	15.01	13.52	*
UML	17.47	5.58	11.90	24.03	*
PG	11.44	7.22	4.22	28.00	*
UL	11.28	7.56	3.72	28.61	*
UMWr	21.10	4.48	16.62	13.21	*
UMK	13.30	7.18	6.12	25.54	*
UMB	17.05	5.92	11.13	35.08	*
PSl	10.92	5.27	5.65	28.51	*
UG	13.20	8.44	4.76	20.79	*
PP	16.34	8.35	7.99	19.68	*
Total for 24 universities	14.03	7.34	6.69	–	–

Source of data: Scopus database.

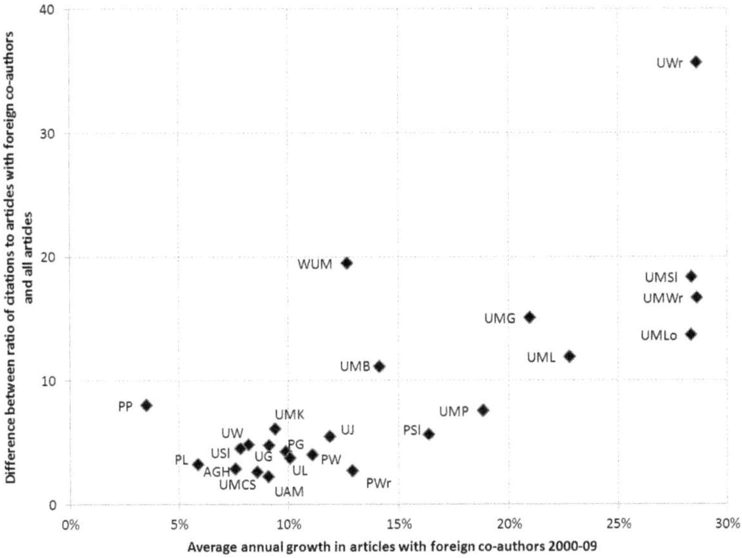

Figure 7.5 Effects of foreign research collaboration on citations to articles of twenty-four Polish universities and average annual growth in the collaboration, 2000–09.

Source of data: Scopus database.

6 Main findings and conclusions

The research project documented in this chapter established the main findings, which are summarized below:

1. The years 2000–09 were a period of systematic increase in scientific output, with impact measured by forward citations and the scale of international research collaboration for each of the top twenty-four Polish universities.
2. Only a limited number of universities account for the majority of scientific production in Poland.
3. The two largest universities, in Warsaw and Kraków, correspond to nearly 19 per cent of all publications from the twenty-four Polish universities and over 26 per cent of forward citations, with the remaining institutions accounting for 2–5.1 per cent of publications and 1.8–5.3 per cent citations each.
4. Jagiellonian University, Kraków, leads in the total number of publications and citations, thus being the most productive university, but the University of Warsaw has the highest ratio of forward citations per published article, which corresponds to the highest impact of its research in the scientific community.
5. Other high impact universities include: the University of Wrocław, the Medical University of Gdańsk, the University of Gdańsk, Poznań University of Technology, the Medical University of Warsaw and the University of Łódź.

6 The leaders of annual growth in the numbers of articles and forward citations are: the Medical University of Wrocław, Silesian University of Technology and Silesian University of Medicine.
7 The most widely internationalized research is carried out by the University of Warsaw, followed by Jagiellonian University, the University of Gdańsk, AGH University of Science and Technology in Kraków and Adam Mickiewicz University from Poznań.
8 For all of the analysed universities, international cooperation has a positive, statistically significant influence on the number of forward citations to articles and has thus led to increased international recognition and perceived scientific impact of research.
9 The most substantial benefits from international cooperation, measured by an increase in citation counts, have been derived by: the University of Wrocław, the Medical Universities of Warsaw, Silesia, Gdańsk, Wrocław, Łódź, Lublin and Białystok. Most of the above-mentioned universities are also leaders in terms of the annual increase in the internationalization of research.

The study demonstrates positive changes in Polish higher education institutions in the last decade. The increase of scientific production has been accompanied by a heightened impact of Polish research and growing international collaboration. The empirical analysis helps identify leaders and laggards, as well as institutions fighting to catch up. This scientific catch-up scenario involves a significant reliance on co-authorship with researchers from foreign institutions. The strengthening of publication-yielding collaborations with universities abroad can thus be recommended both for management of individual institutions and for national scientific and innovation policies.

The chapter proposed also a novel approach to analysing relations between the international collaboration and the perceived importance of scientists and their institutions. While previous studies focused on comparisons between countries or research disciplines, the present findings indicate that the variance between specific academic institutions deserve a closer examination. Differences in international co-authorship tendencies might be linked to the size of an academic institution, scope of its scientific production, experiences in international collaboration and its prestige in the academic world. The observed diversity among the leading national universities not only reveals the stratification of Polish scientific institutions, but also encourages further investigation of co-authorship and citation patterns.

References

Anderson, C. M. (1997) 'Computer mediated communication, the invisible college and visiting scholars: An exploratory study', *Journal of Computing in Higher Education*, 9(1), 27–48.

Bar-Ilan, J., Levene, M. and Lin, A. (2007) 'Some measures for comparing citation databases', *Journal of Informetrics*, 1, 26–34.

Basu, A. and Aggarwal, R. (2001) 'International collaboration in science in India and its impact on institutional performance', *Scientometrics*, 52(3), 379–394.

Bell, G. and Callon, M. (1994) 'Techno-economic networks and science and technology policy', *OECD STI Review*, 14, 59–118.

Boyack, K. W. and Klavans, R. (2010) 'Co-citation analysis, bibliographic coupling, and direct citation: Which citation approach represents the research front most accurately?', *Journal of the American Society for Information Science and Technology*, 61(12), 2389–2404.

Cockburn, I. M. and Henderson, R. M. (1998) 'Absorptive capacity, coauthoring behavior, and the organization of research in drug discovery', *The Journal of Industrial Economics*, 46(2), 157–182.

Cummings, J. N. and Kiesler, S. (2005) 'Collaborative research across disciplinary and organizational boundaries', *Social Studies of Science*, 35(5), 703–722.

De Solla Price, D. J. and Beaver, D. de B. (1966) 'Collaboration in an invisible college', *American Psychologist*, 21(11), 1011–1018.

Ding, Y., Yan, E., Frazho, A. and Caverlee, J. (2009) 'PageRank for ranking authors in co-citation networks', *Journal of the American Society for Information Science and Technology*, 60(11), 2229–2243.

Duque, R. B., Ynalvez, M., Sooryamoorthy, R., Mbatia, P., Dzorgbo, D.-B. S. and Shrum, W. (2005) 'Collaboration paradox: Scientific productivity, the Internet, and problems of research in developing areas', *Social Studies of Science*, 35(5), 755–785.

Etzkowitz, H. and Leydesdorff, L. (1995) 'The triple helix – university–industry–government relations: A laboratory for knowledge based economic development', *EASST Review*, 14(1), 14–19.

Granovetter, M. (1974) *Getting a Job: A study of contracts and careers*. Cambridge, MA: Harvard University Press.

Guo, Y., Huang, L. and Porter, A. L. (2010) 'The research profiling method applied to nano-enhanced, thin-film solar cells', *R&D Management*, 40(2), 195–208.

Hewitt-Dundas, N. (2012) 'Research intensity and knowledge transfer activity in UK universities', *Research Policy*, 41, 262–275.

Inzelt, A., Schubert, A. and Schumert, M. (2009) 'Incremental citation impact due to international co-authorship in Hungarian higher education institutions', *Scientometrics*, 78(1), 37–43.

Katz, J. S. and Martin, B. R. (1997) 'What is research collaboration?', *Research Policy*, 26, 1–18.

Klincewicz, K. (2008) *Polska innowacyjność: Analiza bibliometryczna*. Warszawa: Wydawnictwo WZUW.

Landry, R. and Amara, N. (1998) 'The impact of transaction costs on the institutional structuration of collaborative academic research', *Research Policy*, 27, 901–913.

Lebeau, A. (2001) 'Scientific organizations and European unification: A personal view', *Technology in Society*, 23, 349–359.

Lee, S. and Bozeman, B. (2005) 'The impact of research collaboration on scientific productivity', *Social Studies of Science*, 35(5), 673–702.

Leydesdorff, L., Moya-Anegón, F. and Guerrero-Bote, V. P. (2010) 'Journal maps on the basis of Scopus data: A comparison with the Journal Citation Reports of the ISI', *Journal of the American Society for Information Science and Technology*, 61(2), 352–369.

MacRoberts, M. H. and MacRoberts, B. R. (1989) 'Problems of citation analysis: A critical review', *Journal of the American Society for Information Science and Technology*, 40(5), 342–349.

MacRoberts, M. H. and MacRoberts, B. R. (2010) 'Problems of citation analysis: A study of uncited and seldom-cited influences', *Journal of the American Society for Information Science and Technology*, 61(1), 1–13.

Mahroum, S. (2000) 'Scientists and global spaces', *Technology in Society*, 22, 513–523.

Melin, G. (2000) 'Pragmatism and self-organization: Research collaboration on the individual level', *Research Policy*, 29, 31–40.

Moed, H. F. (2005) *Citation Analysis in Research Evaluation*. Dordrecht: Springer.

Nagpaul, P. S. (2003) 'Exploring a pseudo-regression model of transnational cooperation in science', *Scientometrics*, 56(3), 403–416.

National Science Board (2010) *Science and Engineering Indicators 2010*. Arlington, VA: National Science Foundation.

Okubo, Y. (1997) 'Bibliometric indicators and analysis of research systems: Methods and examples', OECD STI Working Papers, OECD/GD(97)41. Available online at www.oecd-ilibrary.org/science-and-technology/bibliometric-indicators-and-analysis-of-research-systems_208277770603 (accessed 25 July 2011).

Olechnicka, A., Płoszaj, A., Smętkowski, M. and Wojnar, K. (2010) 'Warszawa innowacyjna. Diagnoza potencjału'. Available online at www.euroreg.uw.edu.pl/index.php/dokumenty/doc_download/608-opracowaniewarszawa-innowacyjna-2010.html (accessed 25 July 2011).

Olechnicka, A. and Płoszaj, A. (2008) 'Polska nauka w sieci? Przestrzeń nauki i innowacyjności: Raport z badań'. Available online at www.espon.pl/files/10_2/2/Polska%20nauka%20w%20sieci.pdf (accessed 25 July 2011).

OST (2008) *Indicateurs de sciences et de technologies*. Rapport de l'Observatoire des Sciences et des Techniques. Édition 2008, Paris: Economica.

Parteka, A. and Wolszczak-Derlacz, J. (2011) 'Dynamics of productivity in higher education. Cross-European evidence based on bootstrapped Malmquist indices', Universitá degli Studi di Milano, Working Paper 2011–10. Available online at www.economia.unimi.it/uploads/wp/DEAS-2011_10wp.pdf (accessed 25 July 2011).

Raymond, S. U. (1997) 'Global cooperation in science, engineering, and medicine: An overview of the issues', *Technology in Society*, 19(1), 7–16.

Royle, J., Coles, L., Williams, D. and Evans, P. (2007) 'Publishing in international journals: An examination of trends in Chinese co-authorship', *Scientometrics*, 71(1), 59–86.

Skolnikoff, E. B. (2001) 'The political role of scientific cooperation', *Technology in Society*, 23, 461–471.

Small, H. (1973) Co-citation in the scientific literature: A new measure of the relationship between two documents', *Journal of the American Society for Information Science*, 24(4), 265–269.

van Raan, A. F. J. (1998) 'The influence of international collaboration on the impact of research results', *Scientometrics*, 42(3), 423–428.

Wolszczak-Derlacz, J. and Parteka, A. (2010) *Produktywność naukowa wyższych szkół publicznych w Polsce: Bibliometryczna analiza porównawcza*. Warszawa: Ernst & Young Polska.

Zhou, P. and Leydesdorff, L. (2006) 'The emergence of China as a leading nation in science', *Research Policy*, 35, 83–104.

Zitt, M., and Bassecoulard, E. (2004)[[AQ]] 'Internationalisation in science in the prism of bibliometric indicators: Journals, collaboration, and geographic distribution', in H. F. Moed, W. Glänzel and U. Schmoch (eds), *Handbook of Quantitative Science and Technology Research: The use of publication and patent statistics in studies of S&T systems*. Amsterdam: Kluwer Academic Publisher.

Part II

Universities, entrepreneurship and innovation-driven development

8 The research university, entrepreneurship and regional development

Research propositions and current evidence

Helen Lawton Smith and Sharmistha Bagchi-Sen

1 Introduction

Identifying the formal role of universities in stimulating regional technological and economic development is relatively new (Youtie and Shapira 2008). As recently as the 1980s, factors other than universities, such as the importance of government spending on defence, were considered as the main stimuli to high-tech development in North America (Christy and Ironside 1987) and in the UK (Breheny and McQuaid 1987). An early exception to this was a report in the UK, *The Cambridge Phenomenon* (Segal Quince and Partners 1985), which argued that Cambridge University exerted a direct and indirect influence on the genesis of the high-tech economy. In the US, Annalee Saxenian's 1983 article on *The Genesis of Silicon Valley* appeared in the same year as Etzkowitz's paper on 'the entrepreneurial university'. In the 1990s notable works were by Luger and Goldstein (1991), which reflectedf the growing number of science and research parks, and by Anselin *et al.* (1997) on local geographical spillovers from university research.

By the 2000s, the wealth creation perspective had become pervasive in academic research and public policy, and the regional role was well established. Universities are now variously described as 'engines of growth' and numerous other epithets are ascribed relating to their roles as machines, or as biological agents relating to incubating new firms or as central nodes in networks of innovators (Arbo and Benneworth 2007). These roles are associated with the 'third stream' activities, that is, targeted engagement with external organizations, outreach, enterprise formation, and so on (PACEC 2009).

The basis for the argument that universities stimulate economic development over and above general economic impacts through local spending on wages and services is that university research promotes local knowledge spillovers that lead to regional innovation processes (see Goldstein and Renault 2004; Arbo and Benneworth 2007; Geuna and Musico 2009). Studies that test the hypothesized techno-economic impact of universities have examined human capital, research (basic and collaborative), technology assistance, technology transfer (including spin-offs, patenting and licensing and consultancy). They have used a wide variety of methodologies (see Goldstein 2009 for a review).

Lendel (2010) demonstrates that quantitative studies focused on one or a small number of factors produce mixed results; for example, Beeson and Montgomery's (1993), Link and Rees' (1990) and Gottlieb's (2001) studies of the impact of universities on local labour markets. Qualitative studies have explored culture and strategy to assess differences in impact. For example, Saxenian (1994) found that Stanford University and its faculty were much more engaged than MIT in their respective region's development. Feldman (2001) recorded a limited role in technology transfer in the Capitol region of its research universities (Johns Hopkins, the University of Maryland, Georgetown and George Washington). Methodological problems remain rife. The multifaceted or 'multi-product pathways' by which technology is put into use (Hill and Lendel 2007) make it difficult to assess the impact of each component of these pathways, especially taking into account the specifics of regional path dependencies and business and economic cycles (Lendel 2010).

This chapter focuses on research universities, that is, those characterized by research intensity to examine issues relating to their regional impacts. The US-based Carnegie Foundation classification of thirty-four types of higher education research provides a useful guideline for defining research universities. That classifies research universities as: very high research activity, high research activity and doctoral research universities. Here 'research universities' refer to the first two. There are few such universities and their distribution is highly skewed towards the US and the UK. The first two account for less than 6 per cent of the 4,400 US degree awarding institutions. In both the QS and THE rankings of universities, the UK has four of the world's top ten research institutions: Oxford, Cambridge, Imperial College and University College-London, the rest being in the US.

Against this background, the objective of this chapter is to answer the research question: Under what conditions do research universities actively engage in local and regional economies becoming more than a 'latent asset' (Power and Malmberg 2008)? While research universities have distinctive characteristics, they are far from homogenous in their engagement and impact in their geographical hinterlands. The University of Oxford and changing relationships with its home county, Oxfordshire, is used as an exemplar of these complexities. The discussion draws on published data from the University of Oxford and from a series of studies on the Oxfordshire high-tech economy.

The remainder of the chapter has three sections. The first establishes the rationale for four propositions that form a framework which is used to examine the drivers, evidence and impact of those relationships. In the second, the propositions are examined in the case of Oxford University and Oxfordshire. In the final section, conclusions are presented along with suggestions of possibilities for future research.

2 Research context and propositions

Four propositions are now presented in order to provide a better understanding of the ways in which research universities intentionally contribute to regional development. These are that the extent of regional impact depends on (i) the internal characteristics of the university, (ii) how the university responds to exogenous shocks (Feldman and Francis 2006), (iii) the nature of funding for higher education institutions that accompanies' agenda-setting by national and regional political bodies in relation to wealth creation, and (iv) the attributes of the regional economy (Varga 2001; Lendel 2010). Each of these is examined with respect to drivers of change, evidence of change, and evidence of impact. Note that evidence of change is not necessarily evidence of impact.

2.1 Proposition 1: It depends on the internal characteristics of the university

Drivers of a university's intent to become more than a 'latent asset' in a regional economy are affected by its response to external and internal pressures to improve its economic impact. It can do this by becoming 'entrepreneurial' (Clark, 1998; Siegel *et al.* 2004; Bercovitz and Feldman 2006), varieties of being entrepreneurial: 'commercial' or 'connected' (Kitson *et al.* 2009), or all three. Kitson *et al.* describe the 'connected university' as engaging in building clusters, connecting to national and international economies, and bringing together thinking, practice and finance for regional development.

Clark's (1998) take on the entrepreneurial university is that it is one where the ethos of entrepreneurship is pervasive. It has a strengthened steering core, formal means of university–economy–community interaction and diverse sources of funding that underpin interaction and implementation (Rinne and Koivula 2005; see also Siegel *et al.* 2003). In some cases, motivation driving universities to become entrepreneurial is external such as national legislation and public policy. In other cases, it is internal pressures – both can result in new opportunities, creation of incentives and rewards for technology transfer, and increased industrial funding (Geuna and Muscio 2009).

In the US, drivers of change and the rise of entrepreneurial scientists and the entrepreneurial university have been identified as the outcome of a revolutionary process with big science programmes launched after the Second World War (Etzkowitz 1983). Franzoni and Lissoni (2009), however, suggest an alternative perspective, that is, the American model of entrepreneurial university appears more deeply rooted in the gradual evolution of US universities from teaching colleges to modern research institutions. Moreover, the regional role of land-grant universities, which include some of the top research universities (e.g. University of California and MIT), was enshrined in their establishing principles.

Another debate in the US is the impact of national regulatory frameworks on technology transfer, in particular the 1980 US Bayh-Dole Act, which allowed universities to patent the results of publicly funded research. It has been suggested that this did not have the intended positive impact on commercialization of federally funded research results (Litan *et al.* 2007) and that the current university ownership model might not be optimal in encouraging entrepreneurship (Mowery and Sampat 2005; Kenney and Patton 2009).

In the UK, the first country to develop a national university commercialization policy (Geuna and Muscio 2009), a similar regulatory change half a decade later was enacted but without the same contentious outcomes. In 1985, the British Technology Board (BTG) lost its monopoly access to intellectual property arising from universities and public sector research institutions from Research Council-funded projects. The intention was that universities would (a) give the fullest opportunity and scope to researchers to assume responsibility for exploiting their research findings, (b) encourage researchers to assume this responsibility, and (c) provide guidance and help for those researchers (Harvey 1992).

Whether a university becomes an active regional player results from how a university manages its portfolio of products as 'defined by its mission statements and expressed through its function and policies' (Lendel 2010: 212). Of particular relevance is the interpretation and adoption of policies derived from the interpretation and adoption of national rules relating to IP governance. Such policy decisions are especially influenced by the scope of the university (e.g. research-intensive, teaching intensive) (Andersen and Rossi 2010) and by universities' capacity and/or willingness to exploit opportunities presented by research and other creative activities. Fulfilling the entrepreneurial role, moreover, requires 'a strong, positive knowledge exchange culture at senior management levels' (PACEC 2009: 7).

The implementation of a business-like style of technology transfer whereby the productivity of technology transfer offices (TTOs) is improved is also identified as being a prerequisite for success in commercializing university intellectual property (Kinsella and McBrierty 1997; Lockett *et al.* 2005; Moray and Clarysse 2005; Wright *et al.* 2007). The formulation of a selective incubation strategy for spin-offs (Degroof and Roberts 2004; Clarysse *et al.* 2005) and the development of relationships among academics, TTO personnel and investors (Wright *et al.* 2007) are aspects of this model. These ideal circumstances, however, rarely exist.

The extent of impact of TTOs has been found to be subject to human factors varying from management decisions on which technologies to commercialize (Siegal *et al.* 2007), the extent to which profit maximization is the driving motivation (Muscio 2010), and the likelihood of academics revealing their inventions to the TTO (Siegal *et al.* 2004). Markman *et al.* (2005) found the speed of commercialization and higher revenues was higher when inventors collaborate actively with TTOs. Muscio (2010) found in Italy that research performance drives the academic use of TTOs. Higher levels of research are associated with more contact. Companies are more likely to contact TTOs in universities with good research rankings.

Assessment of impact is also complicated by differences in interpretations of indicators of TTO performance (Lockett and Wright 2005; Chapple *et al.* 2005; Rothaermel *et al.* 2007; Huggins 2008), timescales used and types of universities being compared. Immediate impact is difficult to assess when indicators, such as patents and university spin-offs, are used (Geuna and Muscio 2009) given that revenues from paid consultancy can be higher even in the most elite universities (HE-BCI 2010).

Impact as measured by commercialization of university IP is sometimes but not always associated with research intensity and reputation as much as by the tech-transfer model adopted. In the UK, the most research-intensive universities produce a wide variety of research outputs and hence generate a large variety of IP (Anderson and Rossi 2010: 43). The top four UK universities have the highest number of university spin-offs (HE-BCI 2010). US universities with the most spin-offs in 1995–2001 were flagship entrepreneurial universities (MIT, the UC system, Stanford and Cal Tech) (O'Shea *et al.* 2005) while other top-ranked research universities (Harvard, Yale and Princeton) have not been noted for their spin-offs (Lawton Smith 2006). Research reputation may have an indirect influence on levels of academic entrepreneurship as it may signal the quality of new ventures to both investors and potential partners (Mustar *et al.* 2006). This may help spin-off firms obtain access to external financing and recruit high quality personnel (see also Shane and Stuart 2002; Binks *et al.* 2005; Degroof and Roberts 2004).

At the regional level, a university's impact is multifaceted. Hill and Lendel (2007) identify seven possible products: (1) education, (2) contract research, (3) cultural products, (4) trained labour, (5) technology diffusion, (6) new knowledge creation and (7) new products and industries. Each has its own market niche and can be an asset used by the regional economy or 'exported' outside. Assessment of separate outcomes, they argue, is impossible owing to the interrelationships between university products and the bundled nature of their effect on regional economies.

Evidence from a study by Martinelli *et al.* (2008) of Sussex University in the UK (a high research university) illustrates a slightly different point: that even becoming more entrepreneurial does not necessarily mean a stronger regional impact. They show that, 'the university is at the centre of a dense network of relations with non-academic partners'. Such faculty relationships are described as indicators of entrepreneurialism. They also find that the form of entrepreneurial activity is related to academic discipline, with differing attitudes towards technology transfer. Life sciences had mainly collaborative research agreements, while science and technology faculty preferred consultancy agreements. In general links were more likely to be contract research, collaborative research and consultancy rather than patenting, licensing and spin-offs. Sussex University as a type is more 'entrepreneurial' and 'commercial' rather than 'connected': its links are mainly outside the region. Hence, Martinelli *et al.* conclude knowledge exchange processes may not have the kind of regional impact that stakeholders would wish.

2.2 Proposition 2: It is related to exogenous shocks relating to technological opportunities

Exogenous shocks, as drivers of economic change, encompass economic and business cycles as well as Schumpeterian patterns of technological change (Feldman 2001). Entrepreneurship is about timing. Drivers of change, as is the case for start-ups in new industries generally, can be seen in the Schumpeterian tradition (Schumpeter 1934). A particular strand in the academic literature of the 1980s concerned techno-economic changes (Freeman and Perez 1988) associated with the fifth Kondratieff wave, where entrepreneurship was a key component. The importance of scale economies was reduced in the fifth wave and new industries with new organizational forms emerged offering associated niches for small firms and limited consolidation through vertical integration or mergers (Radosevic 2007).

The rate at which academics become entrepreneurs is associated with such techno-economic changes. Academic entrepreneurs, 'start firms that capitalize on technological opportunities, adapting scientific breakthroughs and generic technologies to create new product markets and reorganize economic activity' (Feldman and Francis 2006, 115). Most are found in 'emerging sectors' such as biotech (Kenney 1986) and information technology (Feldman 2001, PACEC 2003), ones that have grown rapidly since the 1980s. Around the same time, public and private spending on university science parks and incubators increased to capitalize on university research (Phan *et al.* 2005): in the United States, there were twelve business incubators in 1980 and 950 in 2002, and in the UK twenty-five in 1997 and 250 in 2002. The number of science parks increased but at a slower rate.

A further dimension of a regional impact is the extent to which research universities can ameliorate the effects of down-turns in the economy and boost upturns. Lendel (2010) shows that the impact of research universities is very strong during the expansion stages of business cycles but only the very top research universities have positive impacts on regional economies in periods of economic decline. Their consistently high levels of research income, stable flows of students and high academic salaries allow for consistent outcomes.

2.3 Proposition 3: Specific regional direct and indirect effects depend on the nature of funding

This driver of change needs to be considered from the perspective of the balance between national and regional policy (Rip 2002) with respect to the activities that universities 'should' be doing and how funding is allocated. Types of funding for 'third stream' activity include: (i) non-spatial research grants with conditions relating to projections of impact, for example those funded under the seven UK research councils[1] (ii) funding programmes specifically designed to have commercial outcomes (e.g. spin-offs) and (iii) funding that has regional/local engagement or governance built in. Funding can have a beneficial indirect impact (Huggins 2008: 197), that is, 'the probability of investment from the private sector increases with the amount of public funding secured'.

In the US, state-level policy makers drive change by pressing universities to articulate how they can contribute to the economic well-being of the state. Some research universities have adapted to this priority believing that it can result in a more favourable financial treatment by the legislators (Powers 2003). An example is the Georgia Institute of Technology, an entrepreneurial research university, used to spearhead an innovation-driven economic development strategy (Youtie and Shapira 2008). The authors show that state funding has enabled the university to have a local impact by becoming a 'knowledge hub' employing the kinds of activities listed in proposition 1. These have resulted in 'boundary-spanning' roles that 'mediate among academic, educational, entrepreneurial, venture capital, industrial and public spheres' (p. 1188).

National UK policy and funding have had an impact on universities' perceptions of their regional role (PACEC 2009). This is illustrated by the 2008–2009 HE-BCI survey of university–business interaction. This shows that over a third of universities see meeting regional skill needs as a priority right after collaboration with industry and improving access to education. Evidence of the recent importance of UK universities in regional governance was inclusion of universities by regional development agencies (disbanded in 2010) in economic development strategies (Benneworth 2010). Other initiatives involving universities include 'Science Cities', six of which were established by central government in 2005, 'to lead the development of deeper links between business and the science base and ensure that science, technology and innovation succeed in becoming the engine of economic growth'.[2]

2.4 Proposition 4: Impact depends on the attributes of the region

A match between the assets of the university and regional conditions is needed to experience regional impact, but this is contingent on factors such as diversity of university-regions and type of academic research. Boucher *et al.* (2003) offer a taxonomy of based on the functional role of U.S. universities in territorial development: single player universities in peripheral regions, multiplayer universities in peripheral regions, traditional universities in core regions, newer technologically-oriented universities in core regions. Rutten and Boekema (2009) distinguish the impact of different types of academic research in regional economic development as follows: knowledge dissemination (Mode 1) depends on the absorptive capacity of actors in the university's home region and the degree to which actors with high levels of absorptive capacity are regionally networked. The more networked actors, the greater the chances of local direct and indirect knowledge transfer or exchange. Mode 2 (interactive, multidisciplinary form of knowledge production) requires face-to-face contact with practitioners and is often locationally bound. The notion of 'strategic science' (Rip 2002: 125) offers a complementary perspective to this, combining relevance (to specific contexts, possibly local) and excellence (advancement of science as such). Strategic science is basic research carried out with the expectation of producing a broad base of knowledge likely to form the background to the solution of recognized current or future practical problems (Irvine and Martin 1984, in Rip 2002: 125), a modified version of division of labour between research and uptake of research. It always has a local or contextualized component, and contributes to the 'glocalization' of the region with the effect of universities becoming internationally recognized attraction poles (Rip 2002).

Drawing on these ideas and adapting Berglund and Clarke's (2000) seven elements of a technology-based economy, preconditions that would enable universities to create regional competitive advantage (Lendel 2010), we suggest five main regional conditions: (i) local absorptive capacity; that is, local firms that are able to engage with university research (Rip 2002; Chapple *et al.* 2005; Lester 2005; Feldman and Francis 2006), (ii) a nascent or competitive infrastructure (e.g. physical, service), (iii) a technical and professional labour market, (iv) an entrepreneurial culture with coexisting entrepreneurial activity, and (v) quality of life from residential amenities.

Picking up on two of these, infrastructure extends beyond science parks and incubators. A study by Warren *et al.* (2008) on the availability of venture capital, an important element of infrastructure required for incubation, finds that universities isolated from supportive innovation systems has reduced possibilities of tech transfer. The authors suggest that areas where there are weak entrepreneurship communities, early stage innovators rely on the universities for early stage financing and facilities. Hence, close relationships are those with the university not with other actors. Other studies have suggested that in some places, the impact of university spin-offs staying local includes stimulating business support services and infrastructure (Zang 2008), other start-ups (Lockett *et al.* 2005), knowledge spillovers (Acs 2009), as well as jobs. The third, recruitment by local firms, is critical as it is the dynamism (increasing stocks and flows) of the local labour market that accounts for dynamism in clusters (Power and Malmberg 2008). However, not all regions can absorb graduates and skilled labour often leaves in an absence of jobs or entrepreneurial opportunities (McCann and Shephard 2001).

Moreover, Lendel (2010) as well as Harrison and Leitch (2010) show that the local impact of research universities is limited to a small number of places. Lendel (2010) highlights the importance of regional demand for university products: consumers of university products are not necessarily local. Larger metropolitan regions with large companies are more able to absorb the products of universities. Hence demand and impact are stronger than in smaller regions. In the US, top universities are usually located in metropolitan areas with local markets for certain outputs from universities. Two of the top ten UK universities are located in London. The two others, Oxford and Cambridge, are located in relatively small cities, hence where local demand might be expected to be lower but other impacts might be relatively higher for reasons given above.

Such caveats on the universality of university's regional impact are supported by Goldstein (2009: 28), who identifies measureable effects on the impact of technology transfer of universities by distance, types of research and kinds of universities. He finds spillovers from basic research to be less localized than for applied research with spillovers from highly ranked research universities more geographically widespread. In the UK, there are regional differences in the relative importance of revenue from IP and university spin-off – South East England is one of only a few regions where income from spin-offs realization and IP is above average (Harrison and Leitch 2010).

Overall, factors that appear to facilitate change and induce impact are the university strategies and research reputation, the national context, the facility to capitalize on and drive technological change, and the characteristics of the region. There is a difference between being represented in the regional systems by coincidence and being active in forwarding regional agenda. Regional leadership by universities, however, may take the form of direct connections with other organizations, and co-opting regional governance actors rather than vice versa. A summary of the propositions is found in Table 8.1.

Table 8.1 Four propositions setting out reasons why universities might change from being latent assets to active agents

Propositions	*Drivers of change*	*Evidence of change*	*Evidence of impact*
1 It depends on the internal characteristics of university	Recognition of opportunities, rewards and incentives for change (i) for the individual (ii) for the university Ability to capitalize on those opportunities	Institutional change Self interest in exploiting assets, setting terms by which university IP is exploited Establishment and expansion of TTOs' activities	Income from wealth creation activities e.g. spin-offs, patents, licenses and consulting
2 It is related to responses to exogenous shocks	Technological change. Schumpeterian patterns of entrepreneurship Regulatory change	Development of new industries and technologies localized in particular places Rates of university spin-offs and growth Science parks and incubators Economically resilient economies Regional outcomes of national policy change or events	Rising numbers of innovative firms including university spin-offs Investment portfolios of university-associated business angels Tenants in science parks Amelioration of negative effects of recessions Increased wealth creation activities by universities observable in the region.
3 Political agenda	national and regional policies and funding to encourage exploitation of university	Competition for funding from universities 'regional' facing universities Entrepreneurship education & awareness raising	wealth creation and outreach working in practice, alignment of interests Increasing number of student and other entrepreneurs
4 Impact depends on the attributes of the region	Regional absorptive capacity, associated with innovative firms and high quality local labour markets	Changes in the regions (e.g. increasing size of high-tech economy), local measures targeted at developing entrepreneurial culture Local recruitment from the university Involvement in local governance	Networking activity producing tangible and intangible outcomes such as business opportunities Increasing local recruitment from the university Tangible changes in policy towards innovation-led economic development

3 Oxford University and Oxfordshire

The chapter now turns to the changing relationship between the University of Oxford, its departments and some of its thirty-eight independent colleges and six private halls, and the county's high-tech economy. Although activity was in evidence before the 1980s (indeed academic entrepreneurship can be dated to the 1940s), it was then that academic entrepreneurship started to take off, national public policy encompassed both university entrepreneurship and science exploitation, and Oxford University took the first formal steps towards creating a more supportive environment for academic entrepreneurship. The timeline of changes in Oxford University referred to in the text is shown in Figure 8.1.

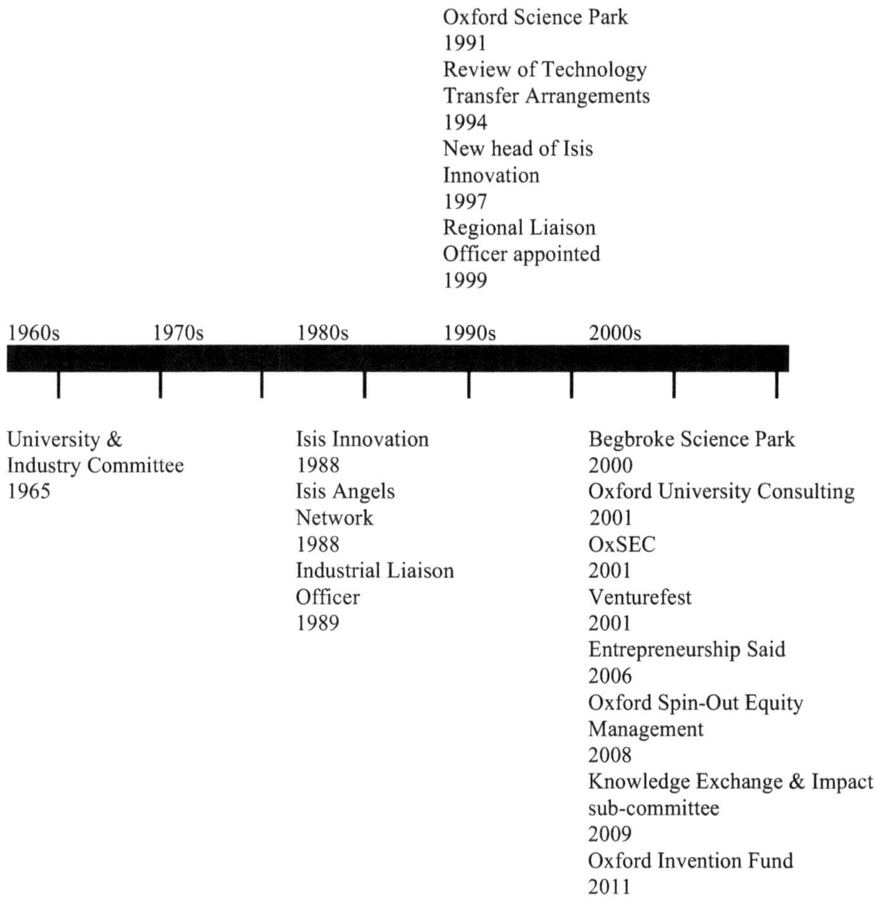

Figure 8.1 Timeline of major Oxford University 'third stream' initiatives.

3.1 Proposition 1: It depends very much on the university

3.1.1 Drivers of change

In the 1960s, early university efforts to improve links with industry over and above the long-standing collaborations by individual academics and departments included the University and Industry Committee (established in 1965, disbanded in 1987). In the 1970s, some departments established consortia with other universities to work on industries' problems. For example, Oxford was a founder member of the UCINA consortium formed in 1979 by a group of applied mathematics departments in five universities to address practical problems in industry as well as to stimulate new and relevant research in numerical analysis (Rip 2002).

In the 1980s, commercializing research became a priority but this still did not include enterprise formation. The driver was internal pressure for the university to benefit from its academics' commercial activities. The outcome was Isis Innovation, the university's wholly-owned technology transfer company, established in 1987 originally as Oxford University Research and Development Ltd. and renamed Isis Innovation in 1988. Initially, its main activity was networking. It established Isis Angels Network connecting private individuals and companies with potential interest in investing in spin-out companies from Oxford University. In 1990, the Oxford Innovation Society was set up to run networking events for scientists and potential funders/acquirers of university research. In 1989 the University took its first major shareholding in a company commercially developing and manufacturing products resulting from scientific breakthroughs of its own scientists: Oxford Glycosystems (later Oxford Glycosciences). By 1997, three companies had been established through Isis Innovation but Isis had not provided the support to help academic entrepreneurs to establish their businesses. In 1989, a full-time industrial liaison officer was appointed, but the role was not intended to have a local remit (Lawton Smith 1998).

3.1.2 Evidence of change

The failure of the university to capitalize on the profits made by spin-off firms formed by academics, technicians and sometimes students independently of the university was the reason why there was the *Review of Technology Transfer Arrangements* in 1994 (Report published 1995). The Report recommended that the University assert its rights to the intellectual property of its academics, students, and visitors (Lawton Smith 2003). A consequence was that a more professional commercialization organization was required to improve returns from university-owned intellectual property.

The Review made it possible for the appointment in 1997 of an entrepreneur as CEO of Isis Innovation. Over time, Isis has added extra functions to market its 'products' (Hill and Lendel 2007). It set up Oxford University Consulting (OUC) in 2001 with the University Research Support Office to facilitate academics in selling their time and helping industries find experts within the university. It is now part of Isis Innovation. In 2008, it established Oxford Spin-out Equity

Management to manage the University's interests in its spin-out companies. In 2011 a new fund, The Oxford Invention Fund, was launched to overcome the 'innovation gap' at early stages of an invention, and is funded by donors to the university. By the mid 2000s, Isis Innovation was widely regarded as 'one of the best technology transfer offices in the country' (HM Treasury 2003) and had grown to having the largest number of commercialization staff of any UK university (Minshall and Wicksteed 2005).

3.1.3 Evidence of impact

The impact of Isis Innovation can be seen in the rate at which research is commercialized through spin-offs, the robustness of those firms, the amount of resources attracted for the spin-off process and the value realized to the university. Since 1997, Isis Innovation has on average created a new spin-off company every two months and has been responsible for creating fifty-nine spin-out companies based on academic research generated within and owned by the University of Oxford. Nearly 90 per cent survive. It has raised an additional £280 million in follow-on venture funding rounds. By 2007, the combined value of Oxford's companies, of which four firms have been launched on AIM (the Alternative Investment Market, London Stock Exchange's international market for smaller growing companies), had reached £2 billion using quoted market capitalization and investor valuations for unquoted companies (£30 million in seed capital/angel and £252 million in venture/institutional funding). Some £4 million has been invested in equity stakes in twenty-one spin-offs, which have attracted £40 million in seed/venture investment (www.isis-innovation.com). In 2008, Isis returned £2.4m on a turnover of £4.7m to the University for distribution to university researchers, departments, and central funds, an increase of 30 per cent over 2006 (www.isis-innovation.com/documents/Isisannualreport2008.pdf (accessed August 11 2009).

3.2 Proposition 2: It is related to a response to exogenous shocks

3.2.1 Drivers of change

Exogenous shocks here refer to technological evolution which gave rise to high-tech entrepreneurial activity from the 1960s onwards in particular locations, for example, Silicon Valley and Route 128 in the US (Saxenian 1994) and the 'Western Crescent' around London, from Cambridge to Berkshire by way of Oxfordshire in the UK (Brehney and McQuaid 1987). These were places dominated by research in the public and private sectors, but very different in scale, location and focus of research. Berkshire and the M4 corridor is characterized by big firms and defence (electronics and big pharma in high tech), while Oxford's and Cambridge's high-tech firms generally were small and their concentrations of public sector, private sector and charitable trust-funded research were civil rather than defence (Garnsey and Lawton Smith 1998).

The increasing number of university-based firms has been a driver in the growth of Oxfordshire's high-tech economy. By the mid 1980s, there were 182 R&D-intensive advanced technology firms in the county which employed 10,659 people. The majority were in manufacturing (125), followed by R&D/consultancy (32) and software (25). These figures reflect national trends. Between 1979 and 1986, while UK manufacturing as a whole was in steep decline, employment in high-tech manufacturing fell less sharply and output increased rapidly. Computer sectors (both hardware and software) showed significant job and firm formation over a similar period (Keeble and Kelly 1986).

More recent data show that growth of entrepreneurship in Oxfordshire has accelerated. In the absence of local data, the Office of National Statistics data show that in 2008 Oxfordshire had some 3,500 high-tech firms employing 45,000 people. Around 14 per cent of the county's workforce[3] is in 12 per cent of businesses in Oxfordshire. Using these figures, the county has the third largest high-tech employment among UK counties (high-tech as a percentage of total employment). The county is characterized by 'diverse specialization' with leading sectors being high-tech services, including software consultancies and biotechnology. Larger businesses are in high-tech manufacturing, including pharmaceuticals, medical instruments and computers.

3.2.2 Evidence of change

In the mid 1980s, the rise in the high-tech economy was matched by a recognition by some of the colleges, but not yet the university, of the possibilities for commercial opportunities presented by the growing number of local high-tech firms including university spin-offs. Magdalen College, established the Oxford Science Park in 1991, succeeding where other colleges such as Wolfson, Exeter and Corpus Christi had failed. Other organizations were more successful in doing so. The provision of space on science parks and incubators is now a defining feature of Oxfordshire's high-tech entrepreneurship infrastructure. By 2008, the county had more incubators and science parks than any other – this is based on the United Kingdom Science Park Association (UKSPA) members and technology-based incubators. It had four UKSPA members and ten incubators.

Oxford University's own Science Park at Begbroke, established in 2000, is a science park incorporating an incubator with space for university spin-offs and local start-up companies. It therefore scores in each UKSPA category. Both the Science Park and the Centre for Innovation and Enterprise are engaged in third stream activities. These include the development of new courses based on applied research with the Department of Continuing Professional Development (CPD).

3.2.3 Evidence of impact

Impact is associated with Schumpeterian patterns of entrepreneurship in emerging technology-based industries. Oxford University was an important source of high-tech firms and employment by the mid1980s. By then some forty firms had been established by academics, technicians and former students, here generally classified as spin-offs. Most originated in the departments of Engineering Science and Metallurgy, but already the medical research units and life sciences had produced a number of firms. Examples include Penlon (medical products) (established 1943), Littlemore Scientific Engineering (1954) and Oxford Instruments (scientific instruments) (1959), which has had the greatest impact through expansion of its own operations and an incubator of firms formed by former employees. These are all still in existence, although Littlemore Scientific Engineering has moved out of the county. By the mid 2000s, Oxford University spin-offs contributed to nearly 3 per cent of the county's total workforce (Lawton Smith and Ho 2006), thereby contributing to path-dependent processes (Martin and Sunley 2006) of growth in the high-tech economy.

3.3 Proposition 3: Specific regional direct and indirect effects depend on the nature of funding

3.3.1 Drivers of change

It was in 1964 that Prime Minister Harold Wilson set his vision for modernizing the UK economy 'in the white heat of technology'. Successive Conservative governments (1979–97) established the paradigm of universities as wealth creators, continued by successive Labour governments (1997–2010), albeit with greater funding for research in universities, for entrepreneurship, and for university–industry interaction.

The Higher Education Innovation Fund (HEIF) has been the most comprehensive measure, set up in 2001 as a single, long-term commitment to 'support universities' potential to act as drivers of growth in the knowledge economy'. The Science Enterprise Fund (SEC) and the Universities Challenge Fund (UCF) were set up as separate funds under HEIF 1. The 2003 White Paper *The Future of Higher Education* (DfES 2003) introduced a spatial dimension as Regional Development Agencies were given a stronger role in steering the Higher Education Innovation Fund. Under HEIF 3, support would be given for 'less research-intensive' university departments. Similarly, in line with the Sainsbury Review, recommendations that more funding be directed towards business-facing institutions, HEIF 4 increased to £150 million in 2010–11, its final round, and is intended to redistribute funding from the richer to poorer universities. An upper limit was set on the amount received by each university, including those in the larger research universities (Gill 2008).

3.3.2 Evidence of change

National funding has enabled Oxford University to expand its 'third stream' activities. It has won funds from each round of HEIF as well as awards from all of the other outreach programmes. These include the University Challenge Fund (UCF) (1999) which granted seed funding to help universities make the most of research funding through support for early stages of commercial exploitation of new products and processes. Under HEIF 2, the university was awarded £2.36 million to increase access to the university's innovation system by funding Isis Enterprises to make technology transfer expertise available to other universities, science entrepreneurship business courses available to scientists from regional companies, and by expanding the Begbroke Science Park Enterprise Fellows scheme, which supports scientists moving out of academia, enabling them to translate academic innovation into commercial applications.

In 1999 a Regional Liaison Director was appointed. He was initially on secondment from Barclays Bank. Funding from HEIF 1 enabled Oxford to establish the Regional Liaison Office, thus giving a more permanent base, albeit one still reliant on HEIF funding. The appointment, 'underlines the University's commitment to play a more prominent part in local economic development'.[4] The Director is actively involved in local governance, being the university's representative on a number of local bodies including the board of the Oxfordshire Economic Partnership (OEP), which represented public and private stakeholders in the Oxfordshire economy.[5] The Office was absorbed into Isis Innovation in 2008.

In 2001, Oxford University was awarded SEC funding to establish OxSEC in the Said Business School with the central tenet of promoting entrepreneurship within the university. OxSEC has sought to add value through entrepreneurship education in three specific areas, namely: Teaching Entrepreneurship, Knowledge Transfer and Links to Business. From the outset, a strategic decision was made that all programmes were to be open to Oxford-based entrepreneurs, high-tech companies and to university members. Oxford Entrepreneurs was launched in 2003. This is a network for students and alumni of the University of Oxford aiming to encourage and support student entrepreneurship. In 2006, *Entrepreneurship Said* was set up as the focal point for entrepreneurship among academics, students and the business community. Subsequently, funded by HEIF 2 funding, HEIF 3 funding and competitive research income, OxSEC has grown to five full-time staff. It has well established connections with other parts of Oxford University such as Isis Innovation and the Begbroke Science Park and with local organizations such as The Oxford Trust and Oxford Innovation.

Other external pressures are driving change. In 2009, in response to the 'impact' agenda relating to measuring outcomes of university research as part of the next round of the research assessment process (the Research Assessment Framework, 2013), Oxford University established the Knowledge Exchange and Impact Sub-Committee, which reports to the planning and resource allocation committee. Part of its remit is advising on, 'Recognition for the wider application of KE (knowledge exchange) and ways to enable development in areas such as teaching and learning, local, regional and national engagement, including public, commercial and industrial engagement'. This is evidence of an institutionalization of a regional economic and societal role within the university system, albeit at a sub-committee level, and is therefore institutionally not analogous to Georgia Tech (Youtie and Shapira 2008).

3.3.3 Evidence of impact

Evidence of impact of the institutionalization of measures to support entrepreneurship within Oxford University is attendance at entrepreneurship education courses run by OxSEC. That is, there is a spatial dimension to this non-spatial policy initiative. Since their launch in 2001, over 9,000 people have attended an OxSEC training course, networking event or seminar with annual attendance reaching in 2005–06 just over 3,200. Its main teaching programme around science entrepreneurship is attended by an average of over 300 people per session (Lawton Smith *et al.* 2007). The number of attendees has risen dramatically. For instance attendance in 2008–09 was double that in 2005–06. This suggests that changes within the university have had the impact of sensitizing large numbers of university and non-university people to opportunities for entrepreneurship.

3.4 Proposition 4: Impact depends on the attributes of the region

3.4.1 Drivers of change

This driver for the change on the part of Oxford University to become more than a latent asset is the increasing absorptive capacity of local region (Rip 2002). This is indicated by the size and entrepreneurial culture of the local high-tech economy, the associated labour market and the presence of networks that provide support for its spin-offs, particularly for finance (Warren *et al.* 2008). Combined, local engagement by Oxford University with regional organizations over time became a viable proposition, with a higher demand for more of the university products (Hill and Lendel 2007) at the local level.

Oxford University indirectly contributed to the establishment of norms of high-tech entrepreneurship in the county as well as the physical and social infrastructure, through networks and networking and labour markets (elements in Berglund and Clarke's 2000 technology-based economies). In 1985, the founders of Oxford Instruments established The Oxford Trust. The Trust set up the county's first incubator units for high-tech firms near Oxford Instruments with six units and networking events for entrepreneurs, would-be-entrepreneurs, professionals with business interest in high-tech entrepreneurship (bankers and lawyers) and local policy makers. The Trust's activities include networking events, an online business network and the Oxford Investment Opportunity Network (OION) (1995), which matches investors to firms. Oxford Innovation, the Trust's spin-off company, has set up four more specialist investment networks for early stage finance. The county now has some sixty-five business networks including ones dedicated to biotech/medical instruments sectors. The Said Business School is an important host for many networking events open to university and non-university people (Lawton Smith 2010).

3.4.2 Evidence of change

The growing convergence of interest by the university in the local high-tech economy and the impact it has on delivering support for enterprise is illustrated by

Venturefest. In 1999 it was launched by an academic in St Catherine's College as a two-day annual event with the objective of bringing together entrepreneurs and venture capitalists. It attracts about 1,500 attendees from Oxfordshire and adjacent regions. Over time, it has been formally supported by the University. Its management team is drawn from representatives of the universities, the local research laboratories, local policy makers, and business. Thus Oxford University was able to be part of the process of building on local physical and network innovation and entrepreneurship infrastructures already in place.

3.4.3 Evidence of impact

The growth in the number of high-tech firms is a possible explanation for increasing impact through local graduate recruitment, suggesting greater alignment between the economy and the university. The effect is twofold. The first is more local recruitment. Higher Education Statistics Agency (HESA) 2004 data suggest high levels of local retention from both of Oxford's universities. About a third of Oxford University (32.62 per cent) and Oxford Brookes University (34.6 per cent) students stay in the county after their first degree. These figures need to be treated with caution. They may also reflect the much greater propensity for Oxford University graduates to stay on for further degrees: a quarter of all Oxford undergraduates go on to further study (Waters and Lawton Smith 2008). The second is the probability of high calibre people in the university and in local organizations working together on a range of initiatives as well as through local recruitment, that is, a 'Mode 1' knowledge dissemination effect (Rutten and Boekema 2009). In 2008, well over a third (38.6 per cent) of Oxfordshire residents were qualified to degree level so that the county ranked as the third most qualified county in England and Wales.

4 Conclusions

The purpose of this chapter is to explore the conditions under which research universities become more than latent assets in their regional economies (Feldman and Francis 2006; Power and Malmberg 2008). Four propositions, which provide the basis for evaluation of regional development impact, are offered and tested using Oxford University as an exemplar: the regional impact of universities depends on (i) the internal characteristics of the university; (ii) how the university responds to exogenous shocks (Feldman and Francis 2006); (iii) the nature of funding for higher education institutions which accompanies agenda-setting by national and regional political bodies; and (iv) the attributes of the regional economy (Varga 2001; Lendel 2010).

With respect to the first, there is an analogy between Meyer's (2003) distinction between 'entrepreneurial academics' and 'academic entrepreneurs'. Some universities have elements of entrepreneurship (academics who undertake commercial activities) and some have a strategic commitment to be entrepreneurial (academics who run businesses). Oxford University, one of the world's top research universities has both. It has a high rate of spin-offs and other indicators of commercialization. Strategic commitment in Oxford University is devolved to three poles: Isis Innovation, the Said Business School and the Begbroke Science Park. While the university benefits from the kudos associated with these successes, it is more a 'connected' university (Kitson *et al.* 2009) than an entrepreneurial university in that it lacks a central steering core (Clark 1998). Of the three poles, Isis Innovation is the lead organization. Its impact has grown broader and stronger as it has captured more of the wealth creating and outreach roles working closely with the other two and with the research support office.

Drivers have been both internal and external. External drivers are derived from national policy. Internal pressures to capture the financial rewards from the IP of staff and students have resulted in a range of extensive wealth creation activities. While the university has produced spin-offs for over sixty years, the evidence suggests that the last twelve have seen Isis Innovation manage the process more effectively. It has been successful in commercializing research through robust spin-offs, patents and licences, thereby increasing the value realized to the university of its IP and in the amount of internal and external resources it has attracted for the spin-off process. On these criteria, it has a very effective business model for technology transfer (Mustar *et al.* 2006; Wright *et al.* 2007; Huggins 2008; Muscio 2010).

The rate of spin-offs, however, did not occur independently of broader changes in the economy, the kind of exogenous shocks suggested in Proposition 2, which led to the rise of high-tech economies in a few favourable locations. The increasing rate of university spin-offs is especially in sectors associated with the fifth Kondratieff wave (Freeman and Perez 1988) – biotech and ICT. The impact is twofold. First, spin-offs contribute to path dependent economic development processes through the formation of new firms. Second, growth in numbers of new firms formed by both university and non-university personnel has stimulated Oxford University's colleges and later the university itself to expand the physical infrastructure and associated support structures for local high tech industry (Phan *et al.* 2005).

Proposition 3 states that national public policy drives change and has a direct and indirect impact on a university's incentive to become entrepreneurial and regionally engaged. Oxford University has been extremely successful in winning funding from all post-1997 third-stream policy measures. It is very unlikely to have become 'connected' (Kitson *et al.* 2009) and as entrepreneurial without central government funding. The university now has an 'innovation structure' that comprises awareness raising, education and training, supporting services, as well as wealth creation strategies. Evidence includes an increasing number of participants in entrepreneurship courses and collaboration with local organizations. How embedded these activities are will in part depend on the nature of future funding under HEIF and other outreach programmes, and in part on what follows from the decision by the Coalition Government elected in 2010 to abolish 'the region' as a concept and scale of policy delivery.

The last proposition is that transformation depends on the attributes of region: that is, the latent or nascent assets of the university and the region become increasingly interconnected. Assets are drivers of change because of the possibilities they create, and include the absorptive capacity of actors (firms, the public sector, local governance bodies) (Rip 2002; Rutten and Boekema 2009) in the university's home region. In turn, these relate to quantity and quality of skills in the local labour market and the hosting and engagement of business networks in local systems of governance. Oxford University has become active in both.

As Lendel (2010) points out, the bundled nature of university production, and the complexity and interdependency of external and internal pressures, makes it difficult to assess which factors are the most important in facilitating regional impact. The evidence suggests that all are required. However, as Martinelli *et al.* (2008) show, technology transfer can be effective, but it is likely to be discipline-related and its impact not necessarily regional. As Rothaermel *et al.* (2007) point out, there is a great need for longitudinal data to establish more clearly the degrees of association between now observed and causal processes at work in influencing the evolution of academic entrepreneurship and their connection to local entrepreneurial activities and regional development. Finally, many governments seem to be attempting to establish further elite, research universities. This study suggests that this may or may not have desirable regional effects right away (Arbo and Benneworth 2007).

Acknowledgement

An earlier version of this chapter appeared in Lawton Smith, H. and Bagchi-Sen, S. (2012) 'The research university, entrepreneurship and regional development: research propositions and current evidence', *Entrepreneurship & Regional Development*, 24, 5–6: 383–404

Notes

1 www.rcuk.ac.uk/Pages/Home.aspx (accessed 21 February 2011).
2 www.timeshighereducation.co.uk/story.asp?storyCode=198532&sectioncode=26 (accessed 9 February 2011).
3 ONS uses data from the annual Business Inquiry which uses a more comprehensive dataset with a less restrictive definition of high-tech and includes higher enumerate numbers of businesses in IT and computer-related services.
4 www.ox.ac.uk/gazette/1999-00/weekly/071099/news/story_4.htm (accessed 24 July 2009).
5 www.oep.org.uk/ (accessed 29 April 2010).

References

Acs, Z. (2009) 'Jaffe-Feldman-Varga: The search for knowledge spillovers', in A. Varga (ed.) *Universities, Knowledge Transfer and Regional Development*. Cheltenham: Edward Elgar, pp. 36–56.

Andersen, B. and Rossi, F. (2010) 'The flow of knowledge from the academic research base into the economy: The use and effectiveness of formal IPRs and "soft IP" in UK universities'. Study commissioned by the Strategy Advisory Board for Intellectual Property Policy (SABIP).

Annual Report (2008). Available online at www.admin.ox.ac.uk/pras/committees/research/knowledge/ (accessed 20 June 2011).

Anselin, L., Varga, A. and Acs, Z. J. (1997) 'Local geographical spillovers between university research and high technology innovations', *Journal of Urban Economics*, 42, 422–448.

Arbo, P. and Benneworth, P. (2007) *Understanding the Regional Contribution of Higher Education Institutions: A literature review*. Paris: OECD. Available online at www.oecd.org/dataoecd/55/7/37006775.pdf (accessed 10 November 2010).

Beeson, P. and Montgomery, E. (1993) 'The effects of colleges and universities on local labor markets', *Review of Economics and Statistics*, 75(4), 753–761.

Benneworth, P. (2010) *Universities & Regional Engagement*. Working Paper 6 'University learning with excluded communities', project CHEPS, University of Twente.

Bercovitz, J. and Feldman, M. (2006) 'Entrepreneurial universities and technology transfer: A conceptual framework for understanding knowledge-based economic development', *Journal of Technology Transfer*, 31, 175–188.

Berglund, D. and Clarke, M. (2000) *Using Research and Development to Grow State Economies*. Washington, DC: National Governors' Association.

Binks, M., Wright, M., Lockett, A. and Vohora, A. (2005) 'The creation of spin-off firms at public research institutions: Managerial and policy implications', *Research Policy*, 34, 981–993.

Boucher, G., Conway, C. and Van der Meer, E. (2003) 'Tiers of engagement by universities in their region's development', *Regional Studies*, 37, 887–897.

Breheny, M. J and McQuaid, R. W. (1987) 'H.T.U.K.: The development of the United Kingdom's major centre of high technology industry', in M. J. Breheny and R.W. McQuaid (eds), *The Development of High Technology Industries*. London: Croom Helm, pp. 297–354.

Chapple, W., Lockett, A., Siegel, D. and Wright, M. (2005) 'Assessing the relative performance of U.K. university technology transfer offices: Parametric and non-parametric evidence', *Research Policy*, 34(3), 369–384.

Christy, C. V. and Ironside, R. G. (1987) 'Promoting high technology industry: Location factors and public policy', in K. Chapman and G. Humphreys (eds), *Technical Change and Industrial Policies*. Oxford: Basil Blackwell, pp. 232–252.

Clark, B. R. (1998) *Creating Entrepreneurial Universities: Organizational pathways of transformation*. New York: Elsevier.

Clarysse, B., Wright, M., Lockett, A., Van de Velde, E. and Vohora, A. (2005) 'Spinning out new ventures: A typology of incubation strategies from European research institutions', *Journal of Business Venturing*, 20, 183–216.

Degroof, J-J. and Roberts, E. B. (2004) 'Overcoming weak entrepreneurial infrastructures for academic spin-off ventures', *Journal of Technology Transfer*, 29, 327–352.

DfES (2003) *The Future of Higher Education*. London: DfES. Available online at www.bis.gov.uk/assets/biscore/corporate/migratedd/publications/f/future_of_he.pdf (accessed 20 June 2011).

Etzkowitz, H. (1983) 'Entrepreneurial scientists and entrepreneurial universities in American academic science', *Minerva*, 21, 198–233.

Feldman, M. P. (2001) 'The entrepreneurial event revisited: Firm formation in a regional context', *Industrial and Corporate Change,* 10(4), 861–891.

Feldman, M. and Francis, J. L. (2006) 'Entrepreneurs as agents in the formation of industrial clusters', in B. Asheim, P. Cooke and R. Martin (eds), *Clusters and Regional Development*. Abingdon: Routledge, pp.115–136.

Franzoni, C. and Lissoni, F. (2009) 'Academic entrepreneurs: Critical issues and lessons for Europe', in A. Varga (ed.), *Universities, Knowledge Transfer and Regional Development*. Chelteham: Edward Elgar, pp. 163–190.

Freeman, C. C. and Perez, C. (1988) 'Structural crisis of adjustment, business cycles and investment behaviour', in G. Dosi, C. Freeman, R. Nelson, G. Silverberg and L. Soete (eds) *Technical Change and Economic Theory*. London: Pinter, pp. 38–66.

Garnsey, E. and Lawton Smith, H. (1998) 'Proximity and complexity in the emergence of high technology industry: The Oxbridge comparison', *Geoforum*, 29(4), 433–450.

Geuna, A. and Muscio, A. (2009) 'The governance of university knowledge transfer', *Minerva*, 47(1), 93–114.

Gill, J. (2008) 'Third stream cash is capped'. Available online at www.timeshighereducation.co.uk/story.asp?storyCode=401043&sectioncode=26 (accessed 14 August 2009).

Goldstein, H. A. (2009) 'What we know and what we don't know about the regional economic impacts of universities', in A. Varga (ed.) *Universities, Knowledge Transfer and Regional Development*. Chelteham: Edward Elgar, pp. 11–35.

Gottlieb, P. (2001) *The Problem of Brain Drain in Ohio and Northeastern Ohio: What is it? How severe is it? What should we do about it?* Cleveland, OH: Case Western Reserve University, Center for Regional Economic Issues, Weatherhead School of Management.

Goldstein, H. A. and Renault, C. S. (2004) 'Contributions of universities to regional economic development: A quasi-experimental approach', *Regional Studies*, 38, 733–746.

Harrison, R. and Leitch, C. (2010) 'Spin-off companies, the entrepreneurial system and regional development in the UK', *Regional Studies*, 44(9), 1241–1262.

Harvey, K. (1992) 'Managing the exploitation of intellectual property: An analysis of policy and practice in nine UK universities'. Available online at https://dspace.stir.ac.uk/dspace/handle/1893/2116 (accessed 19 January 2011).

HE-BCI (2010) Higher Education – Business and Community Interaction Survey: 2008–09. Available online at http://www.hefce.ac.uk/pubs/year/2010/201014/ (accessed 23 July 2012).

Hill, E. and Lendel, I. (2007) 'The impact of the reputation of bio-life science and engineering doctoral programs on regional economic development', *Economic Development Quarterly*, 21, 223–243.

HM Treasury (2003) Lambert Review of Business–University Collaboration (the Lambert Report). London: HM Treasury.

Huggins, R. (2008) 'Universities and knowledge-based venturing: Finance, management and networks in London', *Entrepreneurship and Regional Development*, 20(2), 185–206.

Keeble, D. and Kelly, T. (1986) 'New firms and high technology industry in the United Kingdom: The case of computer electronics', in D. Keeble and E. Weaver (eds) *New Firms and Regional Development in Europe*. London: Croom Helm, pp. 75–104.

Kenney, M. (1986) *Biotechnology: The university–industry complex*. New Haven, CT: Yale University Press.

Kenney, M. and Patton, D. (2009) 'Reconsidering the Bayh-Dole Act and the current university invention', *Research Policy*, 38, 1407–1422.

Kinsella, R. and McBrierty, V. (1997) 'Campus companies and the emerging techno-academic paradigm: The Irish experience', *Technovation*, 17, 245–251.

Kitson, M., Howells, J., Braham, R. and Westlake, S. (2009) *The Connected University: Driving recovery and growth in the UK economy*. London: NESTA. Available online at www.nesta.org.uk/library/documents/Report%2023%20-%20The%20Connected%20Uni%20v4.pdf (accessed 20 June 20 2011).

Lawton Smith, H. (1998) '"Barriers" to technology transfer: Local impediments in Oxfordshire', *Environment and Planning C*, 16, 433–448.

Lawton Smith, H. (2003) 'Knowledge organisations and local economic development: The cases of Oxford and Grenoble', *Regional Studies*, 37, 899–909.

Lawton Smith, H. (2006) *Universities, Innovation and the Economy*. Abingdon: Routledge.

Lawton Smith, H. (2010) 'Formal networks and regional contexts: The case of the Oxford to Cambridge Arc', *Regions*, (March), 5–9.

Lawton Smith, H. and Ho, K. W. (2006) 'Measuring the performance of Oxfordshire's spin-off companies', *Research Policy*, 35, 1554–1568.

Lawton Smith, H., Glasson, J. and Chadwick, A. (2007) *Enterprising Oxford: The Oxfordshire model*. Oxford: Oxfordshire Economic Observatory.

Lendel, I. (2010) 'The impact of research universities on regional economies: The concept of university products', *Economic Development Quarterly*, 24(3), 210–230.

Lester, R. (2005) *Universities, Innovation, and the Competitiveness of Local Economies: Summary report from the local innovation project (MIT-IPC-05–010)*. Cambridge, MA: MIT, Industrial Performance Center.

Link, A. L. and Rees, J. (1990) 'Firm size, university based research, and the return to R&D', *Small Business Economics*, 2, 25–31.

Litan, R. E., Mitchell, L. and Reedy, E. J. (2007) 'The university as innovator: Bumps in the road', *Issues in Science and Technology*, (Summer), 57–66.

Lockett, A., Siegel, D., Wright, M. and Ensley, M. D. (2005) 'The creation of spin-off firms at public research institutions: Managerial and policy implications', *Research Policy*, 34, 981–993.

Lockett, A. and Wright, M. (2005) 'Resources, capabilities, risk capital and the creation of university spin-out companies', *Research Policy*, 34, 1043–1057.

Luger, M. I. and Goldstein, H. A. (1991) *Technology in the Garden*. Chapel Hill, NC: University of North Carolina Press.

Markman, G. D., Phan, P. H., Balkin, D. B. and Gianiodis, P.T. (2005) 'Entrepreneurship and university-based technology transfer', *Journal of Business Venturing*, 20(2), 241–263.

Martin, R. and Sunley, P. (2006) 'Path dependence and regional economic evolution', *Journal of Economic Geography* 6(4), 395–437.

Martinelli, A., Meyer, M. and Von Tunzelmann, N. (2008) 'Becoming an entrepreneurial university? A case study of knowledge exchange relationships and faculty in a medium-sized, research-oriented university', *Journal of Technology Transfer*, 33, 259–283.

McCann, P. and Sheppard, S. (2001) 'Public investment and regional labour markets: The role of UK higher Education', in D.Felsenstein, R. McQuaid, P. McCann and D. Shefer (eds), *Public Investment and Regional Economic Development*. Cheltenham: Edward Elgar, pp. 135–153.

Meyer, M. (2003) 'Academic entrepreneurs or entrepreneurial academics? Research-based ventures and public support mechanisms', *R&D Management*, 33, 107–115.

Minshall, T. H. W. and Wicksteed, W. (2005) 'University spin-out companies: Starting to fill the evidence gap'. A report for The Gatsby Charitable Foundation, online. Available online at www.sqw.co.uk/file_download/56 (accessed 20 June 2011).

Moray, N. and Clarysse, B. (2005) 'Institutional change and resource endowments to science-based entrepreneurial firms', *Research Policy* 34, 1010–1027.

Mowery, D. C. and Sampat, B. N. (2005) 'The Bayh-Dole Act of 1980 and university industry technology transer: A model for other OECD governments?' *Journal of Technology Transfer*, 30(1/2), 115–127.

Muscio, A. (2010) 'What drives the university use of technology transfer offices?' *Journal of Technology Transfer*, 35(2), 181–202.

Mustar, P., Renault, M., Colombo, M. G., Piva, E., Fontes, M., Lockett, A., Wright, M., Clarysse, B. and Moray, N. (2006) 'Conceptualising the heterogeneity of research-based spin-offs: A multi-dimensional taxonomy', *Research Policy*, 35, 289–308.

O'Shea, R. P., Allen, T. J., Chevalier, A., and Roche, F. (2005) 'Entrepreneurial orientation, technology transfer, and spin-off performance of U.S. universities', *Research Policy*, 34, 994–1009.

PACEC (2003) *The Cambridge Phenomenon: Fulfilling the potential*. Cambridge: PACEC.

PACEC (2009) *Evaluation of the Effectiveness and Role of HEFCE/OSI Third Stream Funding: Culture change and embedding capacity in the higher education sector towards economic impact*. Cambridge: PACEC.

Phan, P. Siegel, D. and Wright, M. (2005) 'Science Parks and Incubators: Observations, synthesis and future research agenda', *Journal of Business Venturing*, 20, 165–182.

Power, D. and Malmberg, A. (2008) 'The contribution of universities to innovation and economic development: In what sense a regional problem?' *Cambridge Journal of Regions, Economy and Society*, 1(2), 233–246.

Powers, J. B. (2003) 'Commercializing Academic Research: Resource Effects on Performance of University Technology Transfer', *Journal of Higher Education*, 74(1), 26–50.

Radosevic, S. (2007) 'National systems of innovation and entrepreneurship: In search of a missing link', London:UCL-SSEES Economics Working Paper No. 73.

Rinne, R. and Koivula, J. (2005) 'The changing place of the university and a clash of values: The entrepreneurial university in the European knowledge society. A review of the literature', *Higher Education Management and Policy*, 17(3), 87–122.

Rip, A. (2002) 'Regional innovation systems and the advent of strategic science', *Journal of Technology Transfer*, 27(1), 123–131.

Rothaermel, F.T., Agung, S. and Jiang, L. (2007) 'University entrepreneurship: A taxonomy of the literature', *Industrial and Corporate Change*, 16, 691–791.

Rutten, R. and Boekema, R. (2009) 'Universities and regional development', Book review essay, *Regional Studies*, 43(5), 771–775.

Saxenian, A. (1983) 'The genesis of Silicon Valley', in P.Hall and A. Markusen (eds), *Silicon Landscapes*. Boston, MA: Allen & Unwin, pp. 20–34.

Saxenian, A. (1994) *Regional advantage: Culture and competition in Silicon Valley and Route 128*. Cambridge, MA: Harvard University Press.

Schumpeter, Joseph A. (1934) *The Theory of Economic Development*. Cambridge: Harvard University Press (New York: Oxford University Press, 1961). First published in German, 1912.

Segal Quince and Partners (1985) *The Cambridge Phenomenon*. Cambridge: Segal Quince and Partners.

Shane, S. and Stuart, T. (2002) 'Encouraging entrepreneurship? The effects of the Bayh-Dole Act on university patenting in the United States', *Journal of Business Venturing*, 19, 127–151.

Shapero, A. (1984) 'The entrepreneurial event', in C. A. Kent (ed.), *The Environment for Entrepreneurship*. Lexington, MA: Lexington Books, pp. 21–40.

Siegel, D. S., Waldman, D. A., Atwater, L. E. and Link, A. N. (2003) 'Commercial knowledge transfers from universities to firms: Improving the effectiveness of university-industry collaboration', *Journal of High Technology Management Research*, 14, 111–133.

Siegel, D.S., Waldman, D.A., Atwater, L.E. and Link, A.N. (2004) 'Toward a model of the effective transfer of scientific knowledge from academicians to practitioners: Qualitative evidence from the commercialization of university technologies', *Journal of Engineering and Technology Management*, 21, 115–142.

Siegel, D. S., Wright, M. and Lockett, A. (2007) 'The rise of entrepreneurial activity at universities: Organizational and societal implications', *Industrial and Corporate Change*, 16, 489–504.

THES (2009) *Times Higher Education-QS World University Rankings 2009*. Available online at www.timeshighereducation.co.uk/Rankings2009-Top200.html (accessed 26 July 2010).

Varga, A. (2001) 'Universities and regional economic development: Does agglomeration matter?' in B. Johannsson, C. Karlsson and R.Stough (eds), *Theories of Endogenous Growth: Lessons for regional policies*. Berlin: Springer, pp. 345–367.

Warren, A., Hanke, R. and Trozer, D. (2008) 'Models for university technology transfer: Resolving conflicts between mission and methods and the dependency on geographic location', *Cambridge Journal of Regions, Economy and Society*, 1(2), 219–232.

Waters, R. and Lawton Smith, H. (2008) 'Social networks in high technology local economies: The cases of Oxfordshire and Cambridgeshire', *European Urban and Regional Studies*, 15, 21–37.

Wright, M., Clarysse, B., Mustar, P. and Lockett, A. (2007) *Academic Entrepreneurship in Europe*. Cheltenham: Edward Elgar.

Youtie, J. and Shapira, P. (2008) 'Building an innovation hub: A case study of the transformation of university roles in regional technological and economic development', *Research Policy*, 37, 1188–1204.

Zang J. (2008) 'The performance of university spin-offs: An exploratory analysis using venture capital data', *Journal of Technology Transfer*, 34, 255–285.

9 The second coming of the Triple Helix and the emergence of hybrid innovation environments

Martin Meyer, Kevin Grant and Jari Kuusisto

1 Introduction[1]

This chapter reflects on and attempts to make sense of recent developments of the relevance of the Triple Helix to university industry government relations and examines their impact on economic development in a national and regional context. Sense making seeks to structure the unknown by placing policy formulation into a frame of reference using the theories of policy networks and policy communities, thus allowing one to 'comprehend, understand, explain, attribute, extrapolate and predict' (Starbuck and Milliken 1988: 51), in this case national examples of economic and innovation development.

At the heart of the Triple Helix notion lies the assumption that the spheres of academia, industry and government are moving closer to each other and boundaries are getting blurred. Leydesdorff (2003) developed a formal model that allows us to test the overlay of the various strands of the Triple Helix. One of the salient features of Leydesdorff's model is that it allows for 'negative overlay', or more specifically, a decrease in the overlap between the three different strands. This approach provides inroads to identify policy, structural and focus-related change, or departure to and in the Triple Helix relationship, which has tended to have the 'University' at its heart. Recent research has shown that university patenting – arguably a key indicator in the Triple Helix context – has become static, particularly in the USA (Leydesdorff and Meyer 2006, 2010) and early indicators may suggest it is in decline. This finding suggests blurring of boundaries at the edges of the Triple Helix, coupled with key actors in the Helix system, principally in academe, tending to re-entrench themselves to focus on their 'core business' of academic research (Leydesdorff and Meyer 2010). It also may suggest that the centre of the Helix may be shifting to a different locality: conceptually, intellectually, methodologically and practically.

One could argue that this refocusing on core activities (research and teaching) may actually increase the need for collaboration in what are pre-competitive or non-core areas of academic activity, in different ways to support and enable the research and teaching agenda. Newman writes, when exploring the idea of the academy, that other institutions are far better suited to act as instruments 'extending the boundaries of our knowledge than a university' (Newman 1852: Preface).

Garrod's view complements Newman's work. Garrod suggests that the role of the academy is to promote in others the '*love of truth above the display of mere ingenuity*' (Garrod 1946: 41). Scott develops the idea that a contemporary university should behave like a finishing school for intellectual development and creativity; it should act like the last stage of general education and it should produce a professional elite. It should facilitate the production of scientific knowledge for economic wealth, rather than create it, and it should encourage and promote a cultural ideology that expresses our sense of well-being (Scott *et al.* 1990).

This in turn, necessitates the emergence of new institutional arrangements and forms to enable this to happen and for the economic development of nations and between nations in an organised, managed and focused way to ensure appropriate and relevant economic development.

Such new forms of collaboration allow the various actors and agencies in a Triple Helix to interconnect, collide and create new value propositions, adopting new business models in a collaborative and dynamic way. By drawing on a range of international case studies, we will argue that there is growing evidence of the emergence of hybrid Triple Helix organisations that are better suited to collaboration and co-creation of value, than existing agencies, and indeed there is no longer just the Triple Helix, but several Helix variants that are more suited to different and differing climates, landscapes and intentions. We call this the 'second coming' and this premise is in keeping with the idea of structural innovation, as proposed by Howells and Edler (2011) and it questions to a lesser extent why universities are at the heart of any Triple Helix (or variants) as opposed to other types of agency. This may explain why the earlier findings of Leydesdorff and Meyer (2010) have indicated that universities are retrenching themselves, perhaps into what Schils (1983) claimed, as the purpose of a university has a distinctive task: it is the methodological *discovery* and the *teaching* of truths about serious and important things, not necessarily being the 'power house' behind a or any Triple Helix.

2 The Triple Helix: old structures for new environments?

The 'Triple Helix III' in its '*final*' form is concerned with 'generating a knowledge infrastructure in terms of overlapping institutional spheres, with each taking the role of the other and with hybrid organizations emerging at the interfaces' (Etzkowitz and Leydesdorff 2000: 111; see Figure 9.1). Research has dealt with tri-lateral networks. Far less work has been concerned with 'hybrid organisations' at the touch points. We argue that these are coming now and should be more in the foreground of our thinking and operationalisation of the Triple Helix concept. Research by Etzkowitz (2003) and others (e.g., Tuunainen 2005) have pointed to tensions in boundary-specific organisations that were created with a different mindset in a different time for a different landscape.

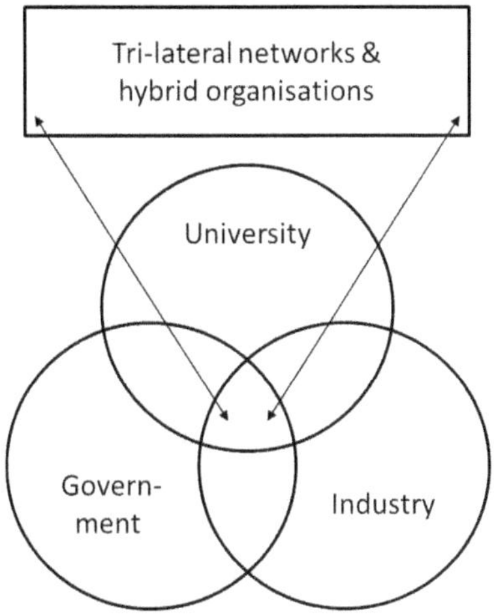

Figure 9.1 The Triple Helix III.

Source: adapted from Etzkowitz and Leydesdorff (2000).

A way of resolving these inherent tensions is via trilateral relationships in the form of new organisational entities that address the need for a plexus between academe, industry and government, which are neither research organisations, nor for-profit businesses, but are often aligned towards policy interventions that share collective ambitions, desires and commitments.

We will argue that these new organisational forms have been created to institutionalise innovation, managed and directed as separate entities with actors from all three Triple Helix strands as stakeholders (if not shareholders). These new forms may resemble 'research in an application context' in how they operationalise what they do and how they do things. This is similar to what Gibbons *et al.* (1994) and Nowotny *et al.* (2001) described as Mode 2 knowledge production (see also Hessels and van Lente 2009) and in what Stokes (1997) characterised as user-inspired basic research located in the 'Pasteur quadrant'. However, the authors mentioned earlier concentrate on knowledge production, not transfer, renewal, consumption and disposal, something these new organisational forms could fulfil.

Often, policy interventions aimed at this sector of innovation-directed research sought to produce knowledge that combines both relevance and scientific excellence (Rip 2004; Hessels and van Lente 2009). The rise and growing spread of 'centres for excellence and relevance', 'collaborative research centres', or 'centres of competence' can be taken as indicators for this new regime and the beginning institutionalisation, or rather 'formalisation', of innovation-directed collaborative research (Rip 2004; Thune and Gulbrandsen 2011; Turpin and Fernández-Esquinas 2011).

As Leydesdorff and Meyer (2010) argue, 'neo-institutional arrangements' are emerging while universities appear to be returning to their core missions.[2]

3 What shape can 'neo-institutional arrangements' take?

While Etzkowitz (2008) describes these spaces from a predominantly regional perspective, they are clearly also relevant in our broader context. We reconceptualise these spaces more generally. One could argue that in 'neo-institutional arrangements', a Triple Helix organisation serves as a platform, generating these Triple Helix spaces. Given this task, these 'hybrid organisations' would need to have a coordinating function for the tri-lateral networks which involves strategic planning of research as well as intermediation and mechanisms for the diffusion and commercialisation of innovation. The drive towards more user involvement in open innovation activities and the focus on use-inspired research would suggest a governance structure that is inclusive, giving prominence to users and their representatives; the structure would also need to offer 'rules of engagement' to guide interactions and a set of shared values. Related to this is the question of how to define or scope areas of use-inspired research. How this will impact on universities is less well understood, but we suggest here that changing the way universities think about generalizations, changing the institutional publication practices, focusing more on understanding the Triple Helix organizational stakeholders, and developing new knowledge creation and transformation networks are needed.

3.1 Knowledge spaces: concentration of R&D activities and resources

Knowledge generation spaces, often organised as (university-based) centres (or units), bring together several research groups around a shared theme for various reasons, such as attracting a greater amount of funding than any single group could; building or acquiring a new facility/infrastructure, and undertaking larger-scale projects (Etzkowitz and Kemelgor 1998; Etzkowitz 2008). Within the Triple Helix context, centres need to be sufficiently broad to generate interest beyond the immediate research, being more widely socially relevant. Topical breadth and relevance will therefore be key descriptors for Triple Helix organisations, as well as sufficiently sizeable funding bases to build a variety of activities on. One could argue that the breadth of the research agenda may reach to the level of a cluster or industry.

3.2 Consensus spaces: strategic research planning and governance

Consensus space can be considered a neutral ground where actors from different organisational backgrounds and perspectives can come together to 'generate and gain acceptability and support for new ideas… knowledge spaces are often transformed from potential to actual sources of economic and social development' (Etzkowitz 2008: 78). Consensus spaces are the place for strategy review and formulation where actors from different strands of the Triple Helix are brought together. This can be associated with a number of intermediary roles and functions (summarised in Table 9.1) based on earlier work by Howells (2006). A Triple Helix organisation could be expected to carry out such functions.

Table 9.1 Consensus spaces – intermediation functions

Type	*Function*
Foresight and diagnostics	(a) Technology foresight and forecasting
	(b) Articulation of needs and requirements
Scanning and information processing	(a) Scanning and technology intelligence – Information scanning and technology intelligence information gathering and identification of potential collaborative partners
	(b) Scoping and filtering (Selection and clearing function, selection of collaborative partners)
Knowledge processing and combination/recombination	(a) Combinatorial [Helping to combine knowledge of two or more partners];
	(b) Generation and recombination
Gatekeeping and brokering	(a) Matchmaking and brokering
	(b) Contractual advice

Source: adapted from Howells (2006).

3.3 Innovation spaces: innovation and intermediation functions

In Etzkowitz's (2008) framework, innovation spaces are new organisational mechanisms that are concerned with realising the goals identified in strategies developed in the consensus space. The Helix framework conceptualises innovation as a multilayered process integrating linear, reverse linear and non-linear processes into a complex adaptive web of relationships and interactions (that collide at times) which is reflected in a broad range of intermediation functions. Table 9.2 summarises some of these functions based on Howells' (2006) earlier work. A Triple Helix-based organisation arguably should be able to facilitate a broad range of these functions to its members or partners, whether this occurs in-house or through networks. Some universities have put frameworks in place that address these issues to some extent already as part of larger networks (e.g. Debackere 2000).

Etzkowitz's framework views resource allocation from the perspective of the applicant rather than the funder. However, a hybrid organisation could see this from a user or consumer's perspective. While allocating public funding is one of the pre-eminent roles of government, typically exercised through research councils or dedicated agencies responsible to government ministries, the rise of public–private partnerships in the 1990s and early 2000s points to another perspective.[3] This development coincides with 'devolution from the centre' in a number of countries

Table 9.2 Innovation spaces – innovation functions

Type	*Function*
Testing and validation	(a) Testing, diagnostics, analysis and inspection (b) Prototyping and pilot facilities (c) Scale-up including manufacturing modelling to overcome bottlenecks (d) Validation, e.g. of analytic methods (e) Training, joint training in use of new technologies
Accreditation	(a) Specification setter or providing standards advice (b) Formal standards setting and verification (c) Voluntary and de facto standards setter
Validation and regulation	(a) Regulation (b) Self-regulation (c) Informal regulation and arbitration (for example, between consumers and producers)
Protecting the results	(a) Intellectual property (IP) rights advice (b) IP management for clients
Commercialisation	(a) Market research and business planning (b) Sales network and selling (c) Finding potential capital funding and organising funding or offerings/early stage capital (d) Venture capital (e) Initial Public Offering
Evaluation of outcomes	(a) Technology assessment (b) Technology evaluation

Source: adapted from Howells (2006).

(Etzkowitz 2008).[4] In some countries, learned societies or research institutes have taken on a 'sovereign' function(s) and role(s) in administering government R&D programmes, supporting applicants in preparing their proposals and informing the decision-making process of the respective ministry or agency (e.g. the so-called 'Projektträger' in Germany; see e.g. Edler and Kuhlmann 2008). In others, ministries delegate responsibilities, including resource allocation, to agencies (e.g. Braun 2008; Pelkonen *et al.* 2008). It may be possible to take this type of delegation a stage further. This would be in line with the Triple

Helix framework, which often emphasises the blurring of boundaries or the relocation of the centre.

These organisations could be envisaged to exist in a different organisational context than the partners in academe, industry and government (which may include users, consumers, beneficiaries and society in general). In a sense, they could be linking structures and entities not based in universities, nor a government agency and not a business either – a 'true' Triple Helix organisation at the centre of trilateral network(s) being and belonging. One could argue that the scope of such an organisation would be broader and more dynamic than that of a typical collaborative research centre. Work on intermediaries specifies a set of functions associated with research and technology organisations (RTOs); typically, no single institute covers these functions in their totality (see e.g. Arnold *et al.* 1998; Howells 2006).

5 Observations from practice

Having outlined the 'ideal type' of a Triple Helix organisation we now explore 'hybrid organisations' and their activities in practice. We focus on centres in the following countries: Canada, Finland, Norway and Sweden. This purposeful sample is appropriate, as each of these nations have well documented, discussed and formalised networks at the science/industry interface that provides an infrastructure or service offering in terms of the three Triple Helix spaces presented earlier. We draw on the three Triple Helix spaces as a framework for reporting our observations.

5.1 Data collection and analysis

Information on the organisations was collected from twelve elite interviews that were conducted face to face or by telephone. Standardised open-ended interviews were chosen as the primary data collection method since they can facilitate the capturing of qualitative data and enabling comparisons to happen between case entities. A semi-structured interview schedule allowed the incorporation of interviewees' own views but at the same time, all interviewees were asked the same basic questions. Hence, data collection was systematic, ensuring that the collected data are complete for each person on the topics addressed. Interviewed experts included research councils, innovation agencies, industry organisations and competence centre management. Interviews focused on the themes outlined in sections 3 and 4. Interview materialwere analysed using thematic analysis, reflected against the earlier literature analysis and existing frameworks (Denzin and Lincoln 2003).

5.2 Overview and comparison

Our international case studies illustrate the wide range of organisations that operate in a Triple Helix environment. Table 9.3 compares the different centres in terms of key Triple Helix-related functions and dimensions. The Appendix to this chapter, offers a more detailed description of the different organisations.

All of them have very specific characteristics and differ considerably in how they operationalise their strategic intent. Some of them cover a very broad range of activities across knowledge, consensus, and innovation spaces, supporting our notion that new institutional arrangements may have led to structural innovations.

All tend to have an inclusive approach and have partners from university companies, public actors and research institutes. The number of core actors varies from five to ten in VINN Excellence Centres to more than thirty shareholders or 100 programme and project partners in the case of the Finnish Strategic Centres for Science, Technology and Innovation (SHOKs). Similarly, they vary substantially in the amount of funding that is associated to them, from not quite EUR 1 million p.a. (VINN Excellence) to some EUR 60 million p.a. (SHOKs).

Only very few resemble the characteristics of a Triple Helix innovation environment. Taking some common themes:

- *Stakeholder participation*: While many centres involve a wide range of stakeholders, only very few centres bring together a broad range of stakeholders also as shareholders in a venture.
- *Organisational framework & governance*: Most of the competence/capability or collaborative research centres are located in a host institution, often a university or a public research institute. Few centres assume an independent legal identity. We have observed a trend towards more formalisation even in centres that are not incorporated as separate legal entities.
- *Coordination function*: Most centres have assumed a coordinating function for the collaborative research programme with which they are associated. This often includes active management of university/industry networks. However, only a few centres have the remit to actively define strategic research agendas for an industry or sector and run several research programmes (rather than a set of projects). Even fewer have the (delegated) authority to disburse substantial amounts of research funding. This is another characteristic that sets hybrid research and innovation environments apart from more traditional collaborative research centres.
- *Intermediary functions*: Centres tend to have a variety of 'intermediary' tasks. In many instances they cover a wide but not the full range of functions as described by Howells (2006).

Arguably, the Finnish SHOK, Strategic Centre for Science Technology and Innovation, comes closest to the 'idea typical' Triple Helix organisation.

Table 9.3 CRCs and their research governance and intermediary innovation functions

Centre	*VINN Excellence Centre*	*CRI Centre for Research-based Innovation*	*BL-NCE Business-led Networks of Centers of Excellence*	*SHOK Strategic Centres for Science, Technology and Innovation*
Country	*Sweden*	*Norway*	*Canada*	*Finland*
'Knowledge generation spaces'				
1 Research agenda for sector or cluster				✓
2 Delegated authority to allocate substantial amounts of research funding				✓
3 Annual budget scale	+	++	++	+++
4 Scope of topics	Thematic	Thematic	Challenge driven	Cluster
'Consensus spaces'				
Organisational form	Hosted	Hosted	Independent	Independent
Industry role	+	++	+++	+++
Foresight and diagnostics			✓	✓
Scanning and information processing			✓	✓
Knowledge processing and combination/recombination	✓	✓	✓	✓
Gatekeeping and brokering			✓	✓

Continued

Centre	*VINN Excellence Centre*	*CRI Centre for Research-based Innovation*	*BL-NCE Business-led Networks of Centers of Excellence*	*SHOK Strategic Centres for Science, Technology and Innovation*
Country	*Sweden*	*Norway*	*Canada*	*Finland*
'Innovation spaces'				
Testing and validation	✓			
Accreditation				
Validation and regulation			✓	✓
Protecting the results			✓	✓
Commercialisation			✓*	✓*
Evaluation of outcomes	✓		✓	✓

Code chart: a(indicates existence of feature), + (indicates relative strength of feature, from + 'existing' to +++ 'very strong/high').

Notes: *By participating companies governed by the CC strategy/network agreement.

5.3 SHOK Strategic Centres as an example of a foremost Triple Helix-type initiative

In Finland, when seeking more effective ways to undertake R&D&I (Research and Development and Innovation) projects, Strategic Centres of Excellence (SHOK) were created, a collaborative venture between the Ministry of Employment and Economy (MEE), the Finnish Funding Agency for Technology and Innovation (Tekes), the Academy of Finland, and the Confederation of Finnish Industries (EK). The first centres were operational in 2009. Key businesses were involved which drove the agenda of these centres.

A unique aspect of the Finnish Strategic Centres (SHOKs) is that research is organised at cluster level, and they are incorporated as not-for-profit companies that coordinate the helix-type activities. It is estimated that in the Nordic context such limited companies would each require around 10 million euros of public funding per year. The funding base of SHOKs and their networks is considerably larger than in most other countries (around thirty shareholders in one instance; an average SHOK centre receives some EUR 50 million initial capital compared to around EUR 20 million for a Swedish CC). The size of these centres and the way they have been organised as a legal entity are certainly the most striking differences to most other comparable Triple Helix-type activities.

The Finnish SHOKs are broadly similar to the Joint Technology Initiatives that have been planned under the EU FP7 programme. They are new types of partnerships between publicly and privately funded organisations involved in research. Their tasks are somewhat different because they cover entire industries: facilitating industrial renewal and fostering radical breakthrough innovations in their industry cluster, including the allocation of substantial amounts of programme funds. Subsequently, the shareholders representing the industry cluster also need to agree with the participating academic actors on strategic research agendas as well as governance structure and procedures for resource allocation, thus ensuring a collective ambition in the success of the project(s) initiated.

Competence Centres are organised as limited companies, which has clear benefits and limitations. The governing structure, responsibilities and principles are clear and are based on company law. However, given the nature of the project work, these centres tend to have a limited life span. Another key consideration has to do with the initial capital requirement for the limited company. In the case of a Finnish Strategic Centre of Excellence (SHOK) the minimum capital requirement is estimated to be round 2 million euros. The legal incorporation of the SHOK is plagued with contractual issues, such as intellectual property, but the issue is avoided if the centre is a linear extension to an existing business that adopts a 'smart source' for its R&D&I.

The Finnish experience demonstrates that businesses have a wide range of needs reflecting differences across industries. SHOKs are a new type of platform seeking to ensure that the knowledge created is available and taken further by several waves of businesses. New knowledge created in the competence centre is initially taken up by the front-running, typically large firms, and followed by the second and third waves of businesses, SMEs in particular.

6 Discussion and conclusions

Recent research raised the issue of whether the blurring of boundaries between the different Triple Helix spheres has come to a halt and actors begin to focus again on their respective 'core business', especially universities on teaching and research. While we cannot offer any evidence that would contradict the trend, we have noted the rise of a new organisational form or configuration that has the potential to step in and merits description as a systemic Triple Helix organisation. Historically, 'hybrid organisations' have been anchored in one of the Triple Helix spheres fulfilling a number of the functions associated. Often the hybrid organisations have been university based or hosted, which perhaps carries too much academic baggage, ways of working and a particular mindset that may not be conducive to contemporary national and international needs and expectations, given the age of austerity. There have so far been very few centres that transcend knowledge, consensus and innovation spaces. The Finnish SHOK centres can be seen as the case that goes furthest in embracing activities across all the Triple Helix spaces. The Canadian model has a long history of success in stimulating and driving collaborative research and its 'business-led' extension offers a more naturally evolved alternative that addresses a wide range of downstream innovation activities. Yet it does not have the reach or even the richness of Finnish SHOKs in defining research and innovation agendas for entire clusters or industries. The other Nordic countries provide examples of focused involvement or engagement, but they have not yet made the 'quantum' leap towards being a 'genuine' Triple Helix organisation. The Finnish SHOKs appear to be a research and innovation environment that resembles the ideological heart of the Triple Helix III. Their strength is that they attempt to integrate all Triple Helix spaces and related functions into a formalised organised framework. In that sense, it can be considered a structural innovation (Howells and Edler 2011). It represents both a new organisational form and a new configurative arrangement.

This new type of establishment allows for open, crowd, service and user/consumer innovation to be embedded in national innovation activities for the well-being of society. Given the world is more dynamic and discontinuous change requires more contemporary ways of enabling innovation, this reconfiguration will allow for a cycle of reflective and reflexive innovation to drive new processes, services and products, as well as new businesses.

The other element that sets the SHOK initiative apart is the delegated authority to develop and approve substantial technology development programmes and projects following non contagious, mutually agreed research agendas by all relevant stake- and shareholders (Gustafsson and Järvenpää, 2011). It is of benefit, as it introduces a new dynamic into innovation, a desire to share power between participants, and fulfils both open and social innovation criteria. Also, its scale is considerably larger than that of other centres; its scope ensures reach and richness at the cluster level. As the contribution illustrates, strategic research agendas alongside a number of programmes are defined by stakeholders participating in the centre. Our interviews reinforce that a balance for stakeholders needs to be struck between basic, applied and strategic research. Importantly, it offers the opportunity to combine challenging basic research questions with the need for user relevance.

Of course, these new forms of innovative organisations will develop, but national governments will need to ensure specific checks and balances are put into place to ensure these hybrids are ambidextrous in responding effectively to the needs of both stake- and shareholders and to find ways to guarantee the needs of many – those who are not geared up or do not feel able to take advantage of what these new dynamic organisations can offer – are represented, rather than satisfying the needs of the keen activists.

In addition, as these 'hybrid organisations' grow out of existing established industry structures, they are very much innovation by design. This raises additional questions: when does an industrial sector become established and when would be the right time to incorporate a hybrid organisation into this sector to stimulate development? Too much pace will fatigue the cluster, too little may dampen its potential. Finally, on policy issues, how do national governments manage the innovation portfolio, given the mix of more orthodox innovation agencies and hybrid organisations operating at the same time?

Since this chapter has established the importance of 'hybrid triple helix organisations', other research questions have arisen. If other nations adopt or create 'true' hybrids, then would they still be hybrids or would they become the established norms? If they remain true to their Triple Helix core, this may lead to a relocation of contract research, which is almost by definition demand-driven, away from universities to different structures or other 'hybrid organisations' that bring together coalitions of actors across the entire spectrum of the innovation system, across all Triple Helix spaces. In light of this, one might argue that the corporate strand of the Helix is now becoming more active in generating the 'overlay', amply supported by the government strand of the Helix, which is creating 'markets' for research through generous innovation support measures and in some instances is even actively engaged in building new institutions for research. So, rather than the end, we may witness a second coming of the Triple Helix and the need for more structural innovation.

Notes

1 This chapter is based on a talk at the UNICA Conference entitled 'The second coming of the Triple Helix: The rise of the corporate strand and the emergence of hybrid innovation environments'. It also draws heavily on related presentations given at the 9th International Triple Helix Conference, Stanford, and a Triple Helix workshop held at Birkbeck College, London, which will be summarised in a contribution for the journal *R&D Management.*

2 One should note that this trend is not limited to universities but encompasses also research institutes, national research centres and research transfer organisations.

3 A case in point for this emergent public–private partnership may be the European Commission's *Joint Technology Initiatives,* which are a major new element of the EU's 7th Research Framework Programme. They provide a way of creating new partnerships between publicly and privately funded organisations involved in research, focusing on areas where research and technological development can contribute to European competitiveness and quality of life. Commission expects this new model of public–private partnership to stimulate additional European research investment, build critical mass by uniting currently fragmented efforts, and ensure effective and efficient programme management (http://europa.eu/rapid/pressReleasesAction.do?reference=MEMO/07/191).

4 Here, power has been moved away from central government to regional and local authorities or organisations that are at arm's length from government.

References

Arnold, E., Rush, H., Bessant, J. and Hobday, M. (1998) 'Strategic planning in Research and Technology Institutes', *R&D Management*, 28(2), 89.

Braun, D. (2008) 'Lessons on the political coordination of knowledge and innovation policies', *Science and Public Policy*, 35(4), 289–298.

Debackere, K. (2000) 'Managing academic R&D as a business at K.U. Leuven: Context, structure and process', *R&D Management*, 30(4), 323–328.

Denzin, N. K. and Lincoln, Y. S. (eds) (2003) *Handbook of Qualitative Research.* Third edition. Thousand Oaks, CA: Sage.

Edler, J. and Kuhlmann, S. (2008) 'Coordination within fragmentation: Governance in knowledge policy in the German federal system', *Science and Public Policy*, 35(4), 265–276.

Etzkowitz, H. (2003) 'Research groups as "quasi-firms": The invention of the entrepreneurial university', *Research Policy*, 32, 109–121

Etzkowitz, H. (2008) *The Triple Helix: University–industry–government. Innovation in Action*. London: Routledge.

Etzkowitz, H. and Kemelgor, C. (1998) 'The role of centres in the collectivisation of academic science', *Minerva*, 36(3), 231–268.

Etzkowitz, H. and Leydesdorff, L. (2000) 'The dynamics of innovation: from National Systems and "Mode 2" to a Triple Helix of university–industry–government relations', *Research Policy*, 29(2), 109–123.

Garrod, H. W. (1946) *Scholarship, its Meaning and Value*. Cambridge, Cambridge University Press.

Gibbons, M., Limoges, C., Nowotny, H., Schwartzman, S., Scott, P. and Trow, M. (1994) *The New Production of Knowledge*. London: Sage.

Government of Canada (2009) *Networks of Centres of Excellence 20th Anniversary Report*. Ottawa, Canada.

Gustafsson, R. and Järvenpää, S. (2011) 'Organising for collaboration and competition in an innovation network: The case of the Finnish metals and engineering strategic network'. Paper presented at Triple Helix Conference, Stanford, July 2011.

Hessels, L. K. and van Lente, H. (2009) 'Re-thinking new knowledge production: A literature review and a research agenda', *Research Policy*, 37, 740–760.

Howells, J. (2006) 'Intermediation and the role of intermediaries in innovation', *Research Policy*, 35, 715–728.

Howells, J. and Edler, J. (2011) 'Structural innovations: Towards a unified perspective?' *Science and Public Policy*, 38(2), 157–167.

Leydesdorff, L. (2003), The mutual information of university-industry-government relations: an indicator of the Triple Helix dynamics. *Scientometrics*, 58(2), 445–467.

Leydesdorff, L. and Meyer, M. (2003) The Triple Helix of university-industry-government relations: a model for innovation in the 'knowledge-based' economy. *Scientometrics*, 58(2), 191–203.

Leydesdorff, L. and Meyer, M. (2010) 'The decline of university patenting and the end of the Bayh-Dole effect', *Scientometrics*, 83(2), 355–362.

Newman, J. H. (1852) *The Idea of a University*. London: Longman, Green, 1929 edition (originally published in 1852) quoted in Hear, A. O. (1989) 'The university as a civilizing force', in C. Ball and H. Eggins (eds) (1989) *Higher Education into the 1990s: New dimensions, the society for research in higher education*. Milton Keynes: Open University Press, pp. 17–28.

Nowotny, H., Scott, P. and Gibbons, M. (2001) *Re-Thinking Science: Knowledge and the public in an age of uncertainty*. Cambridge: Polity Press.

Pelkonen, A., Teräväinen, T. and Waltari, S. (2008) 'Assessing policy coordination capacity: Higher education, science, and technology policies in Finland', *Science and Public Policy*, 35(4), 241–252.

Rip, A. (2004) 'Strategic research, post-modern universities and research training', *Higher Education Policy*, 17(2), 153–166.

Schils, E., (1992) 'Points of departure: the idea of the university: obstacles and ppportunities in contemporary societies', *Minerva*, 30(2), 301–309.

Scott, P., Richards, E., Martin, B. (1990) 'Captives of Controversy: The Myth of the Neutral Social Researcher in Contemporary Scientific Controversies', *Science Technology Human Values*, October, 15(4), 474–494.

Starbuck, W. H. and Milliken, F. J. (1988) 'Executives' perceptual filters: What they notice and they make sense', in D. C. Hamerick (ed.), *The Executive Effect: Concepts and methods for studying top executives*. Greenwich, CT: JAI.

Stokes, D. E. (1997) *Pasteur's Quadrant: Basic science and technological innovation*. Washington, DC: Brookings Institution Press.

Thune, T. and Gulbrandsen, M. (2011) 'Institutionalization of university–industry interaction: An empirical study of the impact of formal structures on collaboration patterns', *Science and Public Policy*, 38(2), 99–107.

Turpin, T. and Fernández-Esquinas, M. (2011) 'Introduction to special issue: The policy rationale for cross-sector research collaboration and contemporary consequences', *Science and Public Policy*, 38(2), 82–86.

Tuunainen, J. (2005) 'Hybrid practices? Contributions to the debate on the mutation of science and university', *Higher Education*, 50, 275–298.

Appendix: Overview of research and innovation environments

Country	*Initiative*	*Brief description*
Norway	CRI – Centres for Research based Innovation (October 2006 – December 2014)	• The objectives of the programme are to: (1) encourage enterprises to innovate by placing stronger emphasis on long-term research; (2) facilitate active alliances between innovative enterprises and prominent research groups; (3) promote the development of industrially oriented research groups that are on the cutting edge of international research and are part of strong international networks; (4) stimulate researcher training in fields of importance to the business community, and the transfer of research-based knowledge and technology; • Activities covered with this programme: basic research, applied research, training and technology transfer • Currently 14 centres established, which receive an annual budget of around NOK 140m (or EUR 17.5m) in public funding with a total budget of NOK 300m per year; total investment over entire 8 year period: NOK 2.4 billion, or EUR 300m. – Maximum amount of funding per project: 50%, at least 25% of the funding is to come from the business partners (NGO) – Minimum/Maximum project duration: 5 + 3 years, after mid-term evaluation
Sweden	VINN Excellence Centres (and similar CRCs)	• The objective of the programme is to create new internationally competitive concentrations of highly qualified experts with the task of conducting research that is problem-oriented and – as a rule – multidisciplinary and generating knowledge and technology that will lead to new products, processes and services. • A VINN Excellence Centre seen as a strong research environment positioned in strong innovative surroundings, typically academic led with industry involvement. Participating actors are universities, companies, public actors, research institutes and other research-performing organisations. • Activities covered with this programme: basic research, applied research. Ideas outside the core actions of the participating actors can also be utilised and further developed, e.g. by the set-up and development of new high-tech and research-based companies. • Typical programme budget: – EUR 20m p.a., max. amount of funding per project: EUR 800k, a third of which to be financed by partners, typically 5–10 members per consortium – 20 centres supported by VINNOVA and 4 other funding agencies, with a total funding of EUR 650m over 5–10 years, incl. EUR 300m investment from industry and others

Country	*Initiative*	*Brief description*
Canada	Business-led Networks of Centres of Excellence Program (BL-NCE)	• In 2007, the Government launched a strategy called Mobilizing Science and Technology to Canada's Advantage to foster three distinct Canadian S&T advantages: (1) Entrepreneurial Advantage, (2) Knowledge Advantage, (3) People Advantage. • BL-NCEs will foster a competitive and dynamic business environment to encourage S&T investments and create an 'Entrepreneurial Advantage': 'The private sector will identify and lead new research networks that address their priorities under the Networks of Centres of Excellence Program.' • Goals: (1) fund large-scale collaborative networks to perform research and commercialization; (2) enhance private sector innovation; (3) deliver economic, health, social and environmental benefits to Canadians and to encourage an Entrepreneurial Advantage. • Initially 4 BL-NCE centres set up, compared to 39 Networks of Centres of Excellence across Canada
Finland	SHOK – Strategic Centres	• 6 new public–private partnerships for speeding up innovation processes. • Main goals: – renew established industry clusters – create radical innovations • SHOKs seen as a permanent co-operation and interaction forum • Centres develop and apply new methods for cooperation, co-creation and interaction • The centre will consist of the coordinating function jointly owned by the parties, and a virtual research organisation – A non-profit limited company will be responsible for the centre's operations; – Shareholders prepare a strategic research agenda for the centre – Large research programmes created for achieving world class expertise – Parties who are not shareholders can also participate • Programme level funding: EUR 300m p.a., EUR 50m per centre

10 Bringing university knowledge to market

Experiences in the Netherlands

Marina van Geenhuizen

1 Knowledge commercialization in an open innovation context

Bringing new knowledge to market is increasingly recognized as the third mission of universities today. Accordingly, universities are seen as creators of new knowledge while being involved in contract research commissioned by the business sector, in collaborative research projects with business partners, in the licensing of patents, in the creation of spin-off firms, etc. (Shane 2004; van Geenhuizen and Nijkamp 2006; Huggins and Johnston 2009). In Europe, this new role of universities started to grow in the early 1980s (Charles and Howells 1992) and has now fully entered the research policy of modern universities (e.g. Rasmussen *et al.* 2006; Etzkowitz 2008; Hussler *et al.* 2010).

Knowledge commercialization is broadly defined in this chapter as 'the process of creation of value from knowledge, by adapting it and/or making it available for economic/societal use and transform it into competing products, services, processes and new economic activity' (Innovation Platform 2009: 8). It is a complex and iterative process with interaction between knowledge institutes and the business world as key in all stages. Knowledge valorization encompasses a chain of processes (partly cycles) that starts with first thoughts about market introduction, eventually together with a firm, and about steps to be taken to reach this through various channels (Bekkers and Bodas Freitas 2008).

Since the early 2000s, the demand for more rapid market introduction and cost reduction in a globalized economy has urged many universities, technology institutes and firms to adopt models of open innovation. The underlying assumption of open innovation is that actors in innovation can and should use external ideas as well as internal ideas, and internal and external paths to market, as they look to advance their technology. Taking the line of Chesbrough (2003) and Chesbrough (2006), open innovation among firms can be defined as: the systematic encouraging and exploration of a wide range of internal and external sources of innovation opportunities, consciously integrating that exploration with firm capabilities and resources, and broadly exploiting those opportunities through multiple channels. This development, it seems, has encouraged collaboration not only between firms but also between firms and knowledge institutes. Of the many channels used in this collaboration, such as selling/obtaining licences, collaborative research projects and spin-off firms that transform inventions into marketable products, processes, etc., the collaboration between universities and firms in research projects has received small attention to date (D'Este and Patel 2007). This also means that there is a lack of knowledge on how such projects develop in reaching the market and on the factors determining the outcomes. For example: is it the type of invention, radical versus incremental, or is it the difference in orientation between universities and industry that causes problems in collaboration and delay in bringing new knowledge to market?

In addition to this gap in understanding, policies to take advantage of the role of universities seldom differentiate between regional systems in core economic areas (usually seen as more competitive) and other areas (non-core). This is despite the fact that increasing evidence of differences is being found between such systems in European countries (such as the UK) (Huggins and Johnston 2009; see also Tödling and Trippl 2005 and Lagendijk and Lorentzen 2007). This chapter responds to the above two gaps in understanding, by addressing the following questions while focusing on new technology:

1 To what extent are technology projects at universities successful in reaching market introduction? How does this vary between different regions?
2 Which factors hamper effective and rapid commercialization? In which respects are these factors related to the invention, the organizational context at universities and university–industry interaction, and to what extent are these related to regional characteristics?

This chapter takes the Netherlands as an example. Despite its small size, the country is faced with different regional innovation systems, potentially causing different opportunities and barriers to knowledge commercialization. The following two regions will be considered: 1) the metropolitan Core area in the western part of the country (Randstad) which has developed as a service-based economy including university cities such as Amsterdam, Leiden and Delft, and 2) the Southeast, stretching from Eindhoven to Maastricht in the South and to Nijmegen in the East, all three cities being university cities. This region has a relatively strong focus on high-technology manufacturing.

The chapter has the following structure. Differences between the regional innovation systems in the Netherlands are presented first (Section 2). Theoretical perspectives are examined in Section 3. This is followed by a discussion of the methodology of the empirical study and an examination of its results focusing on market introduction, delay and failure (Section 4). Barriers and hampering factors in processes of knowledge commercialization are presented in Section 5. The chapter concludes with some policy challenges and three avenues for future research.

2 Regional innovation

The Netherlands is the home of fourteen government approved research universities, thirteen of which focus on research as well as education (offering bachelor's, master's and PhD degrees). Among the research universities there are three universities of technology, in the cities of Delft, Eindhoven and Enschede. Aside from research universities, there are universities of applied sciences (not covered in this chapter). The Netherlands is also home to many non-university research institutes, most of them publicly funded, such as the institutes connected with the Netherlands Organization for Scientific Research (NWO) and those connected to the Royal Netherlands Academy of Arts and Sciences (KNAW).

Government policy for research and innovation at the national level is the joint responsibility of the Ministry of Economics, Agriculture and Innovation and the Ministry of Education, Culture and Science. The national government has established the so-called Innovation Platform to aid in formulating an ambitious strategy on research and innovation. This has led to a new overall programme focusing on acceleration in bringing the Netherlands back into the top 5 (Innovation Platform 2010). Since 2008, the Innovation Platform has been actively involved in knowledge commercialization (Innovation Platform 2009). Knowledge valorization has also been the major focus in the latest economic policy document (MEAI 2011) on 'Top Economic Sectors'.

Table 10.1 presents a brief characterization of the two broad regions in our analysis of the Netherlands: the Core area and Southeast. The Core area is the metropolitan area that stretches as a Rim city (Randstad) of larger and medium-sized towns, from Amsterdam and Haarlem, through Leiden and The Hague to Delft-Rotterdam, Dordrecht and Utrecht back to Amsterdam. In terms of administration and statistics this area is covered by the Provinces of North-Holland, South-Holland and Utrecht. The Southeast contains a chain of cities including one larger city (Eindhoven) and its region (Brainport) and various medium-sized towns such as Venlo, Sittard-Geleen, Maastricht and Parkstad Heerlen-Kerkrade close to the German border. It is covered by the Provinces of North-Brabant and Limburg. Of course, a characterization of regional systems regarding commercialization and relevant conditions is a study in itself; therefore, we have selected statistics that are ready available. Most of these statistics cover a whole set of provinces, but some data refer to individual provinces (Table 10.1).

The Core is much larger than the Southeast, if we take population size as a criterion, that is 7.4 versus 3.6 million persons. Accordingly, there are also important differences in the size of production as indicated by percentage share in national gross value-added: those in the Core are altogether twice as large as those in the Southeast (47.9 versus 20.4 per cent). However, taking regional product per person, there are differences between Core and Southeast: the three provinces in the Core have all a higher regional product per person than the provinces in the Southeast. This situation tends to be connected with a difference in economic structure, with a somewhat stronger emphasis on manufacturing in the Southeast (29.4 versus 18.1 per cent) and on (advanced) commercial services in the Core (56.6 versus 46.5 per cent). With regard to our indicators of knowledge creation and diffusion, the picture is as follows. In a metropolitan area such as the Core – including the large cities of Amsterdam, Rotterdam, The Hague and Utrecht – we would expect all indicators to be relatively high compared to the less urbanized

Table 10.1 Selected characteristics of the Core and Southeast

	Core (North-Holland, South-Holland, Utrecht)	*Southeast (SE) (North-Brabant, Limburg)*a
Size of population (2010) and production systems (2008)		
– Population	Core: 7.4 million	Southeast: 3.6 million
– Gross value-added (% of national total)	Core: 47.9	Southeast: 20.4
– Gross regional product (per person ×1000 Euro)b	– N-H: 40.7 – S-H: 36.8 – Utrecht: 42.0	– N-B: 35.6 – Limburg: 31.5
Structure of production systems (2008)		
– Manufacturing (gross value-added as % in Core and Southeast)	Core: 18.1	Southeast: 29.4
– Commercial services (gross value-added as % in Core and Southeast)	Core: 56.6	Southeast: 46.5
Knowledge creation–diffusion system		
– Number of research universities	Core: 6 (1 technical)	Southeast: 3 (1 technical)
– HRST (% of active population, 2009)b,c	– N-H: 52.5 – S-H: 46.6 – Utrecht: 52.9	– N-B: 43.0 – Limburg: 41.1
– Firms receiving finance for innovation (% of all firms) (CIS 2008)	Core: 5.8	Southeast: 6.5
– R&D expenditure as average per firm in million euro (CIS 2008)	Core: 0.125	Southeast: 0.239
– GERD in euro per inhabitant (2007)d	Core: 267	Southeast: 704.4
– High-tech patent applications (EPO) per million inhabitants (2007)b	– N-H: 21.63 – Z-H: 27.99 – Utrecht: 26.75	– N-B: 164.25 – Limburg: 44.17

a The Province of Gelderland is not included because the city of Nijmegen is at the southernmost point, bordering the Province of Limburg.

b Individual provinces.

c HRST: human resources in science and technology (derived from education and also working experience).

d GERD: total intramural R&D expenditure.

Source: Eurostat (2011); van Geenhuizen (2012); Statistics Netherlands (2011).

Southeast, including one relatively large city (Eindhoven) and a set of medium-sized towns. A relatively high score for the Core is, however, only true for labour market qualification of the active population (HRST). All other indicators (selected here) point to higher levels in the Southeast, such as investment in R&D and also innovation output (patents), clearly reflecting a strong presence of high-tech manufacturing, in firms such as Philips, ASML, FEI and DSM, including their research departments.

3 Theoretical views on knowledge valorization

The extent to which new university knowledge is brought to market is subject to a myriad of influences according to theoretical studies. We will briefly discuss three perspectives: (1) innovation theory focusing on features of the invention, (2) learning theory focusing on learning as a collaborative activity between organizations, and (3) theory on regional innovation (Geenhuizen 2012).

Innovation studies in general tell us that the speed of market introduction depends on the nature of the invention (radical or incremental), including its costs and the market (regulation, subsidies, market demand) (Utterback 1996; Christensen 2003; Tidd and Bessant 2009). In terms of the radical/incremental character, it seems that radical inventions requiring structural changes in infrastructures – such as the fuelling infrastructure in the case of electric cars – face more obstructions than inventions that are incremental and fit into existing structures (Geels 2004). Of course, the cost level plays a role including the trade-off between the newness/additional benefits of the invention and additional costs compared with existing, traditional, products or processes.

In addition, the way to market introduction may be rather long in the case of heavy regulation, such as in designing new drugs, caused by intensive testing and approval procedures. Conversely, inventions without such regulation and at the same time heavily subsidized by public money reach market by a much shorter road. Such inventions might include solutions for treatment of solid waste and of polluted air and water, and solutions in sustainable energy production and storage.

A further relevant body of theory is concerned with types of learning. A distinction can be made between *science-based learning* including laws of nature (know-why), such as in life sciences and nanotechnology, and *problem-based* and engineering types of learning with new applications or combinations of existing knowledge (know-how), such as in medical instrument development and the automotive industry (Asheim *et al.* 2007; Tidd and Bessant 2009). The idea is that problem-based learning gives greater incentive for collaborative learning between the various regional partners involved and thus proceeds faster, whereas science-based learning takes more time and is more often globally oriented than problem-based learning. Firms, but also universities, may differ in this respect; technical universities in particular place greater emphasis on applied and engineering knowledge compared with the beta-faculties of general universities.

Other strands of learning theory tell us that collaborative learning in innovation is only possible if organizations differ sufficiently in a particular expertise that they can learn from each other. They should, however, not be too different in culture and avoid collaboration for that reason or, if collaborating, face the risk of failure in understanding each other (e.g. Nooteboom 2009). The latter point is important in clarifying university–industry relations. Few studies have investigated empirically the nature of inhibiting or of converging factors in university–industry relations, but attention is increasing (e.g. Bjerregaard 2010; Bruneel *et al.* 2010). Accordingly, many obstacles are related to attitude and IP (intellectual property). This can be understood as follows: owing to different methods of knowledge production, there is generally a weak alignment of attitudes between university researchers and private firms. Time lines in most firms are shorter than in university research, and firms need to adapt quickly to changing circumstances, even sometimes ending collaboration when a new technology enters the firm (following a merger/acquisition) or when reorganizations dictate the closure of the R&D department. Universities, in contrast, can remain quite stable in their choices. In addition, while researchers are keen to disclose information (in journals) as fast as possible, firms often prefer to keep new knowledge secret or to appropriate the new knowledge (Westness and Gjelsvik 2010). IP and university administration related obstacles also include differing capabilities of universities and industry in handling patent applications and licensing, and different strategies for maximizing benefits from patents.

We will now briefly discuss regional innovation theory. This set of theories provides important ideas on factors that enhance or hamper knowledge creation and learning, as well as market introduction of the new products, processes, methods, etc. Agglomeration theory places emphasis on the agglomeration advantages of large cities (Feldman 1999; Capello 2006), for example, knowledge spillovers, the presence of pools of specialized workers and other creative people, test markets, and access to global traffic nodes. From a somewhat different perspective – the global city thesis – the advantages of large cities include proximity to major decision-making units, mainly corporate headquarters, a highly specialized service sector, and a large presence of high level professionals (e.g. Sassen 2005).

Other, related schools in regional innovation focus on benefits from a common regional culture, trust and social networks that facilitate knowledge exchange (Keeble *et al.* 1999; Maskell and Malmberg 1999; Cooke *et al.* 2004), particularly the use of 'open innovation'. If well developed, the 'business ecosystem' – including customers, contractors, competitors and collaborators – provides potential for subcontracting, knowledge collaboration and collective learning in the region and, through 'bridges', with knowledge sources abroad. Other major supportive elements in the region are public research organizations and intermediary and financial institutes. The potential benefits from these elements have been extensively addressed in the literature on regional innovation systems and clusters (Porter 2003; Cooke *et al.* 2004; Bathelt *et al.* 2004) and was taken as a guideline in searching for hampering factors in knowledge commercialization (section 5).

4 Technology projects and the way to market

The analysis of the commercialization results of technology projects in this section involves two steps. First, a broad scan of the outcome of almost 370 projects, financed by Technology Foundation STW (various years), in terms of market introduction, continuation and failure, and second, an in-depth study of approximately fifty-one projects representing these different outcomes.

The label 'market introduction/use in society' (Table 10.2 and Table 10.3) indicates that the projects have led to a product, process or method being brought to market through firms or brought to use in society. The label is also given if software (as a main part of a project) was brought to market. This quite strict definition is adopted to identify the 'success stories'. Failure is defined as 'not having reached the stage of market introduction', including not continued. We also use a separate category 'stagnation or development unknown' for older projects.

It should be noted that projects that could not (yet) be brought to market, including the ones that failed, were often not useless. They may have led to some research contracts with potential users, patent applications, or new products, etc. In a few cases failure happened quite early, with the project ceasing before the official deadline or directly following it. In most cases, however, it took a longer time for commercialization to be terminated. Of course, the knowledge developed in such projects may still be partially used in other ways, such as in particular 'branches' of adjacent research at universities and in research at firms where PhD graduates, who were previously involved in a project, find a job.

In view of the long-term nature of knowledge commercialization, two different periods of project take-off are selected in the current study, namely, the years 1995 to 1997, and the years 2000 to 2002. Another reason for examining two separate periods is the change in economic climate in the early 2000s which makes an aggregate analysis of all years dangerous. It is plausible that in the early 2000s firms have been reluctant to engage themselves in commercialization. In addition, we can expect a difference in awareness of commercialization and requirements for success in the two periods: relatively low at the end of the 1990s and stronger since then.[1] For the first period, STW provides evaluation information after ten years, for the second period it provides evaluation information after five years. In addition, each period covers a sufficient number of projects representing the Core area and Southeast in the Netherlands.[2]

In addition, the analysis of hampering factors in commercialization draws on a selected sample of thirty-five research managers that represent projects reaching market introduction (or use in society) and failure (delay) and cover the two periods and regions. All in all this has yielded data on fifty-one projects. Owing to small numbers, however, it is difficult to identify solid trends for the two regions. Data were collected using in-depth interviews and a semi-structured questionnaire. The research managers with which the in-depth interviews were held were mostly the project leaders of research at university. The selected sample covers a limited number of technology segments, namely, biotechnology (medical and industrial), medical technology (instruments/software), new materials (nanotechnology) and systems for sustainable energy, as well as the automotive industry.

Using data on almost 370 projects in the first part of the study, the following outcomes could be identified: *failure* is faced by a minority of the projects, 26 per cent in both sets of cohorts (Table 10.2 and Table 10.3). On the other hand, *market introduction* is also achieved by a minority of the projects, that is, 22 per cent among the older projects and 15 per cent among the younger ones. Aside from the downturn in the economy, the last results also stem from the short time period in consideration.

With regard to the two types of region, the results for the first period reveal a trend of a higher failure rate as well as a higher rate of market introduction in the Core area compared with the Southeast (Table 10.2), witnessed by shares of 28 per cent versus 18 per cent and 25 versus 16 per cent respectively. These outcomes support the idea of higher levels of newness and creativity (breakthroughs) as well as higher levels of risk-taking in core metropolitan areas, particularly Amsterdam and Utrecht, but also favourable conditions in bringing projects to market and use in society, such as tighter connections with public decision making in the national government in The Hague. Delft, with its university of technology, is a city that borders The Hague's suburbs. In addition, by comparing the two regions using only two categories of project outcomes, namely, projects that failed, etc. and successful projects, the differences between the regions turn out to be significant at the 0.01 level (Appendix 1).

In the second period, the Core has remained quite high in failure (29 versus 20 per cent) without a high level of market introduction (14 per cent). This pattern suggests that Core regions have tended to remain 'stuck' with a high failure rate more recently. The in-depth interviews indicated that causes of this change may be, first, a continued level of high risk-taking such as in projects concerning genetic engineering (biology) and medical life sciences (in Amsterdam and Leiden); second, the early adoption of the principle of 'open innovation' in high-tech manufacturing in the Southeast, particularly the region of Eindhoven (Brainport), thereby improving university–industry interaction and reducing failure. In addition, by comparing the two regions using only two categories of project outcomes, the differences between the regions turn out to be significant at the 0.01 level only for the category projects failed (Appendix 1).

Table 10.2 Project outcomes in two types of region (take-off in 1995–97)

Outcome	*Core*		*Southeast*		*Total*	
1 Failure after 5/10 years	45	28%	9	18%	54	26%
2 Stagnation, or development unknown after 10 years	27	17%	15	29%	42	20%
3 Continuation after 10 years	47	30%	19	37%	66	32%
4 Market introduction/use in society	39	25%	8	16%	47	22%
Total	158	100%	51	100%	209	100%

Source: calculated based on information from Technology Foundation STW (Reports on Utilization) and website information (adapted from Geenhuizen 2012).

Table 10.3 Project outcomes in two types of region (take-off in 2000–02)

Outcome	*Core*		*Southeast*		*Total*	
1 Failure after 5 years	31	29%	10	20%	41	26%
2 Continuation (incl. delay)	62	57%	32	64%	94	59%
3 Market introduction/use in society	15	14%	8	16%	23	15%
Total	108	100%	50	100%	158	100%

Source: calculated based on information from Technology Foundation STW (reports on Utilization) and website information (adapted from Geenhuizen 2012).

If we focus on the time needed for new technology to reach market introduction or use in society, it appears that commercialization is a long-term process. In almost a third of the projects that took off in 1995–97 (32 per cent) R&D is still continued after ten years (Table 10.2). The in-depth interviews indicate that in medical life sciences a total of fifteen to twenty years is not an exception. In medical technology (instruments and software) the time is shorter if the invention deals with product improvement that can avoid time-consuming approval procedures. In general, time tends to be shorter if the R&D is driven by demand from the business sector and takes place in close collaboration with a firm, and if the invention is a new application of existing (proven) knowledge. Commercialization of knowledge in projects of these kinds tends to take only five to eight years.

Tentative employment effects of technology projects were calculated from direct effects using the in-depth interviews, and from estimations of various indirect effects. Thus, employment effects include direct employment in the project – often a PhD student and supporting technical workers at take-off, and – as indirect employment – jobs of suppliers to the project (research results, equipment etc.) as well as spin-off firms using the knowledge from the project. The employment effects calculated in this analysis are limited to those in the *region*. We could observe a quite broad range of employment effects (2009) for projects that took off in 1995–97: from 1.3 to 145 full time equivalents (fte). The mean is 20 fte per project, but this value reflects the influence of some large spin-off firms that grew from particular projects. For projects that took-off in 2000–02, the differences are smaller of course, owing to a smaller time period. The range is between 1.3 and 8 fte, and the average is 4 fte. It may be concluded that employment effects in the region tend to be relatively modest, although there are various positive outliers.

5 Hampering factors

In the analysis of barriers in commercialization we distinguish between: (1) factors at university, namely, invention, project/team, organizational circumstances, as well as university–industry links, (2) factors typical for the region, and (3) national factors. The results are as follows (Table 10.4): factors at universities and in university–industry links are most important, almost 60 per cent of all barriers mentioned for older projects. This is followed by the region, with 29 per cent of all these barriers. With regard to younger projects, the first have decreased, and the latter increased. This trend is most probably related to an increased awareness of commercialization at universities and of enabling factors there, and with an increased use of open innovation models covering the region, in which various research institutes and regional businesses take a major role. A discussion of individual factors follows below.

Shortages in the *organizational situation at universities* – their value system, needs and routines – constitute the single most important hampering factor, at 29 per cent of all factors mentioned for older projects. This concerns not only a personal lack of affinity of some researchers with commercialization, but also a reward structure at university that does not encourage this activity. In addition, in some cases commercialization may suffer from reorganization of faculties leading to the closing down and regrouping of specialized research groups. A too small capacity of technology transfer organizations (TTOs) is also mentioned. However, younger projects show a decreasing trend in the frequency of these barriers.

Next is an *unfavourable profile of the invention*, as witnessed by 17 per cent of all factors among older projects. This includes: no outlook on mass production, too expensive for the extras it brings (no breakthrough), and strong technology risks during the next phases of the research. These circumstances cause a lack of interest or change in the interest/involvement of firms in commercialization, but tend to occur less frequently among younger projects.

Obstacles in *university interaction with firms* are somewhat less often mentioned, at 13 per cent of all factors among older projects. Good results from collaboration may be hampered by turbulence and opportunistic behaviour by firms. Thus, unexpected business events disturb the delicate learning processes between university researchers and firms, such as the sudden wish from firms to cease collaboration owing to an emerging competing technology, the closing down of firms' R&D departments owing to reorganization, and merger and acquisition coming with new strategies. Among the younger projects, the situation tends to remain the same.

With regard to *regional factors*, shortages in financial arrangements tend to hamper commercialization most frequently (11 per cent of all factors, for both older and younger projects). This is concerned with public investment subsidies to attract firms from other regions and from abroad, which is absent in the Core area but available in other regions. This situation causes a lack of a level playing field, with the result that the Core might lose high-technology firms (e.g. university spin-off firms) to other regions and might also face more difficulty than other regions in attracting such firms from abroad. The second financial barrier holds true for all regions, that is, lack of easy access to venture capital and large amounts of such capital. It is particularly the shortage of organizational capacity among actors involved that hampers the release of large sums of venture capital necessary for spin-off firms to cross the 'funding gap'. Large sums of venture capital may be necessary in medical life sciences in the stages following proof of principle and prior to manufacturing small badges under strict medical rules.

Table 10.4 Barriers in valorization of technology projects as forwarded by managers.

Barriers	*1995–97 projects abs.*	*%*	*2000–02 projects abs.*	*%*	*Details (in italics if particular region)*
University, university–industry links					
Profile of invention	11	17.4	3	8.1	Much risk at university *(Core)*; much risk for firms; radicality of invention; no mass-market involved; high price for additional new attributes
Value system, needs and routines at university	18	28.6	8	21.6	Small affinity with and little attention to commercialization (general universities); too small budgets for technology transfer; reorganization of research groups; unfavourable reward structure and career for employees in commercialization
Turbulence in university-business interaction (by firms)	6	9.5	4	10.8	Sudden change in collaboration owing to firm reorganization or merger–acquisition (reduction of R&D; sudden loss of attention to invention owing to new, superior, technology)
Opportunistic behaviour of firms	2	3.2	1	2.7	Gaining intellectual ownership; investment driven by immediate profit-seeking
Total	*37*	*58.7*	*16*	*43.2*	
Region					
Business ecosystem	2	3.2	3	8.1	Small ecosystems (critical mass) (*Southeast*) and difficulty in attracting research departments of MNCs; delay in Triple Helix formation
Financial factors	7	11.1	4	10.8	Limited supply of large amounts of venture capital and lack of organizational capacity to arrange this; lack of investment subsidy (*Core*) enhancing relocation to other regions
Business accommodation	4	6.3	3	8.1	Shortage of low rent incubators and close-to-academic-hospital incubators (*Core*); shortage of follow-up accommodation after incubation

Table 10.4 (continued)

Barriers	*1995–97 projects abs.*	*%*	*2000–02 projects abs.*	*%*	*Details (in italics if particular region)*
Remaining factors	5	7.9	4	10.8	–
Total	*18*	*28.5*	*14*	*37.8*	
National system					
Legislation and regulation	4	6.3	1	2.7	Sudden bans on specific animal experiments (medical biotechnology)
Value system and culture	3	4.8	4	10.8	Lack of entrepreneurial culture and risk-taking
Remaining	1	1.6	2	5.4	–
Total	*8*	*12.7*	*7*	*18.9*	
All factors	*63*	*100*	*37*	*100*	

Source: author's interviews with managers at university (35 interviews, 51 projects) (adapted from Geenhuizen 2012).

Other regional factors that were mentioned by research managers are somewhat less important and also work more indirectly, such as a relatively small business ecosystem and delay in Triple Helix formation between university, industry and government (Etzkowitz 2008). A situation of delay in Triple Helix formation is often connected with fragmentation in the region, either between the municipalities in the area (many municipalities or large economic differences between them) or between the different universities or academic hospitals present in one city. The factor 'shortages in (cheap) incubators for university spin-off firms and follow-up accommodation for them' was also mentioned under regional factors (Core). *National factors* tend to be relatively minor in the overall picture and include very different factors, namely, highly specific regulation changes, which occur mainly in the medical life sciences and pharmaceutical industry (Geenhuizen, 2009).

Overall, based on information on changes in hampering factors between older and younger projects (Appendix 2), the decrease of university and university–business factors turns out to be significant on a modest level ($p<0.1$). In addition, age of project and type of hampering factors turn out to be independent.

6 Policy challenges

The suggestions for policy making in this section originate from discussions in the interviews with research managers and are in line with the barriers identified. Accordingly, they do not cover the whole spectrum of solutions and challenges. The suggestions differ in part between the Core and Southeast of the

Netherlands, and are addressed below to universities and city/regional governments respectively.

The challenge of improving commercialization policy at universities and to better prepare researchers for this work is important in all regions, but less important at technical universities compared with beta-faculties of general universities. The most significant change is that of the reward structure at university: there is a need to reward commercialization in a manner similar to scientific work (including career perspectives). Equally important but working more indirectly, is the challenge to closer align the focus in research programs at university with local and regional-economic cluster policies.

Another challenge to be addressed by universities is to increase knowledge on marketing and the market value of inventions as well as on the strategic behaviour of collaborating firms among researchers. These changes could be realized through extra training of university researchers, but also through detachment or part-time appointment of industry managers at university and – conversely – of university researchers in industry. What also may work is the establishment of special (part-time) chairs at universities covering the commercialization of those fields that are key to regional cluster policy. In this case the city/region could co-finance such chairs. The previous policy also contributes to getting industry involved and committed early on, such as in testing pilot projects/schemes/products potentially spreading financial risk more equally over the partners involved. Other challenges include pursuing and developing more *demand-driven inventions* meaning a stronger emphasis on solving practical problems, like hospital practices, water purification processes, and local traffic issues.

An important challenge is to improve the regional subsidy system, particularly location or investment subsidies to create a level-playing field over all regions. This change in regional economic policy should be addressed at the national policy level. A further challenge is to create easy access to venture capital and sufficient amounts of it in all regions. Pre-seed and seed capital seems to be sufficiently available, but at the stage of commercialization where scaling-up of production for testing is needed while the outcomes are still uncertain (funding gap), large amounts of venture capital are needed. Local and regional governments may be helpful here as they can initiate the establishment of public–private consortia and funds together with local and regional banks and the universities. Such initiatives are now growing, with the Life Sciences Fund Amsterdam as one of the first (AIM 2010).

Regarding the desire in non-core areas to increase the 'critical mass', the idea has arisen to enhance growth in particular niches (clusters) that have strong links with other nearby clusters (synergy). With regard to growth of universities, we propose enhancing alliances with other universities and universities of applied sciences in the region, as well as various models of attracting foreign students (Masters). The latter implies a need to provide extra student housing and to create a quality of life encouraging students to remain in the area.

It is worth noting that a new 'generation' of policy concepts is emerging aimed at an acceleration of knowledge commercialization in open innovation. These concepts put emphasis on the capacity of the region and regional actors to perform as 'testing beds' or true 'Living Labs'. The latter distinguish themselves by using experiments in real-life environments in which inventions are interactively developed (co-designed), validated and tested with customers (Living Labs 2010). The idea is also that commercialization efforts are undertaken in order to respond to real needs for innovations in society. Much knowledge commercialization lends itself to being accelerated in this way, in healthcare, environmental protection, sustainable architecture, local safety, and transport services. However, the aims of 'Living Labs' are still somewhat fluid and different for diverse parties (e.g. Dutilleul *et al.* 2010). This situation means that before cities/regions start to promote them and/or participate in them, it is worth identifying bottlenecks in their performance and disadvantages for particular parties involved.

A final point to be discussed is the role of the city/regional government. Their role seems to be moving on from simply accommodating spin-off firms and creating science or technology parks (spatial planning and financial involvement in spatial projects) to additional new roles that include acting as the *initiator* and *connector of networks*. City/regional governments may play a prominent role in initiating the development of new commercialization tools/incentives, such as local/regional venture capital, the establishment of university–industry bridges by co-financing specific appointments (chairs) in faculties, and the establishment of 'Living Labs'. In addition, city/regional governments may act as connectors between parties in Triple Helix development in a situation of fragmentation.

7 Concluding remarks

This chapter has explored the commercialization of technology projects at various universities in the Netherlands. Drawing on an evaluation of 370 projects and a detailed analysis of some fifty-one projects, the chapter has brought to light preliminary results on the extent to which technology inventions are brought to market and on factors hampering such development. The chapter responds to two knowledge gaps in the current literature: first, a lack of insight into the commercialization of university knowledge on the project level; and, second, a lack of knowledge of differences between various regions in commercialization outcomes.

Failure is faced by a minority of technology projects whereas market introduction is also faced by a minority of such projects. Older projects in the Core area tend to face a relatively high rate of market introduction but also a relatively high failure rate. However, with regard to more recent projects, the differences between the Core and Southeast tend to change as more projects in the Core area seem to stagnate. Indications of factors underlying this pattern were found in the different choice of projects at universities, that is, more science-based with technologies facing greater risk in the Core versus more problem-based or engineering-based projects, allowing for open innovation, in the Southeast. In general, the main hampering factors include circumstances in the organisational context at university and in the interaction between university researchers and firms, as well as a shortage of (regional) financial instruments, particularly a shortage of (large amounts) of venture capital. Regional factors tend to have increased in importance over the last couple of years. Overall, distinguishing between regions turned out to be relevant both in the identification of problematic situations and suggestions for policy solutions. We found a trend that knowledge valorization at the project level does not benefit from circumstances in large cities in the recent past.

Whether the differences also hold true for knowledge commercialization in other countries remains a question. However, we may infer from the Innovation Scoreboard indicators (Pro Inno Europe 2011) that countries such as Austria and Norway, and to some extent the United Kingdom, display similar patterns of commercialization of technology projects in service clusters and high-tech manufacturing clusters. The work in this chapter contains some limitations. The first follows from the use of information from one particular source of subsidies for technology research. It is possible that other financing or subsidy programs in the Netherlands may produce a somewhat different picture. However, the in-depth interviews in the current analysis were structured in such a way that a broader picture of the projects receiving other subsidies could be gained. This suggests that the results on most important barriers to commercialization have larger implications. Second, this study had to be narrowed with regard to technologies and applications. Fields such as micro-electronics and information and communication technology, including multimedia and gaming, creative industries etc., have remained beyond the study.

Given the previous results and implications, the following *three research avenues* are envisaged. First, the next stage of research will focus more on demand-driven projects and add these to the database. Second, the current study was merely descriptive, the next part will allow for testing causal models on the individual project level. Third, one of the conclusions of the current study is that industry needs to be involved in commercialization much earlier than usual. This means a shift to more 'open' forms of commercialization, such as in regional 'test-beds' and 'Living Labs'. The concept of 'Living Labs' is mushrooming today, with participation by universities, customer groups, city/regions, large firms and non-profit organizations. However, there is a need to better clarify the concept, to identify which network structures and forms of steering are the best, which advantages and disadvantages for participants may arise, and what obstacles may prevent the optimal design of 'Living Labs'. Indeed, there is a huge amount of research ahead.

Acknowledgement

The study has benefited from a subsidy by NICIS (Netherlands Institute of City Innovation), four cities (Amsterdam, Delft, Eindhoven and Maastricht), TU Delft and the Netherlands Ministry of Education and Science.

Notes

1 The year 2002 was selected for practical reasons; this was the year of the most up-to-date evaluation information available at the time that the current study commenced (2008).

2 The study is restricted owing to financial support. Therefore, we could not include all relevant research universities. We selected as follows: presence in the Core area and presence outside this area at a substantial distance, in the East and South of the country. We included Delft University of Technology in Delft, the University of Amsterdam and Free University (both in Amsterdam), Leiden University (Leiden), Erasmus University (Rotterdam) and Utrecht University (Utrecht) (all in the Core area), and Eindhoven University of Technology (Eindhoven), Radboud University (Nijmegen), and Maastricht University (Maastricht) (other regions).

References

Amsterdam Innovation Motor (AIM) (2010) Available online at www.aim.nl (accessed 28 September 2010).

Asheim, B., Coenen, L. and Vang, J. (2007) 'Face-to-face, buzz and knowledge bases: sociospatial implications for learning, innovation and innovation policy', *Environment and Planning C: Government and Policy*, 25, 655–670.

Bathelt, H., Malmberg, A. and Maskell, P. (2004) 'Clusters and knowledge: Local buzz, global pipelines and the process of knowledge creation, *Progress in Human Geography*, 28, 31–56.

Bekkers, R. and Freitas, Bodas (2008) 'Analysing knowledge transfer channels between universities and industry: To what degree do sectors also matter?' *Research Policy*, 37, 1837–1853.

Bjerregaard, T. (2010) 'Industry and academia in convergence: Micro-institutional dimensions of R&D collaboration', *Technovation*, 30(2), 100–108.

Bruneel, J., d'Este, P. and Salter, A. (2010) 'Investigating the factors that diminish the barriers to university, industry collaboration', *Research Policy*, 39, 858–868.

Capello, R. (2006) *Regional Economics*. London: Routledge.

Charles, D. and Howells, J. (1992) *Technology Transfer in Europe: Public and private networks*. London: Belhaven Press.

Chesbrough, H. (2003) *Open Innovation: The new imperative for creating and profiting from technology*. Cambridge: Harvard Business School.

Chesbrough, H. W. (2006) *Open Business Models: How to thrive in the new innovation landscape*. Boston, M: Harvard Business School Press.

Christensen, C. (2003) *The Innovators Dilemma*. Cambridge, MA: Harvard University Press.

Cooke, P., Heidenreich, M. and Braczyk, H. J. (2004) *Regional Innovation Systems: The role of governance in a globalised world* (2nd edn). London: Routledge.

D'Este, P. and Patel, P. (2007) 'University–industry linkages in the UK: What are the factors underlying the variety of interactions with industry?' *Research Policy*, 36, 1295–1313.

Dutilleul, B., Birrer, F. A. J., and Mensink, W. (2010) 'Unpacking European living labs: Analysing innovation's social dimension', *Central European Journal of Public Policy*, 4, 60–85.

Etzkowitz, H. (2008) *The Triple Helix: University–industry–government innovation in action*. London: Routledge.

EUROSTAT (2011) *Regional Yearbook 2010*. Online. Available online at http//: epp. eurostat.ec.europe.eu.

Feldman, M. (1999) 'The new economics of innovation, spillovers and agglomeration', *Economics of Innovation and New Technology*, 8, 5–25.

Geels, F. W. (2004) 'From sectoral systems of innovation to socio-technical systems: Insights about dynamics and change from sociology and institutional theory', *Research Policy*, 33, 897–920.

Geenhuizen, M. van (2009) 'Life sciences in the Netherlands: Weakness in a strong national policy', *European Planning Studies*, 17(1), 1–17.

Geenhuizen, M. van (2012, forthcoming) 'From ivory tower to living lab. accelerating the use of university knowledge', *Environment & Planning C, Government & Policy*.

Geenhuizen, M. van and Nijkamp, P. (2006) 'Learning regions in an evolutionary context: Policymaking for high-technology firms', *International Journal for Entrepreneurship and Innovation Management*, 6(3), 265–282.

Huggins, R. and Johnston, A. (2009) 'The economic and innovation contribution of universities: A regional perspective', *Environment and Planning C: Government and Policy*, 27, 1088–1106.

Hussler, C., Picard, F. and Tang, M. F. (2010) 'Taking the ivory from the tower to coat the economic world: Regional strategies to make science useful', *Technovation*, 30, 508–518.

Innovation Platform (2009) *From Aims to a Position Ahead: Enhancing knowledge circulation* (in Dutch). The Hague (NL): Innovation Platform.

Innovation Platform (2010) *The Netherlands back in the Top-5* (in Dutch). The Hague (NL): Innovation Platform.

Keeble, D., Lawson, C., Moore, B. and Wilkinson, F. (1999) 'Collective learning processes, networking and "institutional thickness" in the Cambridge region', *Regional Studies*, 33, 319–332.

Lagendijk, A. and Lorentzen, A. (2007) 'Proximity, knowledge and innovation in peripheral regions: On the intersection beween geographical and organizational proximity', *European Planning Studies*, 15, 457–466.

Living Labs (2010) Available online at www.livinglabs.nl. (accessed 1 October 2010).

Maskell, P. and Malmberg, A. (1999) 'Localised learning and industrial competitiveness', *Cambridge Journal of Economics*, 25, 167–185.

Ministry of Economics, Agriculture & Innovation (MEAI) (2011) *Policy to Boost 'Top Sectors'*. The Hague.

Nooteboom, B. (2009) *A Cognitive Theory of the Firm: Learning, governance and dynamic capabilities*. Cheltenham: Edward Elgar.

Porter, M. (2003) 'The economic performance of regions', *Regional Studies*, 37, 549–578.

Pro Inno Europe (2011) *European Innovation Scoreboard*. Maastricht: MERIT.

Rasmussen, E., Moen, Ø. and Gulbrandsen, M. (2006) 'Initiatives to promote commercialization of university knowledge', *Technovation*, 26 (4), 518–533.

Sassen, S. (2005) The global city: Introduction to a concept. *Brown Journal of World Affairs*, XI(2), 27–43.

Shane, S. (2004) *Academic Entrepreneurship: University spin-offs and wealth creation*. Northhampton, MA: Edward Elgar.

Statistics Netherlands (2011) *Community Innovation Survey 2008*. Available online at www. statline.cbs.nl (accessed 10 November 2011).

Technology Foundation STW (various years) *Reports on Utilisation, 2006–2007–2008* (in Dutch). Utrecht: STW.

Tidd, J. and Bessant, J. (2009) *Managing Innovations: Integrating technological, market and organizational change*, 4th edn. Chichester: John Wiley & Sons.

Tödtling, F. and Tripple, M. (2005) 'One size fits all? Towards a differentiated regional innovation policy approach', *Research Policy*, 38(4): 1203–1219.

Utterback, J. (1996) *Mastering the Dynamics of Innovation*. Boston, MA: Harvard Business School Press.

Westness, P. and Gjelsvik, M. (2010) 'The linkages of firms, universities and research institutes in innovation processes: A conceptual framework', in M. van Geenhuizen, W. Nuttall, D. Gibson and E. Oftedal (eds), *Energy and Innovation: Structural change and policy implications*. West Lafayette: Purdue University Press, pp. 213–232.

Appendix 1

The results of the Goodness of Fit test suggest significant differences between Core and Southeast with regard to both project outcomes in the first period and only with regard to failure, stagnation, etc. in the second period. The outcomes of the independence test, however, suggest that region and project outcome are statistically independent in both periods.

Table 10.A1.1 Regions by project outcomes (take-off in 1995–97) and chi² statistics

Project outcome	*Core* *Abs. %*	*Southeast* *Abs. %*	*Goodness of fit*
Failure, stagnation, continuation (including development unknown)	119 75.3	43 84.3	11.590**
Market introduction/use in society	39 24.7	8 15.7	10.960**
Total (n=209)	158 100	51 100	

Independence test: 1.8; *p<0.05; ** p<0.01.

Table 10.A1.2 Regions by project outcomes (take-off in 2000–02) and chi² statistics

Outcome	*Core* *Abs. %*	*Southeast* *Abs. %*	*Goodness of fit*
Failure, stagnation and continuation	93 86.1	42 84.0	7.869**
Market introduction/use in society	15 13.9	8 16.0	2.475
Total (n=158)	108 100	50 100	

Independence test: 0.1; *p<0.05; **p<0.01.

Appendix 2

Table 10.A2.1 Age of projects by hindering factors and chi² statistics

Hindering factors	*Older (1995–97) projects abs. %*	*Younger (2000–02) projects abs. %*	*Goodness of fit*
University, university–industry links	37 58.7	16 43.2	5.250*
Regions	18 28.5	14 37.8	1.069
National	8 12.7	7 18.9	0.378
Total (n=100)	63 100%	37 100%	

Independence test: 2.3; * p<0.1; degree of freedom=2.

In addition, independence tests using two categories of hindering factors (instead of three) do not produce results indicating dependence between age of project and type of hampering factor.

11 Academic entrepreneurs in post-socialist Central European countries

Evidence from the Hungarian biotechnology sector

Katalin Erdős and Attila Varga

1 Introduction

Spin-off founding is a high-profile entrepreneurial activity that benefits regional economic development through job creation and enhances technological change by bringing research results to the market (Lawton Smith and Glasson 2005). It is estimated that since the 1930s, MIT has spawned 4,000 companies employing more than a million people. The Cambridge Phenomenon constitutes a European success story: Cambridge University in the United Kingdom was identified as being indirectly the origin of virtually all the high technology companies in the Cambridge area (Wicksteed 1985). Lawton Smith and Glasson (2005) investigated 114 high-tech spin-offs originated in the Oxfordshire region's universities and research laboratories and found that the companies had approximately 9,000 employees in 2002, and their total turnover was nearly 1 billion pounds.

As first portrayed by Etzkowitz (1983) the 'academic entrepreneur' is the ultimate driver of university spin-off firm formation. What is a special 'academic' motivation behind university spin-off firm formation is that (alongside with the normal profit incentive) its ultimate aim is to further support academic career. This is because commercialization of research results via spin-offs can enhance scientific activity in the research lab through ensuring additional funding for university research. Also, there is a potential synergy between basic research carried out at the university and applied research that is undertaken in the firm. Additionally, employment in spin-offs can keep talented students near the university.

The 'academic entrepreneur' concept is strongly rooted in the American system of research organization (Etzkowitz 2003; Franzoni and Lissoni 2009). In the US, academics traditionally act like entrepreneurs: besides research they are involved in multiple activities that are typical for entrepreneurial managers. Etzkowitz (1983, 2003) even characterizes research labs as 'quasi-firms' since their existence depends crucially on the ability of securing financial resources and recruiting the most talented people. Academics should acquire funds to set up and maintain their laboratories, hire assistants, and network to have access to additional resources. They 'sell' their products at conferences and journals to enhance their reputation among fellow scientists. Principal investigators should be endowed with skills very much similar to those that characterize successful businessmen who manage their enterprises. Thus, spinning off a company is just a one-step distance from other 'ordinary' entrepreneurial tasks of academics.

Continental Europe performs less successfully than the US in transferring knowledge from university labs to the regional economy via spin-off companies. One potential reason behind this difference might be that the historical evolution of institutions determining the continental European research system (Franzoni and Lissoni 2009; Geuna and Mowery 2007; Lissoni *et al.* 2008) holds back the emergence of the 'classical' (in the above Etzkowitzian sense) academic entrepreneurs even against some success stories (Wright *et al.* 2007). Fundamental differences can be observed in the research system of the US and that of continental Europe[1] regarding the status of researchers, the role of competition and mobility, and the organization of research funding. These can all significantly influence whether a firm is spun off by the 'classical' academic entrepreneur.

Scientific structures of the post-socialist Central European countries have their roots in the continental European tradition, but they inherited many features of the Soviet model as well. The more rigid Central European system thus seems to offer even fewer incentives for academic entrepreneurship, which suggests that the academic entrepreneur might not be a natural ingredient of the Central European university system.

The main research question of this chapter is whether the specific 'academic' drivers behind university spin-off firm formation are present at all within the Central European context. Empirical evidence on university spin-offs in Central and Eastern European countries is very limited so far and usually provides a rather negative view. The literature pays even less attention to the particular 'academic' motivations behind the formation of university spin-offs in the Central European context (Erdős and Varga 2012).

This chapter is based on interviews carried out with university researchers who actively participate in firm formation in Hungary. Hungary is an excellent Central European case since its university system shows German (continental) and Soviet (socialist) features as well. The chapter is structured as follows. The second section first compares the American and Continental European research systems, with an emphasis on Germany as one of the best examples of this system. Then it introduces the Soviet model, which was formerly implemented in Central European countries. The third section presents the results of the empirical research. A summary concludes the chapter.

2 Institutional differences between the US and Central Europe

Differences between the US and continental Europe in the status of researchers, the role of competition and mobility, and the organization of research funding are among those institutions that can significantly influence whether a firm is spun off by the 'classical' academic entrepreneur.

There are major differences in the *status of researchers* between the US and continental Europe. University faculty members in the US are university employees, whereas in continental Europe they are civil servants or state employees (Franzoni and Lissoni 2009). This naturally has an effect on the salary of researchers. In the US system, compensation is determined in a decentralized manner, while the more centralized European systems do not reflect productivity differences directly in salaries (Bonaccorsi 2007).

Related to the status of researchers, there is fierce *competition* among universities for the most talented researchers in the US (Bonaccorsi 2007) as they can attract additional public funding to the universities. Consequently, in the US there is a high *mobility* of researchers and short-term moves between academic and non-academic environments are also permitted. In contrast, the continental European system seems to offer less mobility and competition, which inhibits systematic collaboration across institutional boundaries, impeding the transfer of technologies (Franzoni and Lissoni 2009).

Regarding *research funding*, the US system is a multilayered, decentralized one where financial sources come from several political levels (federal, state, local etc.) and from several types of agencies in terms of governance (public, private, third sector, foundations) and time horizons (short or long term). In contrast, in the more centralized continental European system important funding sources are allocated only at a few levels and there is only a low variety of funds with a limited role of donations. Furthermore, the allocation of R&D resources in the US follows an evidence-based decision-making process while in the continental European system decision making is also affected by a political compromise about equivalently competing claims. The former system leads to the concentration of resources favouring long-term and large-scale funding while the latter usually results in a much more equalized distribution of resources, attracting short-term and limited-size funding (Bonaccorsi 2007).

Knowledge transfers among universities in the US are further supported by one of the important features of US *doctoral education*. Fresh PhDs are not allowed to take their first jobs in the university where their degree was earned but they have to look for academic positions at different institutions. Thus, the graduate labour market significantly contributes to the flow of new knowledge within the university system (Bonaccorsi 2007). Conversely, the continental European system seems to offer less mobility and competition, which results in less intense systematic collaboration within and across institutional boundaries and ultimately less intensive technology transfer (Franzoni and Lissoni 2009).

The Central European model of science can be best understood if we study those systems that have played the deepest influences on its development. Most of these countries have German university traditions with strong impacts by the Soviet model. Germany, as one of the best representatives of the continental European system provides an excellent ground for general comparison with the USA. Its main ideological spirit is rooted in the Humboldtian model developed in the early nineteenth century. The primary notion behind this was the '*Bildung*' or formation of the person. Integration of teaching and research and academic freedom of professors were further inevitable principles of this system (Goldstein 2007).

In Germany there is a more balanced division of public research between universities and other public research organizations (PROs) compared to the USA (Buenstorf 2009). This has many consequences on the possible occurrence of entrepreneurial activities at universities, since PROs usually have a higher autonomy than universities (Franzoni and Lissoni 2009) typically directed by the federal states, whereas PROs include organizations with variable legal statuses and funding schemes (Koschatzky and Hemer 2009).

A further peculiarity of the German system was the so-called Hochschullehrerprivileg (professor's privilege) which meant that university scientists could retain IPR resulting from their research (Buenstorf 2009). This feature of the system contributed to a more active role of PROs in patenting (Bacchiocchi and Montobbio 2009). Though professor's privilege was abolished in 2002 and today universities have the right to patent inventions developed by their employees (Geuna and Rossi 2011), low autonomy and lack of experience in patenting may inhibit the fast spread of university patenting in Germany (Franzoni and Lissoni 2009). Furthermore, one cannot exclude the probability that informal technology transfer mechanisms evolved in the past still remain substantial (Grimpe and Fier 2010).[2]

Bonaccorsi (2007) and Franzoni and Lissoni (2009) highlighted that the civil servant or state employee status of researchers in many continental European countries can be detrimental for spin-off process since they inhibit the integration of entrepreneurial activities into academic reward structures. Salary of German academics is fixed according to an administrative scale (Musselin 2004). Though financial constraints initiated some changes in the funding structure as part of modifications introduced in the legal status of researchers (Geuna and Nesta 2006), integration of entrepreneurial activities into promotion procedures is rather an exception than the rule. Grimpe and Fier (2010) underline the difference between the work contracts of German and American faculty members. Whereas German academics are paid for the whole calendar year, American scientists are typically paid only for nine months, which can provide an incentive to complement their salary with entrepreneurial income.

Bonaccorsi (2007) highlighted the role of competition between universities in trying to attract the most talented researchers and the related public funding. However, mobility of academics in Europe is hindered by strong national features of academic labour markets that include differences, for example, in recruitment procedures, promotion rules, career patterns, just as in salaries and statuses of researchers (Musselin 2005).

As already mentioned, the university system in Hungary, as in several other Central European countries, has been developed under German influences and later on with significant impacts from the Soviet research system. Etzkowitz and Leydesdorff (2000) argue that in the Triple Helix model of the socialist countries nation states encompassed academic and industrial spheres and also directed their relationships. Gaponenko (1995) gives a detailed description of the Soviet model of science that evolved around the 1920s–1930s and was exported to the members of the Soviet Bloc. He argues that the research system was subordinated to bureaucratic and party interests that limited internal democracy and creative freedom. Unlike in the US system, where military inventions are aimed at being converted into civil products creating wealth, if it does not threaten security (Bush report 1945), the overwhelming secrecy of the Soviet model resulted in limited or no civil application of military inventions (Gaponenko 1995).

Somewhat similar to continental European experiences a division of tasks between universities and specialized research organizations existed in the Soviet model as well. Academy of Sciences had the highest prestige and consequently abundant resources (but at the same time with ballasts as well), whereas universities carried out R&D only as a secondary activity, with inadequate finance and equipment (Gaponenko 1995). Fundamental features of this model remained unchanged until the 1990s. This system suffered from many deficiencies, for example, inefficiency weakness in research evaluation (Balazs *et al.* 1995).

There is still only limited empirical evidence on academic spin-off processes in Central and Eastern Europe. Serious economic problems following the collapse of the socialist political system resulted in significant R&D budget cuts that caused the exodus of many excellent scientists. As a result, the system fell into a deepening crisis. Balazs (1995) provided a detailed description of the problems of the first half of the 1990s. A hostile environment for R&D activities was further exacerbated by the lack of science and technology policy.

Spin-off firm formation was found as one of the active adjustment strategies by research organizations. The option of establishing spin-off companies created a symbiotic relationship between these firms and the university department. Many of the new firms preferring to stay close to the university did not target fast growth (Balazs 1995). Inzelt (2002) argued that Hungarian scientific 'backyard farms' have questionable economic development impact. Tchalakov and his co-authors (2010) highlight the importance of path-dependency of the Eastern European spin-off activity. They also point to the differences and asymmetries that evolved between researchers who were involved in contract research and those who were not.

The role of contract research is also highlighted by Balazs (1995). She argued that the necessary skills and abilities on the side of the researchers have been accumulated through contract research work in the former system (Balazs 1995). She also highlights the importance of faculty members by arguing that the professional interest of researchers in R&D gave an impetus towards practical utilization of knowledge. Unfortunately the limited interest of companies was a major obstacle, as also suggested by Inzelt (2004) who found that lack of innovative companies hinders university–industry cooperation.

3 Can we find the 'classical' academic entrepreneur in Central Europe? An empirical analysis[3]

In our study we investigate researchers' scientific motivations behind spinning off a firm and the factors that influence the realization of these motivations. To eliminate the differences among different scientific and entrepreneurial fields we focused on one particular area: biotechnology. Based on websites of the Hungarian Biotechnology Association, the Hungarian Spin-off and Start-up Association and that of the technology transfer offices we prepared a list of companies that was later completed and corrected by suggestions of consultants from different fields. The twenty-two companies chosen cover the majority of companies operating in the country and meet our selection criteria (biotechnology, academic founder). We conducted interviews with eighteen researchers (three faculty members could not participate in the research in the interviewing period owing to international or other programmes and one researcher declined to participate). We interviewed five entrepreneurial researchers in the Central Hungarian region (from Budapest and Gödöllő) and thirteen from the countryside based on a structured scheme. The duration of the interviews varied between thirty and ninety minutes. Researchers in the Central Hungarian region are affiliated with Eötvös Loránd University, Semmelweis University and Szent István University. We met five researchers at the University of Debrecen, five at the University of Pécs, and three at the University of Szeged. All of them held a CEO or CSO (or equal) position in the firm, and all are physicists, chemists, veterinarians and more than third are medical doctors. The oldest firm was established in 1992, the youngest in 2008. Four of the five firms in Pécs were created in the first half of the 1990s, whereas the youngest companies were found in Debrecen where all of them were established in 2005 or later. Fields of the companies are biotechnology and medical devices. It is interesting to note that at least half of the companies predate the legislative changes that promoted university entrepreneurship. This seems to support the view observed in the US and also in Europe that legislative changes may only legitimize already existing academic entrepreneurial processes that are largely an outcome of the rise of biotechnology (Geuna and Nesta 2006; Mowery and Ziedonis 2002). Age of founders is various: one third of them are under forty, eight of them are between forty-one and sixty and four of them are over sixty.

Three of the companies are operating in the field of medical devices (surgery, gastrotonometrics and allergology), one of the companies is active in the field of medical biology, biotechnological research and bioinformatical software development, and one in genomics; three of them develop diagnostic devices, molecules; one of the companies is active in the field of toxicology, two are related to the food industry, six are connected to pharmaceuticals and cancer therapy and one is involved in gamete and embryo manipulation.

Nearly half of the companies have two or fewer employees, which can be related to the facts that either the firms are relatively young and/or they have subcontractors. However, five companies have more than ten employees. It is hard to measure how successful these companies are, especially since some of them are very young. Since biotechnology is not a typical big employer, we can argue that those five companies that have more than ten employees are very successful; furthermore – based on the interviews – we assume that four additional companies are also profitable or stable and self-sustaining. Two of the companies are still in the developmental phase but have really breakthrough ideas in their field of operation.

3.1 The 'classical' academic entrepreneurs

Eight classical academic entrepreneurs are identified in the sample. Most of these researchers have excellent publication and citation records and their scientific work is widely known internationally. They are usually positioned in the upper segment of the academic hierarchy and most of them lead their own research groups. Their aim with spin-off establishments is to further extend scientific activity. Firms might focus on the development of a product or a technology originated in a single idea or on the creation of broader product ranges. Firms might provide services too, but this activity is not their main profile – it targets additional revenue to improve own research performance. Financial motivations do not play a crucial role though one of them mentioned that it is a significant incentive. Monetary rewards are usually considered as a measure of success in the business world.

Academic and firm activities are symbiotic with each other. Almost all of them prefer academic work though the firm is also very important. They enjoy the atmosphere in academia and some of them work at the university even after retirement. Some others assume that they will work in the firm more intensively after retirement.

Though most of them did not have any business education, the majority had prior business experience in forms of contract research with industry or consultancy services. Additionally, all of them have acquired some project management experience, thus they already had an insight into financial administration, and contacts with accountants and lawyers. Usually, a CEO position is held by them or by their MBA partners or both, at least at the beginning, but at a certain stage of firm development they usually extend their management with appropriate experts and sometimes hire professional managers. The most often mentioned reason for this was not the lack of necessary knowledge but the shortage of time that is faced by most scientists after a while. Usually they feel that managing administrative and financial issues of different projects demands too much time and it is wiser to hire a professional management at a certain stage of the development of the company.

Most of the academics in this group studied or even worked abroad for longer or shorter periods and are still today in contact with their foreign partners. Some of them have met role models abroad, and this is even more important if we consider that there are only a few of them in Hungary. However, those few try to help their colleagues and may hold a consultant position in the firm of the younger researcher.

It is important that several members of this group are deeply embedded in international research networks and have regular interactions with their colleagues abroad. These contacts usually originate from visiting positions at such excellent scientific centres as Karolinska Institute in Sweden, University of California in San Francisco, University of Wisconsin-Madison or at firms such as Genentech. One of them had the chance to work with a Nobel laureate.

Academic networks play important roles in the start-up process. Sometimes the idea of establishing a firm occurs during conversations with academic colleagues.

In other cases development of the firm happened through academic networks and contacts. We also observed cases where academic contacts were business contacts as well where the foreign academic is a spin-off founder and cooperates with the company. The intensity and extension of academic contacts often do not decrease, but increase after the establishment of the firm, since this generates cooperative research and co-publications. In some cases business networks resulting from former contract works with industry appear significant in the start-up phase as well.

Publication delay is a natural phenomena occurring regularly in patent protection. Sometimes this is mentioned as a problem, but in most of the cases it is accepted as necessary and not regarded as a big impediment in scientific work. No tensions are reported with respect to secrecy even though in some cases there is a formal secrecy policy at the company. Academics usually share their ideas with their university colleagues (who are often colleagues in the firm as well) since trust-based atmosphere is crucial for scientific research. In some cases they prefer to share their firm-related research only with colleagues working in the firm. However, it is well understood by their university colleagues and does not lead to tension or mistrust. This might be related to the fact that there is a division of labour between applied research carried out at the firm and basic research at the university laboratory; thus, university colleagues do not feel they are in a disadvantageous situation by not being informed about firm-related knowledge. Free flow of information and knowledge exchange are also advantageous to a certain extent for the entrepreneurial scientist since there is a huge accumulated general and specific knowledge at the universities and it can easily happen that a specialist university colleague can solve a question that has arisen in a certain field.

With respect to norms at the university department, academics in this group describe them as favourable for entrepreneurial activities. Some of the colleagues support the company while academic entrepreneurs submit project proposals together with the university.

State supports and grants play significant roles in the establishment of companies. The other important source of income is contract research. Most of the companies would be able to completely exist on sales and service incomes – only two of them need state support as they are not on the market yet with their otherwise promising technologies. These are the only companies in this category where business angel and venture capital (VC) financing are present. For the rest of the academic entrepreneurs, almost all of them are a bit sceptical about VC and they usually try to avoid it. The main reason is that they are afraid of losing control over their firms, which would be harmful for their original academic intentions.

There is sometimes a very limited, but not typical cooperation with the university technology transfer office (TTO) at companies established at the beginning or in the middle of the 2000s. University equity holding is not typical (it happens only in one case), though parent universities theoretically support entrepreneurial activities. It seems that science parks, an often mentioned actor of the technology transfer process, do not play a significant role for academic entrepreneurs as only two researchers mentioned their importance.

3.2 'Unbalanced' academic entrepreneurs

In contrast to 'classical' entrepreneurs where activities in the university laboratory and in the firm are in a close symbiosis, for this type of researchers either research (three cases) or the company (one case) gets the dominant focus, though both of the activities are present.

Half of the researchers in this group limited their activity to science from the beginning and did not participate in the management of their firm, which was established together with a surrogate entrepreneur or with a company. The reason – next to shortage of time – is that they are only interested in the development of the product, which is based on their idea, and not in the business operation of the firm. Sometimes they do believe that more intense, deeper entrepreneurial engagement would be harmful for their scientific activity. One of these researchers admitted that it might result in losing control over the firm but he also believes that it is normal because the industry's task is to develop the product.

The difference here with the first group might be rooted in the fact that three of these companies develop medical devices and the professors are medical doctors in all three cases, with their primary interest in healing. They are very successful and acknowledged scientists – half of them have outstanding publication and citation records. They are placed in the highest levels of the academic hierarchy. Academic entrepreneurial role models were not mentioned. Also, the three researchers attached primarily to academia did not have any formal business education.

Products are clinically tested at university laboratories. Academic network connections are utilized to get feedback about the appropriateness of the product and to gain advice for potential further development. University devices and facilities are used during product development. Research results are often published in scientific articles. The fourth researcher represents the opposite side: he is stronger as an entrepreneur than as an academic. He used to be a faculty member but while working at a company abroad he had already decided to quit his job and start a company. Although he is primarily motivated by business he was already involved in the research line in which the company was operating at the time he held a research position at the university. He still maintains close contacts with universities. Besides his academic network, his business network also played a very important role by giving advice and an interest-free loan to establish the company.

Even though three of the professors are dominantly interested in academia, researchers of this group did not mention either secrecy or publication delay as a problem that would hold them back from being engaged in entrepreneurial activities in the future.

State supports and grants play a very important role by creating the chance to set up a company and carry out the applied research that is needed to develop a product. Lots of them mentioned that calls for university–industry cooperation projects meant a crucial step at the beginning, and such projects are also beneficial later on.

As for entrepreneurs of the first type, the university is supportive not only at the departmental level, but also at a more general policy level as well; however, neither the TTO nor the science parks of VC funds seem to play a dominant role in any steps of the technology transfer process.

3.3 Academic entrepreneurs impeded by environmental factors

Their motivations and characteristics in many respects are the same as those of the 'classical' academic entrepreneurs but for various reasons they cannot fulfil both roles. Their original aim with the firm was to promote their scientific work and to develop and put into practice an invention based on academic research. However, owing to an unfavourable departmental attitude, dislike of their direct superior (maybe jealousy), or lack of resources they cannot realize their original aims. We identified three cases in this group.

The scientific work of the researchers is well acknowledged by the research community. Two of them have very impressive publication and citation records, even though one of them is in a younger age cohort. Despite this, none of them has the chance to set up their own research group within the university. Despite good scientific performance they are positioned at the bottom/middle level of the academic hierarchy, and sometimes have been there for a while, and this is not likely to change significantly in the near future. Thus they do not feel they are appreciated.

Two of these researchers had taken visiting researcher positions abroad where they saw role models and found that the symbiotic relation of the spheres of academia and business is an attractive idea that is worth following. Only one of these researchers had prior business experience before setting up the spin-off firm and neither of them had an existing business network that would have helped in the establishment and development of the spin-off. Even though there is sometimes a strained relation with the university academic networks including colleagues from parent and overseas universities helped two of the researchers to become academic entrepreneurs.

Firms are based on a concrete idea or the recognition of the biotech trend; however, it is not always a breakthrough. Where the idea is a real advancement in the scientific field, the reputation of the scientist increases and the success of the firm and the researcher sometimes awakes envy among colleagues and heads of departments with unfortunate negative consequences. One researcher left the university because his superior did not welcome his entrepreneurial success. But these tensions arising in academia are not rooted in secrecy or publication delay.

So it can happen that the researcher has to choose between academic and entrepreneurial engagements because the university management supports entrepreneurial activities only on the surface. In fact the faculty member is continuously subject to negative discrimination at the promotion/tenure procedure. The supportive attitude of the university policy is indicated by the university equity holding that occurs in one of the companies. Since the decision about tenure and promotion is made at the departmental level, university regulation and the TTO cannot do very much if the head of the department is against entrepreneurship. At this point departmental norms are more important than written laws.

Since in some cases there is a hostile university environment behind the seemingly supportive strategy, most of these companies used the services of the TTOs in a very limited manner or not at all. There are also examples of entrepreneurial performance being under expectation even where it concerns researchers with good scientific results, significant attempts to set up a company and considerable grant experience. In these cases the tension does not seem to be between scientific and entrepreneurial activities but between the company and the framework conditions. The researcher may not have the necessary business knowledge but cannot or does not intend to hire appropriate management. The lack of specialized patent experts can also block spin-off development. Some insist that these experts are only available abroad which involves enormous cost. The problem is more striking in cases of firms located in the countryside since the management always has to travel to the capital to arrange patent issues.

Also, lack of appropriate financial resources can stop companies growing. At the beginning there is usually state and EU support available, which is very helpful but in some stages of the development a more significant amount would be needed that is not covered by these grants. If state support is suspended for a while this can cause considerable problems. They do not have VC in the firm; one of the researchers mentioned that they try to avoid it because of the associated risk of losing control over the firm. One of the companies had a laboratory in an industrial park but in the others scientific parks did not play a role.

3.4 'Externally motivated' academic entrepreneurs

The three entrepreneurs in this category are different from the previous ones in terms of both motivation and characteristics. They are in the bottom/middle segments of the academic hierarchy and they do not necessarily have international experiences. Though one of them visited a non-entrepreneurial university he did not meet academic entrepreneurial role models. On the other hand, he has the highest publication and citation record in this group while the others are at the beginning of their academic careers.

Since the companies operate in the fields of chemistry and genomics where contract research is common, most of the entrepreneurs have business experience but they do not have real business networks that would have facilitated the process of spin-off development. On the other hand, academic networks can be helpful. One of the companies was established by colleagues and in another case it seems that they kind of 'use' academic personnel in the firm. Secrecy does not play a role, academic community and information flows are rather supportive for the research, and also publication delay is acceptable.

The absence of tensions between academic and entrepreneurial activities is perhaps related to the fact that the establishment was mostly initiated by the university and there is university equity holding as well. The motivation to set up the company is rooted in the entrepreneurial strategy and practice of the university or triggered by need for resources. In the university push-driven case a very active TTO executes the aggressive entrepreneurial strategy of the top management of the institution. The TTO is eager to introduce entrepreneurial activities with more and more researchers to gain a deeper insight into the research carried out at the university and it may set up an own company to manage contract research works. Since inventions are filed and sometimes they put up research questions and topics that are likely to generate patents, TTOs are often active initiators of firm establishment, asking the principal investigator (the leading researcher) to become the CEO. It is about companies with considerable, sometimes 100 per cent university equity share. Sometimes there is industrial equity share as well, but with no VC in these firms either.

The spinning-off process can also be initiated by the researcher but with the involvement of the university. Thin basic research sources motivate more and more researchers to obtain money from different types of applied research grants to conduct basic research. This can be particularly true for researchers positioned in the bottom/middle segment of the academic hierarchy as in the competition for grants they start with a significant setback compared to colleagues with high publication and citation records.

There is a 'negative Matthew effect'[4] for researchers at the beginning of their careers as there are no significantly sized grants available. Since these are usually young companies it would be hard to predict what influence these circumstances will have on the growth and success of the firms but it is likely that with modest enthusiasm or unclear vision their growth will be slow or zero. It cannot be excluded that, based on their first entrepreneurial experience, the faculty will decide to establish a real firm to exploit the synergies between university and industry, moving into one of the above-mentioned categories depending on environmental factors.

It is very likely in the case of externally motivated entrepreneurs that they would never have started a company in the absence of a supportive university environment or that of grants. Thus state support and grants play a major role in this group. Firms of externally motivated entrepreneurs might never be born in a hostile university environment. Even firms that aim at the utilization of an idea, there is much more an institutional entrepreneurial than a personal motivation.

These firms typically do not have a professional external management; there is no business angel money or VC in the firms, but there is sometimes an industrial partner and lots of industrial contract research.

One of the companies aims at the development or commercialization of certain research results, while the other two – although they also carry out in-house research – are still searching for their right place and one of them even admitted that they do not expect significant growth. The basic difference between these and the 'classical' academic entrepreneurs is that in the latter case they target profitable, large markets. The externally motivated entrepreneurs are either mainly specialized in services or operate on a 'will see' principle, searching for their place, and are not growth oriented. The firm is rather an alternative commercialization method of an idea as compared to licensing. This can be realted to the fact that they are operating with university equity support.

University pressure and equity lead to an increase in the number of firms but it might create a hostile atmosphere among researchers. The 'everyone is guilty unless proven otherwise' is not a good approach on the part of the TTOs. A personal, trust-based relationship might have resulted in the same number of disclosures but could have created a better university atmosphere, motivating researchers to decide about the firm formation themselves in view of the opportunities. However, it must also be admitted that purely volunteer firm formation is hindered by a lack of successful role models.

Not only university strategy but also the interest of the TTOs might induce proactivity (sometimes seen by the faculty as aggression) of technology transfer offices in spin-off firm formation. This is because TTOs are established and operated by grants, which are needed to sustain their operation. Thus, they have to prove their right to exist and their effectiveness. Since licensing – although it usually means quick direct revenues – is not easy and is unlikely to secure the operation of the TTO, regional employment and development effects of spin-offs can be good arguments while applying for operating expenses.

4 Summary and conclusions

In this chapter we were searching for the 'classical' academic entrepreneur behind university spin-off formation. Specifically, we tested the assumption whether the academic entrepreneur is indeed a typical US phenomenon rooted in the institutional set-up of the American research system. Interviewing university spin-off firm founders in the Hungarian biotechnology sector, we found that the 'classical' academic entrepreneur does exist even within the context of Central European institutions.

Eight out of the eighteen cases clearly show the characteristics of firms established by the 'classical' (in the Etzkowitzian sense) academic entrepreneur. The main motive behind the formation of these firms is to enhance the carrier of the academic researcher by maintaining a synergic relationship between the university lab and the spin-off firm. It was also shown in the chapter that there are important factors that determine whether a firm is spun off by a 'classical' academic entrepreneur or the spinning-off process results in different types of companies.

'Classical' academic entrepreneurial firms are spun off by well-established, internationally recognized scholars with rich academic network connections. The existence of role models appears to be critical in the emergence of these companies. It was also shown that academic and business relationships can significantly increase the success of firm formation. While entrepreneurial policies of the universities do not show sensible impacts on the emergence of these 'classical' academic entrepreneurial firms, supportive departmental norms are crucial to their success. In the absence of a friendly environment at the department or the necessary business knowledge and financial resources, academic entrepreneurs become 'impeded' in the sense that a successful company does not enrich scientific activities at the university laboratory. On the other hand, specificities of the product and missing role models could lead to the formation of the 'unbalanced' type of entrepreneurs where integration of the firm into scientific research becomes limited, as the firm exists 'on the side' of the activities of the academic. Surprisingly we found that aggressive university policies supporting spin-off firm formation does not result in 'classical' academic entrepreneurial firms but rather in companies with limited business–academia synergies.

Our results suggest that, under specific circumstances, the 'classical' academic entrepreneur can indeed emerge and become successful even within the context of the seemingly not supportive Central European research institutions. As such, the power of academic motives in firm formation is clearly shown by the empirical analysis. These findings have important policy consequences. Our results imply that institutional changes in the European research system would increase the participation of academics in university technology transfer via spin-off firm formation. Based on our empirical findings we suggest that increased financial autonomy of universities, real competition among academic institutions in attracting highly qualified researchers and the introduction of a multilayered system of research funding could significantly enhance the success of already existing academic entrepreneurs, providing further incentives to spin-off companies motivated by strong academic goals.

In this study we followed an 'indirect' route to address the research problem, whether academic motivations that rule university spin-off firm formation in the US are present in the Central and Eastern European context. We selected Hungary, a country where institutional traditions are even more disadvantageous with respect to academic entrepreneurship than in other continental countries located in the Western part of Europe. It is a natural question, whether our results can be generalized to other post-socialist countries in Central and Eastern Europe. Universities in Soviet Bloc countries were encompassed by the nation state that directed their relations in a Triple Helix I configuration (Etzkowitz and Leydesdorff 2000). Thus transition economies' university systems have similar roots, suggesting that it is highly likely that the findings of this study hold true for all of the Central European countries. However, the generality of our research results within the Central European context needs further investigation, especially if we consider that the transformation of the research system of the countries belonging to different waves of the EU enlargement happened at a different pace (Tchalakov *et al.* 2010). Owing to the peculiarities of biotechnology and academic entrepreneurial activities in this field the extension of our findings to other sectors would be questionable. Any effort in this direction needs further investigation (Franzoni and Lissoni 2009; Geuna and Nesta 2006; Lissoni *et al.* 2008; Mowery *et al.* 2004; Nerkar and Shane 2003).

Acknowledgements

The research leading to these results has received funding from the European Community's Seventh Framework Programme (FP7/2007–2013) under grant agreement no. 216813. Previous versions of this chapter were presented at the 49th Congress of the European Regional Science Association (Lodz, 2009), at the 3rd Central European Conference in Regional Science (Kosice, 2009), at the Culture of Business – Capital of Culture conference (Pécs, 2009), at the workshop on Cooperation and Participation in Technoscience in the Socialist and Post-Socialist Space (Graz, 2009), at the second IAREG progress meeting (Vienna, 2009) and at the Seminar of the Department of Economics and Regional Studies, Faculty of Business and Economics, University of Pécs (Pécs, 2009). Comments received by participants of the above-mentioned events are gratefully acknowledged. We would like to express our special thanks to the useful comments and suggestions by Zoltán Bajmócy, János Barancsuk, Attila Buday-Sántha, Edward Bergman, Pablo D'Este, Javier Revilla-Diez, Harvey Goldstein, Simona Iammarino, Annamária Inzelt, Jan-Philipp Kramer, Francesco Lissoni, Gunther Maier, Gerd Schienstock, Sabine Sedlacek, Helen Lawton-Smith, Zoltán Szabó, László Szerb, Franz Tödtling, Michaela Trippl, Nick von Tunzelmann, and also for the experts and consultants who offered their help by providing contacts to interviewees: Norbert Buzás, Kata Dobay, Arnold Fehér, Kinga Homolay, Zsolt Makra, Lenke Rónaszegi, Marcell Veidner, and to all of the researchers who agreed to be interviewed. Of course the usual disclaimers apply.

Notes

1 Albeit, no single European model exists (Etzkowitz *et al.* 2000); there are some common characteristics of universities in continental Europe which makes an Anglo-Saxon (US)–continental comparison reasonable. The importance of this comparison is also supported by the strong reliance of the Central and Eastern European research systems on the continental tradition.

2 For comparison, in the US the Bayh-Dole Act created a clear intellectual property right structure already in 1980 (Henderson *et al.* 1998).

3 This section draws on Erdős and Varga (2012).

4 Merton (1988) argued that owing to the Matthew effect already acknowledged scientists are usually better appreciated, which results in a higher level of trust and access to even more funding.

References

Bacchiocchi, E. – Montobbio, F. (2009) 'Knowledge diffusion from university and public research: A comparison between US, Japan and Europe using patent citations', *The Journal of Technology Transfer*, 34(2), 169–181.

Balazs, K. (1995) 'Innovation potential embodied in research organizations in Central and Eastern Europe', *Social Studies of Science*, 25(4), 655–683.

Balazs, K., Faulkner, W. and Schimank, U. (1995) 'Transformation of the research systems of post-communist Central and Eastern Europe: An introduction', *Social Studies of Science*, 25(4), 613–632.

Bonaccorsi, A. (2007) 'Explaining poor performance of European science: Institutions versus policies', *Science and Public Policy*, 34(5), 303–316.

Buenstorf, G. (2009) 'Is commercialization good or bad for science? Individual-level evidence from the Max Planck Society', *Research Policy*, 38(2), 281–292.

Bush, V. (1945), 'Science: The Endless Frontier. Available online at http://www.nsf.gov/od/lpa/nsf50/vbush1945.htm (accessed 31 March 2012).

Erdős, K. and Varga, A. (2012) 'The academic entrepreneur: Myth or reality for increased regional growth in Europe?' in M. van Geenhuizen and P. Nijkamp (eds), *Creative Knowledge Cities: Myths, visions and realities*. Cheltenham and Northampton, MA: Edward Elgar, pp. 157–181.

Etzkowitz, H. (1983) 'Entrepreneurial scientists and entrepreneurial universities in academic science', *Minerva*, 21(2/3), 198–233.

Etzkowitz, H. (2003) 'Research groups as "quasi-firms": The invention of the entrepreneurial university', *Research Policy*, 32(1), 109–121.

Etzkowitz, H. and Leydesdorff, L. (2000) 'The dynamics of innovation: From national systems and "Mode 2" to a Triple Helix of university–industry–government relations', *Research Policy*, 29(2), 109–123.

Etzkowitz, H., Webster, A., Gebhardt, C. and Terra, B. R. C. (2000) 'The future of the university and the university of the future: Evolution of ivory tower to entrepreneurial paradigm', *Research Policy*, 29(2), 313–330.

Franzoni, C. and Lissoni, F. (2009) 'Academic entrepreneurs: Critical issues and lessons for Europe', in A. Varga (ed.), *Universities, Knowledge Transfer and Regional Development: Geography, entrepreneurship and policy*. Cheltenham and Northampton, MA: Edward Elgar, pp. 163–190.

Gaponenko, N. (1995), 'Transformation of the research system in a transitional society: The case of Russia', *Social Studies of Science*, 25(4), 685–703.

Geuna, A. and Mowery, D. (2007) 'Publishing and patenting in US and European universities', *Economics of Innovation & New Technology*, 16(2), 67–70.

Geuna, A. and Nesta, L. J. J. (2006) 'University patenting and its effects on academic research: The emerging European evidence', *Research Policy*, 35(6), 790–807.

Geuna, A. and Rossi, F. (2011) 'Changes to university IPR regulations in Europe and the impact on academic patenting', *Research Policy*, 40(8), 1068–1076.

Goldstein, H. A. (2007) 'Institutions for knowledge generation and knowledge flows: Building innovative capabilities for regions', Paper for the 10th *Uddevalla Symposium*, 14–16 June 2007, Uddevalla, Sweden.

Grimpe, C. and Fier, H. (2010) 'Informal university technology transfer: a comparison between the United States and Germany', *Journal of Technology Transfer*, 35(6), 637–650.

Henderson, R., Jaffe, A. and Trajtenberg, M. (1998) 'Universities as a source of commercial technology: A detailed analysis of university patenting 1965–1988', *Review of Economics and Statistics*, 80(1), 119–127.

Inzelt, A. (2002) 'Restructuring and financing R&D: New partnerships', in A. Varga and L. Szerb (eds), *Innovation, Entrepreneurship, Regions and Economic Development*. Pécs: University of Pécs Press, pp. 27–50.

Inzelt, A. (2004) 'The evolution of university–industry–government relationships during transition', *Research Policy*, 33(6/7), 975–995.

Koschatzky, K. and Hemer, J. (2009) 'Firm formation and economic development: What drives academic spin-offs to success or failure?', in A. Varga (ed.), *Universities, Knowledge Transfer and Regional Development: Geography, entrepreneurship and policy*, Cheltenham and Northampton, MA: Edward Elgar, pp. 191–218.

Lawton Smith, H. and Glasson, J. (2005) *High-tech spin-offs: Measuring performance and growth in Oxfordshire*. England: Oxfordshire Economic Observatory.

Lissoni, F., Llerena, P., Mckelvey, M. and Sanditov, B. (2008) 'Academic patenting in Europe: New evidence from the KEINS database', *Research Evaluation*, 17(2), 87–102.

Merton, R. K. (1988) 'The Matthew effect in science, II: Cumulative advantage and the symbolism of intellectual property', *ISIS*, 79, 606–623.

Mowery, D. C., Nelson, R. R., Sampat, B. N. and Ziedonis, A. A. (2004) *Ivory Tower and Industrial Innovation*. Stanford, CA: Stanford University Press.

Mowery, D. C. and Ziedonis, A. A. (2002) 'Academic patent quality and quantity before and after Bayh-Dole act in the United States', *Research Policy*, 31(3), 399–418.

Musselin, C. (2004) 'Towards a European labour market? Some lessons drawn from empirical studies on academic mobility', *Higher Education*, 48(1), 55–78.

Musselin, C. (2005) 'European labor markets in transition', *Higher Education*, 49(1/2), 135–154.

Nerkar, A. and Shane, S. (2003) 'When do start-ups that exploit patented academic knowledge survive?', *International Journal of Industrial Organization*, 21(9), 1391–1410.

Tchalakov, I., Mitev, T. and Petrov, V. (2010) 'The academic spin-offs as an engine of economic transition in Eastern Europe: A path-dependent approach', *Minerva*, 48(2), 189–217.

Wicksteed, S. (1985) *The Cambridge Phenomenon: The growth of high technology industry in a university town*. Cambridge: Wicksteed.

Wright, M., Clarysse, S., Mustar, P. and Lockett, A. (2007) *Academic Entrepreneurship in Europe*. Cheltenham and Northampton, MA: Edward Elgar.

12 Universities, researchers and their perspective on the fulfilment of the third role

The case of Czechia

Pavla Žížalová and Vladislav Čadil

1 Introduction

The growth in importance of knowledge and knowledge-based activities for economic competitiveness in academic research as well as in political and business circles has led to a shift in the view on the role and function of universities as key players in innovation systems (whether national or regional) (Cooke 2001, 2004). Universities have traditionally performed mainly two functions – research and teaching. In recent years, however, these functions have extended to a new role – in simple terms – that of 'economic development institutions'.

The modern university, therefore, not only focuses on research and teaching, but cultivates an entrepreneurial culture and an openness to change; it is not afraid of science leading to commercial applications, is connected to the business sphere, and contributes to the social and economic development of the region where it is located. All these activities are often described as 'the third role' of universities (e.g. Gibbons *et al.*1994; Chatterton and Goddard, 2000; Etzkowitz and Leydesdorff 2000; Harloe and Perry 2004). It is often argued that the third role is an extremely important mechanism for generating technological spillovers, and significantly contributes to entrepreneurial competitiveness enhancement as well as social returns of research and development (R&D) investment (Martin and Scott 2000; Siegel and Zervos 2002).

Quite a lot of attention has already been paid to this issue, reflecting the increased pressure on universities to adopt the third role, and work actively both in society and the economy. However, the majority of studies focus on evaluating the extent to which universities fulfil their third role, mainly regarding how and how much they cooperate with the private (application) sector, namely, narrowly conceived technology transfer (Anderson *et al.* 2007; Chapple *et al.* 2005). Another line of inquiry focuses on the role of geographical proximity for the cooperation between universities and industry and whether universities are becoming a major engine of regional innovation systems (see also Gulbrandsen *et al.* 2011). Much less attention has been paid to how exactly the universities themselves fulfil this role or why they decided to implement it. Recently, Srinivas and Viljamaa (2008) followed this path, analysing the third role of the Finnish Turku University. They argue that universities cannot be expected to deliver automatically the third role demanded and that it is a slow, demanding process which can hardly be generalized into a 'one-size-fits-all' structure. Rather, they stress it needs to be analysed on a case-to-case basis, also taking into consideration the specific institutional set-up of the regions and national economies where the universities are located.

This chapter aims to take this analysis further, by following the university perspective on the development of the third role in a specific institutional context characterized so far by very limited university–industry links – that of the Czech Republic (see e.g. Blažek and Žížalová 2010; Žížalová 2010). While Finland (and other countries and regions mostly subject to case studies on this topic – e.g. the UK, Germany, the USA etc.) belongs to one of the most innovative countries with above-average intensity of science–industry links (see e.g. the result of the latest Innovation Union Competitiveness Report 2011 where Finland scores among the top countries regarding, for instance, the number of public–private co-authored research publications in the Web of Science database or share of business R&D expenditures financing public research, EC 2011), the Czech Republic (hereafter referred to as Czechia) is almost at the other extreme. According to the European Innovation Scoreboard 2009, it ranks among the group of moderate innovators, with innovation performance below the EU27 average and whose very limited science–industry links is one of the key bottlenecks of the national innovation system (Žížalová 2011a, 2011b; Žížalová 2010).

Besides, the modern innovation policy does not have a long tradition here. Only recently the separate R&D policy and innovation policy were incorporated into the National Research, Development and Innovation Policy 2009–2015, relying heavily on copying foreign best practices, without a comprehensive understanding of underlying preconditions and without the necessary adaptation of such practices to the specific characteristics of particular national and regional innovation systems (Blažek and Žížalová 2010; Blažek *et al.* 2011; Tödtling *et al.* 2010; Žížalová 2010). Only now, more intensive discussions both at the national and regional levels about a need for more strategic and 'fine-tuned' approaches in designing innovation and competitiveness strategies has started, related to the overall reform of the research, development and innovation (R&D&I) and the higher education system. However, the discussion is still ongoing, precluding forecasts of any concrete results and changes.

Thus, this chapter aims to add to the discussion on the adoption of the third role by analysing universities embedded in a less advanced innovation system with a rather unclear and changing institutional setting and innovation policy. Srinivas and Viljamaa (2008) argue that we cannot expect universities to adopt the new roles and functions that society demands automatically and without difficulties, especially in the much less favourable and more unstable environment of the Czech innovation system. Despite the unclear outcomes of the current discussion on the R&D&I system setting in Czechia, it is evident that universities will have to build and somehow incorporate a 'third role'. The first question of the chapter is, therefore, how are the universities motivated to become more involved in cooperation with the application sphere and economic development in general, and how do they adapt to these new roles. The question will be analysed not only at the university level, but also at the level of individual researchers: how, when and why they engage in these activities.

Next, the chapter aims to push this discussion further and evaluate how the adoption of the third role by a university might affect its 'traditional' functions – especially basic research and teaching. The reason lies in a critique of the feasibility of the new role for the universities (or public research in general). In public discussions of the reform of the Czech research system, some researchers have raised worries that a strong emphasis on the third role will mean a departure from basic research towards intensively application-oriented research and development related to the current needs of local businesses. This may restrict open science, where knowledge flowing freely brings greater social benefits than when channelled into technology transfer (Eisenberg and Nelson 2002). On the other hand, it is worth noticing that this is only one side of the coin as also our analysis shows that cooperation between science and business often leads to mutual positive effects. The problem lies in the fact that many researchers do not see these benefits and the government pushing them into cooperation with industry is not capable of explaining and demonstrating them clearly.

Contrary to the majority of current studies, the article does not focus only on 'fashionable', modern disciplines such as biotechnology or other so called high-tech industries (Haeussler and Colyvas, 2011; Hirsch-Kreinsen and Jacobson 2008; Kohler 2008). Rather, our aim was to cover a broad range of scientific disciplines ranging from traditional technical and natural science disciplines to modern ones, such as nanotechnology and biotechnology, as well as to social and human sciences. Particularly, the latter have been largely overlooked despite the growing proclamations of the importance of non-technical or public sector innovations for competitiveness (Gulbrandsen *et al.* 2011).

The chapter will proceed with a description and evaluation of the national context for the development of universities and their functions in Czechia. Next, the methodology, which combines extensive desk research and a large number of semi-structured interviews with various actors, will be described. The fourth and fifth sections will look at how and why the universities engaged in the third role, how they motivate (or not) their researchers to cooperate more intensively with the application sphere and what the key barriers are for the effective adoption of the third role. The chapter concludes with a discussion of the appropriate policy response to help motivate universities to adopt the third role but at the same time sustain the quality and competitiveness of basic research and teaching.

2 The national context for the adoption and development of the third role functions of universities

The role of universities in the economy and society cannot be fully understood without reference to the local and national context (Harloe and Perry 2004; Srinivas and Viljanaa 2007). The political and social shift in Czechoslovakia after 1989 also brought fundamental changes to the higher education system. During the previous communist regime most research was carried out only at the

Academy of Sciences or several branch research state-owned institutions oriented towards applied research. These companies were subsequently privatized and usually, unable to withstand the competitive pressure, gradually left the market (Blažek and Uhlíř 2007). The academic freedom and self-governance of universities were restored and research activities were strengthened when new legislation defined their role as both higher education and scientific institutions. Thanks to their newly gained autonomy, the universities slowly started fulfilling their roles, though without previous links to industry.

Consequently, this also led to the establishment of new faculties and new regional universities, loosening the previously high concentration of universities in the largest metropolitan centres. The first wave of this process occurred in 1991 when four regional universities outside the main centres were established. While some were established almost from scratch (for example, the Silesian University in Opava), others could follow in the tradition of existing higher education establishments. The three youngest public universities are the University of Tomas Bata in Zlin (founded in 2000), College of Polytechnics Jihlava (founded in 2004) and the Technical University of Economics in České Budejovice (established in 2007). Currently all the self-governing regions with the exception of the smallest, Karlovy Vary, have at least one public university or faculty.

One of the most important reasons for establishing regional universities is to encourage the inflow of educated people into the region and prevent a brain drain to traditional higher education centres, thus indirectly stimulating the level of regional culture (OECD 2007). Nevertheless, the establishment of regional universities was not directly supported from the central level, nor were the universities seen as the national level instrument to support socially or economically underdeveloped regions (OECD 2007).

In general, the Higher Education Act (in §1) states that 'colleges are the highest educational system's top centres of education, independent knowledge and creative activity and play a key role in the scientific, cultural, social and economic development of society' but no strategic document obliges them to participate actively in (regional) economic development. Related activities supported by the Ministry of Education are mostly of a recommendatory nature.

The reform of tertiary education and the growth of universities, together with a very low proportion of tertiary educated population and particularly the above-average and growing economic return of higher education (Večerník 2001, 2006) led to a growth in the number of students enrolling for university studies. The growth of student numbers visible in Table 12.1 was also supported by the funding mechanism for higher education (Matějů *et al.* 2006; OECD 2007). The basic part of the budget of public universities is composed of a state subsidy grant for teaching activity, calculated from a formula combining the number of students and the financial assessment of each accredited programme (i.e. a coefficient reflecting the differences between disciplines and fields). No link to the results or quality measures was considered, which led to an expansive growth in the number of university students, without any significant reform of the institutions.

Table 12.1 University development in Czechia – the number of public universities, their students and graduates and R&D expenditures between 1990 and 2010

			School year		
	1990/1991	*1995/1996*	*2000/2001*	*2005/2006*	*2009/2010*
Number of public universities	23	23	23	27	28
Number of faculties	82	110	111	127	134
Number of students enrolled	118 194	173 945	228 423	289 571	389 231
Number of graduates (Ms)	15 693	15 156	20 778	24 200	34 160
			Year		
	1998	*2000*	*2005*	*2010*	
Higher education R&D expenditures as % of GDP	x	0.11%*	0.17%	0.23%	0.28%
Share of higher education R&D expenditures in total R&D expenditures	x	9%*	14%	16%	18%
Share of business sector financing higher education R&D expenditures	x	1.0%*	1.1%	0.8%	1.1%

Note: *indicators only from the year 1998 are available.

Source: CZSO, UIV – Institute for information on education.

The growth of the universities did not only include the increasing number of students and expansion of teaching activities but also involved research and development activities. The research capacity of the universities grew steadily after 1990, closing the gap between higher education and government R&D expenditures (10.6 and 11.4 billion CZK in 2010). While the institutes of AS CR remain focused on basic research, according to official R&D statistics, universities realize more applied and industrial research. Despite this character of the university research, the share of private funding of Czech universities research remains very low. According to CZSO data, industry finances approximately only 1.0 per cent of the total R&D funding at Czech universities. These figures are considerably higher in other countries, for instance 7 per cent in Finland, 6 per cent in Austria and Estonia or 14 per cent in Germany (CZSO 2010).

Even though universities are formally expected to diversify their financial sources, and gain other funding from supplementary activities, they are still highly dependent on state subsidies. According to a recent OECD study, the total share of extra-budgetary revenues varies on average around only 15 per cent. There are three universal sources of these revenues: revenues from services for students and study-related fees, revenues from property and revenues from research and development activities and services (OECD 2007). The latter is still the least important source of revenues for the majority of the Czech universities and faculties (see also Marek and Žížalová, 2011).

Thus, despite the vast expansion of universities, their system of financing and other organizational aspects has not much changed since the first 'reform' in the mid 1990s. On the other hand, the environment in which universities operate has shifted fundamentally, creating both financial and structural deficiencies in the system. According to Matějů *et al.* (2006), the current situation is largely owing to the unwillingness of the universities themselves to change, for instance, allowing competition for students among the various universities to be introduced in the system of financing and evaluation.

The current status quo is, however, being further challenged, not only by the demographic changes and state budget constraints, but also by the construction of new large research infrastructures supported by EU Structural Funds. According to the latest information available (August 2011), eight large research centres and thirty-three regional research centres with a total investment of 42 billion CZK (1.7 billion EUR) are to be built and start operating by 2015. The centres expect to need about 25 billion CZK (1 billion EUR) between 2011 and 2020 for their operation costs, that is, 2.5 billion CZK yearly or in other words 4 per cent of total R&D expenditures in 2010 (between 2009 and 2010 the total R&D expenditures in Czechia grew by 3.6 billion CZK, approximately 6 per cent; however, this was only in the business sector while public expenditures decreased slightly).

The existing pressure on universities to reform and also diversify their financial sources, which has led to several unsuccessful attempts to pass new regulations (for details see Matějů *et al.* 2006), will therefore only intensify and the universities will need to reform and adapt unless they want to lose their competitiveness.

3 Methodology

The empirical analysis is based on a combination of a desk research analysis and field survey. The desk research focused on the analysis and evaluation of available national strategic R&D and innovation policy documents and of university strategic and conception documents, as well as internal regulations. First, publicly available documents were searched and analysed; later, all the public universities' managements were asked to provide their documents related to the key third role functions, namely, university–industry cooperation, knowledge and technology transfer and researchers' motivation schemes.

The desk research was complemented by extensive field research consisting of an electronic survey among university representatives (owing to high autonomy conducted at the faculty level), focused on the availability and content of the university strategies, policies and directives in the field of knowledge transfer and cooperation with industry. All the universities were approached at the faculty level owing to the high level of their autonomy.

The functionality of knowledge transfer strategies and incentive systems cannot be assessed only by their existence, content or settings, but it is necessary to assess how these systems are actually implemented and used in practice, that is, by individual researchers for whom they are indeed aimed. In order to evaluate the functionality of the knowledge transfer system at the university level, its effect in motivating researchers to cooperate with the application sphere, as well as its impact on their research and teaching activities, an extensive interview survey was carried out among leading researchers, covering all the main academic disciplines, public universities and regions. Therefore, the sample is considered highly representative and relevant. The experiences of knowledge transfer offices (KTOs) were also evaluated through direct interviews with their managers.

Altogether, we received eighty-three responses from the faculty management (out of 152 requests sent) and carried out seventy-six interviews with researchers and ten[1] interviews with technology transfer offices (centres) of the major public universities. The structure of the interviews and responses according to the main academic disciplines is described in Table 12.2.

When choosing the researchers to be contacted and interviewed, several criteria were applied. First, the aim was to cover at least one interview at each of the key public universities (in terms of their research activities and results), while covering the widest possible regional pattern. Second, researchers who have achieved important, above-average results were selected. This criterion consisted of three steps: (i) identification of research sub-fields with above-average R&D results (standardized publications record on Web of Science and patent activity observed in the three patent offices: the Czech Office of Industrial Property, the European Patent Office and the US Patent Office) (ii) identification of organizations with above-average contributions to the science sub-fields and (iii) identification of researchers with the strongest results for the two above-mentioned lists. This method of assessment omits social sciences and humanities; therefore, researchers with excellent results from these academic fields were added to the sample. In total, 106 researchers from sixty-one research workplaces (faculties or specific research units at universities) were contacted.

In addition to the tailor-made research for the chapter, the authors built on knowledge accumulated through various forms of personal involvement in the processes of drafting and evaluating R&D and innovation strategies and policies, both at the national and regional level.

Table 12.2 Structure of the responses and interviews

	Natural and agricultural sciences	*Engineering and Technology*	*Medical and Health Sciences*	*Social sciences and humanities*	*Total*
Number of survey responses (no. of faculty responses)	13	29	9	33	84
Number of researchers interviewed	33	31	8	4	76

4 Universities and the third role

4.1 Strategic level and university management

It is widely accepted that knowledge transfer, the scope of science–industry interaction and researchers' behaviour is determined by the specific institutional environment of individual universities and the setting up of the national innovation system (Bercovitz and Feldman, 2003; D'Este and Fontana 2007; D'Este and Patel 2007; Schartinger *et al.* 2001). Therefore the following paragraphs focus on illustrating some fundamental changes in the national R&D&I system and their drivers, and universities' response to them.

During the last decade, the importance of knowledge transfer (and other third role activities) has been increasing for Czech universities (and other research institutes as well) because of several, mostly external, drivers. The first lies in the need to reflect the growing attention paid to more intensive collaboration of R&D institutions with the business sphere in EU strategies, such as the Lisbon Strategy, and in national R&D&I policy documents. Though such a shift is visible in Czech national policies, the reflection remained to a large extent on a rhetorical level rather than in practical implementation of concrete programmes supporting science–industry links and knowledge transfer. Thus the main contribution can be seen in the increased awareness and discussion of the necessity and benefits of cooperation with private enterprises among university management members.

The second driver for the growing attention paid to the third role of universities is the EU Cohesion Policy, where the role of research and innovation activities has fundamentally increased in the last two programming periods (Barca 2009). Czech operational programmes (OPs) co-funded by the EU structural funds partly incorporated this idea and launched some measures that enable universities and public research institutes to establish and further develop their knowledge transfer offices. With some modifications, such support has been provided in the present programming period 2007–13.

Despite this shift in the perception of the third role of universities, not only in the OPs but lately also within the planned reform of tertiary education, neither the existing University Act nor the draft of the act amendment currently being prepared contain any obligations or objectives for the development and employment of the third role of university practice. As a result, support for the development of the third role is still very passive and formal, with two general implications for our chapter. First, there is no motivation or a legislative push for the universities to employ their third role and engage more intensively in cooperation with the application sphere. Second, universities, particularly their management, but also a large share of the researchers, do not consider their third role as important as their research and education activities. As a result, commercialization activities and cooperation with industry are considered only as a minor (supplementary) economic activity and are not systematically and actively promoted.

The absence of legislative regulations and the free choice of universities to engage or not in knowledge transfer and cooperation with industry are also reflected in a very general description of knowledge transfer activities and their goals in the universities' long-term strategies. These strategies are the main strategic documents specifying developmental objectives and measures for individual university activities. Their preparation is obligatory and has to be approved by the Ministry of Education responsible for tertiary education and research activities. The strategies accept the importance of knowledge transfer and science industry links for development of research and teaching activities; nevertheless, the absolute majority of them only adopted general objectives of the long-term Ministerial strategy without specifying their own objectives, particular activities, resources and responsibilities. These general objectives state that universities will support participation in the operations of science and technology parks; build technology transfer centres and business incubators; design study programmes/fields of study with regard to employers' needs; or create conditions for spin-off from the academic community. However, they do not mention how, when and under whose responsibility these activities will be carried out.

Except for the overall long-term strategy, none of the Czech public universities have an additional strategy (or specific documents that could be considered as a strategy) related to third role activities, in particular knowledge transfer and cooperation with industry. The long-term strategy is usually further supplemented only with various technical directives dealing with the regulations and formal procedure of IPR and commercialization. These documents specify (in more or less detail) the process of intellectual property protection and commercialization, and rules and steps that university management, researchers and knowledge transfer offices should follow in this process. Nevertheless, there are several universities without any such directives, although the Act on R&D Support clearly defines them as obligatory for organizations receiving public support for R&D activities. These universities explain the absence of such directives by a long tradition of an intellectual property rights (IPR) regime that does not need to be codified in any formal document. Similar opinions were expressed by the managers of those universities where the technical regulations and directives are very brief, usually not exceeding one or two pages of text. While this is the view of the university management interviewed in our analysis, the researchers hold a distinct opinion as will be explained later in the chapter.

4.2 Knowledge transfer offices

In general, knowledge transfer offices play an important part in the stimulation and deepening of university–industry collaboration, carrying out commercialization of R&D results as well as providing feedback to researchers in terms of the applicability of their knowledge (Friedman and Silberman 2003; Mansfield and Lee 1996). Such importance of knowledge transfer offices has been positively perceived by Czech universities in recent years.

In spite of the above-mentioned slack attitude of the universities towards a strategic framework for knowledge transfer, the majority of them have established or plan to establish their knowledge transfer offices (KTOs) in the near future. The establishment of the KTOs was mainly initiated by the above-mentioned OPs during both programming periods (2004–06 and 2007–13). An overview of the KTO projects supported by EU funds as well as other sources is shown in Table 12.3. Currently, seven out of twenty-six public universities are disposed with specific KTOs and ten others are preparing them.

In several cases, the KTOs supported during the years 2004–06 used EU funds again in the current period for further development, despite the fact that the universities with such projects had to declare they were able to ensure their own financial funds for sustaining the project activities after project completion, or to guarantee sustainability by providing revenues from knowledge commercialization. The reason lies probably in the above-mentioned lack of interest of the university management in real knowledge transfer activities and their active support. As a result, further development of the established, existing KTOs and their services is considerably limited. In order to provide at least elementary knowledge transfer services and be able to keep knowledge transfer specialists, the KTOs themselves apply for publicly funded projects which would compensate for the lack of university financial resources, as the interviews revealed. Unfortunately,

Table 12.3 University knowledge transfer offices supported by public programmes during the years 2004–11

Source (funding)	*Number of KTOs newly established by public universities*
Operational programme Industry and Enterprise	2
Joint programming document 2 (Prague)	1
Operational programme Research and Development for Innovation	10 (applications)
Operational programme Enterprise and Innovation	4
Operational programme Prague Adaptability	1

Source: own analysis.

the majority of the programmes available are not directly focused at the realization of knowledge transfer services themselves (i.e. realization of concrete knowledge transfers or cooperation with industry), but rather on human resource development activities. Consequently, the KTOs carry out plenty of lectures, training courses or workshops related to IPR and commercialization, but few real knowledge transfer activities. These are the activities for which they were (or should have been) established.

5 Third role and university researchers

Contrary to the vague formulation in official university strategic documents, the researchers generally view knowledge transfer and cooperation with the application sphere as positive and beneficial, both for research and teaching. Surprisingly, this finding is in contradiction with researchers' official proclamations given in public discussions mentioned above.

The researchers collaborate with firms for a diverse set of reasons. Similar to the motives identified by D'Este *et al.* (2005), Howell *et al.* (1998) or Meyer-Krahmer and Schmock (1998) in universities from the most developed EU countries, reasons for collaboration include additional research income and the applicability of research results. Nevertheless, other important motives such as access to industry skills and facilities or keeping abreast of industry problems are not considered as important as the motives of foreign researchers. The emphasis on financial motives for cooperation expressed by many interviewed researchers is in line with the shortage of public financial resources and their efforts to seek money from other sources including science–industry collaboration or participation in EU framework programmes.

Most of the interviewed researchers engage in knowledge transfer and collaboration with industry (see Table 12.4). Only fifteen out of seventy-six researchers mentioned that they do not realize these activities and focus strictly on theoretical, non-oriented research and research results. Literature mentions several forms of cooperation and knowledge transfer, the most frequent being meetings and conferences, consultancy and contract research, the creation of physical facilities (spin-off companies, new physical facilities), training (postgraduate training in company, training company employees) and joint research (D'Este and Patel 2007; Faulkner and Senker, 1995; Schartinger *et al.* 2001). Cooperation is often carried out on an individual basis, and stimulated by personal links between researchers and firm representatives, who are, in several cases, former schoolmates or graduates from the same university. The involvement of knowledge transfer offices in the initiation of cooperation has been rare so far.

Although heterogeneity in the form of links between universities (researchers) and the application sphere was identified, the variety of knowledge transfer channels that actually come into action and were mentioned is limited, as is indicated by the literature or foreign practices (see e.g. Larsen 2011). Two large areas encouraging links between universities and the application sphere and engagement in the third role were missing in the responses – (i) activities regarding

Table 12.4 Character of the research–industry cooperation

Type of cooperation	*Number of researchers interviewed – domestic cooperation*	*Number of researchers interviewed – cooperation with subjects located abroad*
Contracted research*	21	9
Join projects (domestic, international)	17	8
Licence purchase	3	8
Consultations, methodologies applied	22	10
Spin-off companies**	8	
No cooperation	15	

Note: Open questions caused various forms of cooperation and knowledge transfer to be mentioned in the interviews. These forms were later standardized by the interviewers into the above-mentioned types; multiple answers remained possible. *Contracted research includes very diverse activities varying from simple measurement to relatively broad applied research projects. **Spin-offs do not include only real spin-off companies with a share of research organizations in their assets, but also companies founded by researchers in order to implement knowledge transfer through them. Only two cases represent the spin-off by definition.
Source: own analysis.

personal mobility (e.g. staff exchange, holding double positions, recruitment policy) and (ii) training opportunities (e.g. research organizations training industry staff, industry funding PhD students). Probably, the reason lies in the very conservative nature of universities, the unclear and insufficient competencies and responsibilities given by the Act, and the lack of state incentives for development and utilization of these channels.

The majority of the researchers who cooperate with industry were also able to explain the benefits of such links and whether and how they affect the research activities. First, it is important to mention that all the researchers indicated basic, theoretical research as their primary goal and mission, prior to knowledge transfer and commercialization of their results. They are propelled by a drive, in some cases accompanied by a desire to see their solutions realized in practice. The latter activities are seen as a potentially fruitful, beneficial and/or inspirational supplement, yet not by all the researchers interviewed. Those researchers who engage in cooperation with industry were able to describe the motives for and benefits of such activities, though some of them were critical towards the character of the cooperation (see below).

First of all, according to the interviews, the links and cooperation with the application sphere represent significant revenues for the university, faculty and/or research team budget and, unlike public funding or grant funds, can be used without restriction. Second, the researchers view the application of their knowledge and cooperation with the application sphere as an important source of new ideas, new lines of questioning or new problems to be solved by their research.

To put it differently, the researchers find the basic research and links to industry complementary.

Satisfaction with the content of these links, however, varies among the addressed researchers. Many of the researchers (approximately one third of the interviewed researchers) do not find current demand from the application sphere challenging enough for a combination of theoretical knowledge and applied research. According to their experience, the majority of the companies with which they cooperate demand only relatively simple, standardized services. In such cases the cooperation does not provide the inspiration sought for further research and the researchers even see it as making use of cheap university services instead of full-price market ones.

Yet, even where cooperation with industry is not fulfilling, the links with the application sphere are seen as an important part of teaching. Students can and very often also do participate in projects with industry (this example was spontaneously mentioned by eleven researchers interviewed). They offer them a chance to test their theoretical knowledge in practice and gain very practical experience, contributing to their future employability. If the university is available to offer such cooperation with industry to its students on a more regular basis and promote it sufficiently, it can also help increase its attractiveness and pull in more new students and talent, a factor that is becoming more and more important owing to the current unfavourable demographic trends. Additionally, most of the surveyed research teams are still gaining new researchers from among their doctoral students. Therefore, the ability to attract the best ones, the most talented ones, comprises an important factor for the long-term development of research teams.

Hence, in general, researchers view the third role of universities positively and emphasize the benefits of university–business collaboration for their research, development and teaching activities. At the same time, they stress the necessity of favourable conditions in the whole knowledge transfer system of universities. They often argue that they would cooperate with the business sphere more intensively on condition that the system provided them effective services and motivated them to cooperate in creating applicable R&D results. The insufficient and ineffective knowledge transfer system leads many researchers to their own commercialization of their knowledge. In several cases individual researchers commercialize their knowledge in an unofficial (and illegal) way without informing the university management of the commercial potential of their knowledge, the commercialization process and revenues. Thus, researchers admit that often the pursuit of transferring their R&D results into application drags them away from the research itself. On the other hand, if they want to stay fully concentrated on the research, they often have to abandon the desire to follow through to practical applications.

Owing to inefficient and deficient knowledge transfer services, researchers have to spend much more time dealing with these activities on their own, at the expense of their primary activities – research and teaching – and without the proper knowledge, skills and experience that 'professionals' would have. This adds to the findings of Thursby and Thursby (2011), that individual researchers' involvement in knowledge transfer activities does not lead them away from basic research. A few examples of excellent researchers with above-average theoretical R&D results and several applications, including a spin-off company in our sample confirm these results. However, in all of them, the above-mentioned conclusion of Thursby and Thursby (2011) is valid, according to our results, only because these researchers are disposed with quality support services, for instance by the KTOs. As such, more intensive emphasis on third role activities could be in conflict with the need to continue the quality of traditional functions, research and teaching, unless a supportive environment is created.

5.1 Motivation to engage in the third role

To minimize the risk of bypassing the official paths, free-riding or selection of non-optimal, effective and in-line-with-the-R&D-organization-strategy option of knowledge transfer, the interviews revealed the need to create a motivating, transparent system for the work of the researcher – their research and teaching, but also knowledge transfer.

The active participation of researchers in strengthening the third role can be stimulated in various ways, with both financial and/or non-financial tools. Examples of such tools are shown in Figure 12.1. The scheme of motivation tools is based on our own analysis of the existing university incentive schemes described or actively used.

Financial motivation means that creation of applicable or commercialized knowledge (R&D results) is rewarded by a share in the revenues from commercialization (percentage or fixed rewards) or merit awards (extra bonuses). Other types of financial incentive can include more indirect support such as ensuring better R&D equipment (modern laboratories, appliances, etc.), reflecting the quality of the R&D results of the researchers and their revenues, profits and/or savings.

Building an academic career is a transitional type of incentive between the two main groups mentioned because it can combine both financial stimuli (higher wages, merit awards) and personal satisfaction from building a career and creating applied R&D results.

In general, non-financial motivation types stem from the ethical, moral and altruistic attitudes of researchers, or reflect traditional and codified academic liberties. Apart from these attitudes, the availability of knowledge transfer services provided to researchers and their IPR awareness seems to be important as well. The lack of information on IPR and the underdeveloped knowledge transfer services can discourage researchers from commercialization of their knowledge, and this was also expressed in the interviews.

Almost all the Czech public universities analysed have some kinds of the above described mechanisms to motivate researchers to create applicable R&D results and cooperate with companies and further subjects such as hospitals, ministries, from the application sphere. Usually, however, the motivation schemes are not

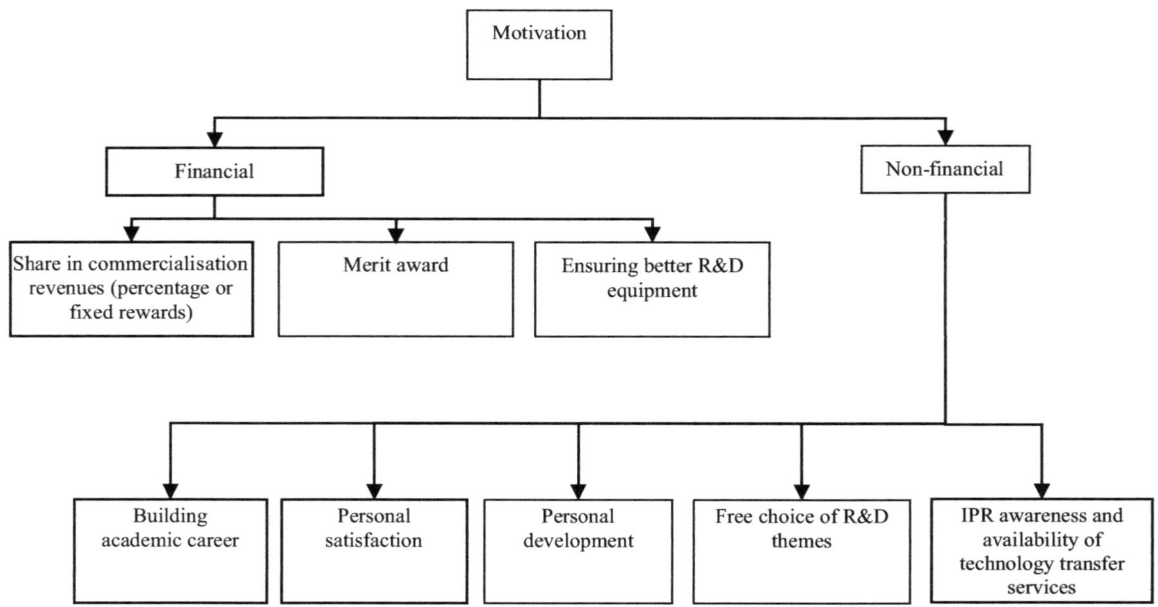

Figure 12.1 Typology of possible motivation schemes for knowledge transfer activities.

Source: own analysis.

codified in one university regulation but are non-governmental organizations, etc. fragmented into several internal university documents or even left informal and uncodified. The key documents and regulations include university strategies, university statutes, wage directives, IPR and knowledge transfer directives, and ethical standards. Exactly how and where they are formulated differs from university to university – with their strong autonomy, it is really up to them how they define the inner regulations. As a result, in some cases these schemes are defined at the university level, usually in case of smaller, regional universities. In case of the largest, traditional universities (such as the Charles University in Prague with more than 600 years' tradition), the possibility of preparing such regulations is left at the department level. So, for instance, the department of inorganic chemistry has a different reward system for its employees than the department of human geography, both being part of the Faculty of Sciences. On the other hand, there is one central KTO for the whole university and some of the regulations are at the same time also centralized. Several researchers also mentioned that the official university, faculty and/or department documents do not actually describe the motivation scheme as a whole, as in part they are left informal or unofficial, leaving space for speculation, subjective evaluation and even potential for envy and bad personal relations in the workplace.

Further, the interviews also revealed low awareness of the existing motivation schemes in general. Through the desk research, we were able to find at least some, though sometimes very formal and short, official documents describing motivation schemes at all the universities. Surprisingly, there were only thirty researchers (less than a half of those interviewed) who mentioned a certain form of motivation used and codified at their universities or faculties. Such low awareness can result from the above-mentioned fragmentation of the regulations, which are described in many varying documents at varying levels of the organizations. Additionally, the internal regulations are very often not promoted or passed over directly to the researchers or the administrative personnel who should be helpful to researchers in such cases, but published only by posting them on the organization's intranet, with often few yearly amendments.

The incentives mentioned during interviews with researchers and KTO managers as being used and known can be summarized as follows (the number of researchers who mentioned such a motivation scheme are indicated in brackets):

- rewards consisting of wage costs planned in individual R&D projects. If the researcher is able to obtain publicly funded R&D projects, project wage costs are considered to be a reward, not a regular wage. The university does not determine the highest level of the reward at all (7 researchers);
- particular additional rewards for specific research and applied results. A part of commercialization revenues is distributed to researchers as a percentage or fixed reward. Excellent and purely R&D results are rewarded as well, often preferentially (12 researchers);

- evaluation of researchers (quality assessment of their results). Activities (mainly learning) and R&D results of individual researchers are evaluated regularly and researchers are rewarded according to the evaluation. The evaluations mostly include traditional university functions – pure R&D results and teaching activities (7 researchers);
- indirect support by ensuring high-quality knowledge transfer services (2 researchers).

Comparison of these stimuli and the motivation scheme and its tools shown in Figure 12.1 indicates that universities, at least according to their researchers, apply few tools from the whole motivation schemes given in their developmental documents. The main weakness of the financial incentives, mentioned by researchers, consists in the decision-making process for distributing financial bonuses among researchers. In many cases the bonuses are assigned by their direct management according to unclear rules. Another important finding is that researchers, with only minor exceptions, do not mention the use of the non-financial tools, especially those developing academic careers and the social importance and responsibility of researchers.

Surprisingly, researchers and even technology transfer specialists do not mention any incentives for KTO specialists who are responsible for carrying out knowledge transfer processes and interaction with the business sphere. Because the quality of knowledge transfer services is perceived as rather low, and researchers note the necessity of this incentive, the absence of motivation for KTO specialists can be seen as one of the major barriers to third role development.

Overall, from the research carried out, it is apparent that Czech universities build their strategy and development around the traditional maximization of their institutional funding (i.e. funds from the state budget), allocated to universities principally according to their publication activity (the number of papers in scientific journals, books, etc.), namely, basic R&D results. The universities create motivation systems leading to a higher number of eligible publications and other types of R&D results, omitting other research and research-related activities. As a result, researchers are under pressure to publish as many papers as possible, while relatively low attention is paid to the development of science–industry links.

6 Conclusion: how to stimulate universities and researchers to commit fully to cooperation with the application sphere?

The results of our research show that researchers and universities differ profoundly in their opinions on university–industry links, knowledge transfer or the third role more generally. Whereas the mindset of researchers at a lower level towards knowledge transfer activities is generally positive and the interviewed researchers were able, in most cases, to clearly describe the benefits of such activities, the university and faculty management is more conservative, seeing the key mission of their organizations in basic, theoretical research and the high university autonomy in the orientation of research topics and study programmes.

In the majority of Czech universities, basic research predominates over applied research, although many members of the academic staff would prefer applied research and more intensive cooperation with the application sphere.

Such an attitude among the university management and also some of the researchers is encouraged by the evaluation methodology for research organization used for the distribution of institutional R&D funds. The methodology is purely quantitative in nature and is based only on the quantification of R&D outputs generated by individual researchers. In general, institutional R&D funds are allocated according to the number of points received on the basis of the quantity of R&D outputs. Although the range of R&D outputs includes outputs typical for applied research as well, scientific papers in journals with impact factors and books received the highest score in the evaluation. Consequently, the application of the methodology, and the university management, force researchers to produce as many papers and books as they are able in order to maximize institutional R&D funds, which create a decisive part of university revenues. As a result, researchers do not have the time and capacity to deal with the generation of applied R&D results.

The prevailing view of the university management leads to a clear distinction between basic research and applications (which include more applied, oriented research, knowledge transfer and cooperation with industry other than pure basic research) rather than a complementarity stated by the researchers interviewed. Many of the university managers addressed in our research consider that basic research and knowledge transfer activities cannot be pursued effectively at the same time, they have to be carried out on an 'either–or' basis. With the existing R&D support system in Czechia in mind, it is obvious for them to concentrate all their effort on basic research and teaching and leave room for more applied oriented activities only where they do not collide.

Many researchers and university managers declare that the excessive stress on traditional university activities results from the failure of the government to create the necessary conditions for the initiation and development of science–industry collaboration. Up to now, policies at all levels do not provide clear and specific priorities for basic or applied research, neither do they contain strong measures to directly promote science–industry links. The policy only declares the necessity of cooperation, without specifying the individual steps leading towards stimulation and further development of university–business collaboration.

Although the policy level plays an important role in the support for the third role, our research confirms that the role of the universities themselves is undeniable. With probably just a little more support and motivation, the interviews have shown that researchers (or at least the majority) are willing to engage in these activities and would actually welcome a more active part played by their organizations. So, what essentially should the universities realize in order to support third role development and at the same time not jeopardize their primary roles?

The fulfilment of the third role demands greater involvement of management and strategic governance than mere research or teaching activities. Up to now (in the case of Czech universities) the universities have been managed by open scientific and academic freedom rules. Together with the high autonomy of faculties and even their departments, both these factors contribute significantly to the low intensity of information transfer from the university management to individual researchers, resulting in their low awareness of the university's strategic visions and goals, as well as personal management tools (e.g. low awareness of the motivation and incentive system for researchers).

The current rapid changes and shift in society bring a new challenge here. The universities now need to grasp their development more firmly and start guiding the organization, not just administering it. Only through such strategic management and guidance will universities be able to achieve the necessary synergies between the various activities they engage in, particularly research, teaching and knowledge transfer, and the expectations of society and the economy. Particularly, the growing heterogeneity of the linkages in the innovation systems calls for stronger coordination and management. Otherwise, ad hoc, solitary solutions will very probably mean the fragmentation of constrained resources, both human and financial. Therefore, the first recommendation arising from our research is for universities to start to think of themselves as any other democratic organization – and thus to leave enough freedom so as not to hinder their creativity and free spirit, but at the same time to define and enforce clear, transparent rules of the game, outlining not only how to play but also the end goal.

In order to carry out third role activities without a negative impact on the productivity of traditional ones, universities should not only provide researchers with clear game rules and motivate them with a joint vision, but also offer them reliable and effective support. More concretely, this means offering them knowledge transfer and project administration services that would relieve them of those activities not directly linked to their formal responsibilities. Thus, instead of having to deal with project administration, promotion, market studies and so on, they can focus their efforts on what they are best at – research and teaching.

Moreover, the engagement of researchers in science–industry collaboration and effective involvement of knowledge transfer specialists should be stimulated and supported by an appropriate incentive scheme providing them with a wide range of suitable tools for both. The researchers should be motivated through a variety of transparent and accepted financial and non-financial tools and evaluation mechanisms, which would take applied R&D results into consideration and would be in line with the goals the university has set up in their strategic plans. The reason for also motivating technology transfer specialists lies in the necessity to cover the whole knowledge commercialization and transfer process, not just its beginning. The third role of universities cannot be seen separately but as an important source of new ideas, information, skills, specialists and finance for teaching and research. As such, it is an integral part of the main activities – teaching and R&D. The third role provides feedback for university activities and helps universities to enhance their position in society and the economy as a whole. Since third role development depends on people and their thinking, it is not a short-term issue but a long-term enterprise.

Note

1 At the time of the research being carried out, there were sixteen technology transfer offices operating in Czechia.

References

Anderson, T. R., Daim, T. U. and Lavoie, F. F. (2007) 'Measuring the efficiency of university technology transfer', *Technovation*, 27, 306–318.

Barca, F. (2009) 'An agenda for a reformed cohesion policy: A place-based approach to meeting European Union challenges and expectations', Independent Report prepared at the request of Danuta Hübner, Commissioner for Regional Policy.

Bertzovitz, J. and Feldman, M. (2003) 'Technology transfer and the academic department: Who participate and why?' Paper presented at the DRUID Summer Conference 2003, Copenhagen, 12–14 June.

Blažek, J., Uhlíř, D. (2007) 'Innovations and innovation policies in the Czech Republic: The case of Bohemian regional innovation strategy', *European Planning Studies*, 15, 871–888.

Blažek, J. and Žížalová, P. (2010) 'The biotechnology industry in the Prague metropolitan region: A cluster within a fragmented innovation system?' *Environment and Planning C: Government and Policy*, 28, 887–904.

Chapple, W., Locket, A., Siegel, D. and Wright, M. (2005) 'Assessing the relative performance of UK university technology transfer offices: Parametric and non-parametric evidence', *Research Policy*, 34, 369–384.

Chatterton, P. and Goddard, J.B. (2000) 'The response of higher education institutions to regional needs', *European Journal of Education*, 35(4), 475–496.

Cooke, P. (2001) 'Regional innovation systems, clusters, and the knowledge economy', *Industrial and Corporate Change*, 10, 945–974.

Cooke, P. (2004) 'Evolution of regional innovation systems: Emergence, theory, challenge for action'. in P. Cooke, M. Heidenreich and H. J. Brraczyk (eds), *Regional Innovation Systems*. 2nd Edn. London: Routledge, pp. 1–18.

CZSO (2010) R&D Indicators in 2009. Available online at www.czso.cz/csu/2010edicniplan.nsf/p/9601-10.

D'Este, P. and Fontana, R. (2007) 'What drives the emergence of entrepreneurial academics? A study on collaborative research partnerships in the UK', *Research Evaluation*, 16, 257–270.

D'Este, P. and Patel, P. (2007) 'University–Industry linkages in the UK: What are the factors determining the variety of interactions with industry?' *Research Policy*, 36, 1295–1313.

D'Este, P., Nesta, L. and Patel, P. (2005) 'Analysis of University–Industry research collaborations in the UK preliminary results of a survey of university researchers'. SPRU Report.

EC (2011) *Innovation Union Competitiveness report 2011*. Brussels: European Commission Directorate-General for Research and Innovation.

Eisenberg, R. S. and Nelson, R. R. (2002) 'Public vs. proprietary science: A fruitful tension?' *Daedalus*, 131(2), 89–101.

Etzkowitz, H. and Leydesdorff, L. (2000) 'The dynamics of innovation: From national systems and "Mode 2" to a Triple Helix of university–industry–government relations', *Research Policy*, 29, pp. 109–123.

Faulkner, W. and Senker, J. (1995) *Knowledge Frontiers: Public sector research and industrial innovation in biotechnology, engineering, ceramics and parallel computing*. New York: Oxford University Press.

Friedman, J. and Silberman, J. (2003) 'University technology transfer: Do incentives, management, and location matter?' *Journal of Technology Transfer*, 28, 17–30.

Gibbons, M., Limoges, C., Nowotny, H., Schwartzman, S., Scott, P. and Trow, M. (1994) *The New Production of Knowledge: The dynamics of science and research in contemporary society*. London, Thousand Oaks, New Dehli: Sage.

Gulbrandsen, M., Mowery, D., Feldman, M. (2011) 'Introduction to the special section: Heterogeneity and university–industry relations', *Research Policy*, 40(1), 1–5.

Harloe, M. and Perry, B. (2004) 'Universities, localities and regional development: The emergence of the 'Mode 2' university?' *International Journal of Urban and Regional Research*, 28(1), 212–223.

Haeussler, C. and Colyvas, J. A. (2011) 'Breaking the ivory tower: Academic entrepreneurship in the life sciences in UK and Germany', *Research Policy*, 40(1), 41–54.

Hirsch-Kreinsen, H., and Jacobson, D. (2008) *Innovation in Low-tech Firms and Industries*. Cheltenham, UK, Northampton, MA: Edward Elgar.

Howell, J., Nedevea, M. and Georghiou, L. (1998) *Industry–Academic Links in the UK*. Report to HEFCE, PREST, University of Manchester.

Kohler, H-D. (2008) 'Profit and innovations strategies in low-tech firms', *Estudios de Economía Aplicada* 3(26), 1–20.

Larsen, M. T. (2011) 'The implications of academic enterprise for public science: An overview of the empirical evidence', *Research Policy*, 40(1), 6–19.

Mansfield, E. and Lee, J. Y. (1996) 'The modern university: Contributor to industrial innovation and recipient of industrial R&D support', *Research Policy*, 25, 1047–1058.

Martin, S. and Scott, J. T. (2000) 'The nature of innovation market failure and the design of public support for private innovation', *Research Policy*, 27, 835–851.

Matějů, P., Zlatuška, J., Bartoš, W., Simonová, N. and Basl, J. (2006) *Tertiary Education in the Czech Republic: Main trends and the causes of the current crisis*. Report, Institute of Sociology. Prague: ISEA.

Meyer-Krahmer, F. and Schmock, U. (1998) 'Science-based technologies: University–industry interactions in four fields', *Research Policy*, 27, 835–851.

OECD (2007) *Country Background Report for Czech Republic*. OECD thematic review of tertiary education.

Schartinger, D., Schibany, A. and Gassler, H. (2001) 'Interactive relations between university and firms: Empirical evidence for Austria', *Journal of Technology Transfer*, 26, 255–268.

Siegel, D. S. and Zervos, V. (2002) 'Strategic research partnership and economic performance: Empirical issues', *Science and Public Policy*, 29, 331–343.

Srinivas, S. and Viljamaa, K. (2008) 'Emergence of economic institutions: Analysing the third role of universities in Turku, Finland', *Regional Studies*, 42(3), 323–341.

TC (2008) Zelená kniha výzkumu, vývoje a inovací v ČR.

Thursby, J. G. and Thursby, M. C. (2011) 'Faculty participation in licensing: Implications for research', *Research Policy*, 40(1), 20–29.

Večerník, J. (2001) 'Earnings disparities in the Czech Republic: Evidence of the past decade and cross-national comparison', *Prague Economic Papers*, 10, 201–222.

Večerník, J. (2006) *Evolution or Revolution? Disparities in earnings and household income in the Czech Republic 1988–2002*. IES Occasional Paper: 4/2006.

Žížalová, P. (2011a) *Mapa výzkumného a aplikačního potenciálu ČR*. Praha: Technologické centrum AV ČR.

Žížalová, P. (2011b) *Vědecká excelence a její přenos k aplikační sféře*. Ergo 03/2011.

Žížalová, P. (2010) 'Geography of knowledge-based collaboration in a post-communist country: Specific experience or generalized pattern?' *European Planning Studies*, 18(5), 791–814.

13 Entrepreneurial universities, entrepreneurial students

Higher education and creation of innovation in the Spanish area

Ana Lanero, José Luis Vázquez and María Purificación García

1 Introduction

Encouragement for entrepreneurship education is currently at the heart of a host of political requirements in the countries of the European Union (EU), in an effort to develop a dynamic enterprising culture and foster new firm creation as a source of sustainable competitiveness in the continent (Council of the European Union 2000; Communication 2010/2020 from the Commission 2010). An outcome of that has been the inclusion of the sense of initiative and entrepreneurship in a *European Framework on Key Competences for Lifelong Learning* (Recommendation 2006/962/EC of the European Parliament and of the Council). In this framework, entrepreneurship competence is defined as "an individuals' ability to turn ideas into action. It includes creativity, innovation and risk-taking, as well as the ability to plan and manage projects in order to achieve objectives" (p. 17). From this view, it is recognized that entrepreneurship acts as a source of personal and professional self-realization, active citizenship and social inclusion for individuals, and that is why entrepreneurial competences should be developed by the end of compulsory school or training, acting as a foundation for further lifelong learning.

In this context, contemporary educational systems are seeing their training missions expanded by the assignment of a further responsibility to provide a socio-economic boost, taking the form of channeling future generations of the working population towards entrepreneurial goals in accordance with the new needs of the productive sector. Among all educational institutions, the universities' response to this aim is of particular relevance, as they are the principal agents for generating and disseminating specialized knowledge in the context of a social reality in which access to higher education is more and more generalized in developed countries. In fact, the advisability of promoting entrepreneurial mindsets has extended to the current Bologna Process aimed at building a modern degree structure adapted to the professional profiles required by the current EU society through the establishment of a common European Higher Education Area (EHEA). In this context, the project *Tuning Educational Structures in Europe* (González and Wagenaar 2003), devoted to the identification of learning results and desirable competences by thematic area, has included entrepreneurship into the group of systemic transversal competences to be trained along all levels of university higher education.

Despite this political commitment, it is estimated that more than half of university students in Europe do not have access to entrepreneurial education, some differences existing by country (Enterprise Directorate General 2008a). Particularly in Spain, most universities have developed and implemented specific extracurricular actions to give support to potential entrepreneurial initiatives of students, in the form of University–Enterprise foundations, business chairs, spin-off programs or specific institutional programs and centers on entrepreneurship (Directorate General of Small and Medium Enterprise Policy 2006; National Agency for Quality Assessment and Accreditation 2007). However, the specific impact of these institutional initiatives on the entrepreneurial projections of Spanish graduates seems to be largely unknown, or somewhat disappointing. In this regard, previous evidence points to the general conclusion that students perceive a scarce consideration of entrepreneurship topics within university programs (Vázquez *et al.* 2010a), and since academic courses focus on the wage-employment paradigm, the transit through university has a poor effect on the entrepreneurial vocations of students (Vázquez *et al.* 2009, 2010b). Furthermore, even if experts acknowledge the importance of adapting entrepreneurship education to the needs of students with different academic backgrounds (Enterprise Directorate General 2008b), the differential effects of institutional programs on the entrepreneurial orientation of students within different academic areas has not been analyzed in previous literature.

To fill this gap, this chapter aims to analyze the impact of various sources of institutional support in two Spanish universities on the entrepreneurial goals of students, and to validate the generalization of such effects by considering the academic background of undergraduates. To this end, the next sections are organized as follows: First, we review previous attempts to clarify the impact of entrepreneurship education in the university. Second, we propose a sequential model to explain the origin of entrepreneurial behaviour in young students and the potential influence of university programs on its various stages. Third, we present an empirical study aimed to test the hypothesis of the model in a representative sample of undergraduate students. Lastly, the main conclusions, implications and limitations of the study are discussed.

2 The effects of entrepreneurship education

As entrepreneurship education has become a widespread phenomenon worldwide, several authors have pled for the establishment of operating mechanisms to assess the impact of educational and institutional programs oriented toward stimulating the enterprising spirit of students in university settings (e.g., Gibb 1987; Johannisson 1991; Vesper and Gartner 1997; Storey 2000; McMullan *et al.* 2001; Henry *et al.* 2005). In this sense, Jack and Anderson (1998) have proposed a five-step model based on a previous version of Block and Stumpf (1992) which stress the assessment of the results of every program over time (Table 13.1). From

Table 13.1 Jack and Anderson five-step framework of program effectiveness

Temporal moment	*Impact criteria*
Ongoing measures	• Student enrolment • Number and type of course • Awareness of entrepreneurship • Entrepreneurial interests
Pre and post program measures	• Entrepreneurial intention • Knowledge gained • Self-perception of learning and competence
0–5 years post program	• Number and type of start-ups • Foothold acquisitions • Entrepreneurial position
3–10 years post program	• Survival and reputation of new start-ups • Change in reputation and innovation level
More than 10 years post program	• Contribution to society and economy • Firm performance • Career satisfaction • Personal self-satisfaction and psychological success

Source: Adapted from Jack and Anderson (1998, p. 10) and Henry *et al.* (2005, p. 161).

that temporal framework, the model considers both immediate impacts—based on the development of entrepreneurial competences, interests and intentions—and long-term criteria—in terms of business start-up and future performance as well as other measures of success.

According to the model proposed, most previous studies devoted to analyzing the impact of entrepreneurship education in the university have been preferably set in a temporal framework close to the implementation of the specific program or course. In these studies it has been frequent practice to use subjective measures of effectiveness, through which participants are asked about the effects caused by the implemented action on various aspects of their entrepreneurial orientation. The results of this line of work confirm, in general terms, that such programs positively influence the perception of acquired skills to start a business (e.g., Galloway *et al.* 2005; Hartshorn and Hannon 2005; Okudan and Rzasa 2006; Izquierdo and Buyens 2008; Matlay 2008), the development of positive attitudes toward self-employment (e.g., Johannisson 1991; Autio *et al.* 1997; Gorman *et al.* 1997; Hegarty 2006; Vázquez *et al.* 2009), and the emergence of enterprising aspirations and intentions (e.g., Kolvereid and Moen 1997; Vesper and Gartner 1997; Galloway and Brown 2002; Vázquez *et al.* 2009; Oosterbeek *et al.* 2010).

While it is true that subjective reports have been found useful in understanding the experience of students when building their enterprising projections (Fayolle 2005), experts warn about the relative validity of such measures as real indexes of program effectiveness, given their frequent limitations in terms of sample representativeness, honesty of responses, and the like (McMullan *et al.* 2001; Westhead *et al.* 2001; Heinonen and Poikkijoki 2006). For this reason, many academics refer to long-term observable impacts as ideal measures of program effectiveness (Vesper and Gartner 1997; McMullan *et al.* 2001), thus calling for a higher number of longitudinal studies that allow the estimation of the real effect of the actions developed on the incidence and evolution of new enterprises actually created (e.g., Garavan and O'Cinneide 1994; Storey 2000).

Even if less frequent, studies following this approach confirm the positive effect of specialized education on start-up behavior (e.g., Varela and Jiménez 2001; Placer and Vázquez 2005; Rasmussen and Sørheim 2006; Matlay 2008; Sierra 2010), just as on other measures of business growth (Galloway and Brown 2002). However, generalization of these findings is limited to the scarcity of studies carried out, given both the technical limitations of follow-up research approaches and the difficulty of isolating the effect of education from other factors potentially affecting entrepreneurial behavior and subsequent performance (Sandberg and Hofer 1987; Shaver and Scott 1991; Pollitis 2005).

Taken together, most previous evidence on the effects of entrepreneurship education in university settings has focused on the impacts of particular programs or initiatives on specific results, while the overall analysis of various types of institutional support on different facets of the entrepreneurial behavior of students over time is quite unknown. In this context, further research should consider the real effectiveness of institutional entrepreneurship programs in the European area, according to the career needs of different groups of students. With this in mind, we next propose a model to test the influence of university programs in the Spanish area on the development of entrepreneurial interests, intentions, and behavior of students, and validate its generalization to different academic fields.

3 A step-based model of entrepreneurship support in the university: Hypotheses.

The effectiveness of entrepreneurship programs in the university depends on their ability to fit the career needs of future graduates at that stage of professional development. In this sense, most previous works explaining the development of entrepreneurial vocations among university students refer to psychological intention-based models derived from Ajzen's (1991) theory of planned behavior by and Shapero's (1982) model of the entrepreneurial event. From these frameworks, it is assumed that the most immediate antecedent of any planned behavior, in this case the creation of a new firm, is the intention to carry out such behavior.

In psychological literature, intentions are assumed to capture the motivational factors that influence the behavior. That is to say, they are indicators of how hard people are willing to try in order to behave in a specific manner to achieve a goal. For this reason, intentions appear as a good predictor of planned behavior (Ajzen 1991), especially if this is difficult to perform and demands a great amount of resources. New businesses emerge over time and involve considerable planning. Nascent entrepreneurs don't launch their new firms in an automatic manner as a conditioned response to a stimulus, but starting a business is a complex career decision reflecting some degree of cognitive processing. Since entrepreneurship clearly represents planned, intentional behavior (Bird 1988; Katz and Gartner 1988), it seems useful to research it by using formal models of intentions (Krueger *et al.* 2000). In conclusion, as entrepreneurial intentions involve a deep commitment to new venture creation, the greater the intention to start a business, the greater the likelihood of performing such behavior.

While it is true that some success has been achieved by this line of research, some authors have questioned the usefulness of intention-based models to predict the future start-up behavior of students (Robinson *et al.* 1991; Soutaris *et al.* 2007). In this sense, research shows that, regardless of the entrepreneurial propensity of undergraduate students, setting up a business is seen as a long-term goal, which is preceded by some leeway devoted to gain experience and gather resources (e.g., Galloway and Brown 2002; Galloway *et al.* 2005).

With this is mind, we propose in this chapter that models derived from vocational literature are more suited to explain the entrepreneurial behavior of students, since they consider the specific academic and professional reality experienced by students at the end of higher education. For instance, Social Cognitive Career Theory (SCCT) by Lent *et al.* (1994) identifies three different stages in the development of career goals in youth, namely interests, intentions, and entry behaviors. As in literature derived from intention-based models, career intentions are supposed to directly influence the implementation of congruent career behavior. However, in this case career behavior is defined as "actions designed to implement the [career] choice (e.g., enrolling in a particular training program or academic major)" (Lent *et al.* 1994, p. 94). Applied to entrepreneurship career decisions, this allows us to conceptualize entrepreneurial behavior in terms of early actions previous to business start-up, in reference to a stage of business gestation or "nascency" that involves observable activities such as investing time and personal effort, developing a business plan, attending training courses, or gathering resources (e.g., Soutaris *et al.* 2007; McGee *et al.* 2009). Lent *et al.* (1994, p. 88) define vocational interests as "patterns of likes, dislikes and indifferences regarding career relevant activities and occupations," and propose them as determinants of both career intentions and behavior. Parallel, entrepreneurial interests have been used to explain career goals in adolescence (e.g., Bonnett and Furnham 1991; Schröder and Schmitt-Rodermund 2006). Nevertheless, the construct and its link to entrepreneurial intentions and nascent behavior have not been analyzed in university samples. To fill this gap, and following the SCCT approach, we propose two hypotheses defining our general step-model of entrepreneurship career choice among undergraduate students (Figure 13.1):

H1: Entrepreneurial intention directly influences nascent behavior.
H2: Entrepreneurial interest directly influences entrepreneurial intention, and indirectly influences nascent behavior by total mediation of entrepreneurial intention.

Around this sequential model, most Spanish universities nowadays have developed and implemented specific actions to give support to the potential entrepreneurial initiatives of students, in the form of University–Enterprise foundations, business chairs, spin-off programs, or specific institutional programs and centers on entrepreneurship (Directorate General of Small and Medium Enterprise Policy 2006; National Agency for Quality Assessment and Accreditation 2007). From this framework, initiatives are generally concerned with supporting and providing accompanying services for potential and nascent entrepreneurs, in order to drive them toward successful start-ups. Revision of several descriptions of best practices on this matter (e.g., Placer and Vázquez 2005; Corduras *et al.* 2008; Sierra 2010) point to three categories of institutional support to promote entrepreneurship in the Spanish area: motivational, formative, and instrumental.

First, motivational support is concerned with actions oriented at raising students' awareness of business start-up as an alternative career choice, thus including motivation and encouragement actions, receptivity to students' interests, awareness campaigns, etc. From this view, the objective of these actions is to stimulate student's vocational interests in entrepreneurship.

Second, formative support embraces actions devoted to improve undergraduates' access to sources of information and training in entrepreneurship competences. It includes information about business creation, counselling services, entrepreneurship training courses, etc. Therefore, these activities are oriented at consolidating a formal entrepreneurial intention in students.

Finally, instrumental support refers to provision of resources for business start-up, including mentorship and monitoring, business plan assistance, financial resources, etc. Hence, the purpose of this stage would be to support observable nascent behaviour among future graduates.

However, the specific impact of those institutional initiatives on the entrepreneurial projections of Spanish graduates seems to be largely unknown, or somewhat disappointing (Vázquez *et al.* 2009, 2010a, 2010b). At the same time, no previous literature has focused on the effects of different sorts of institutional support over the entrepreneurial sequence of interests, intentions, and behavior. In this context, we complement our model with the following three hypotheses:

H3: Motivational support directly influences entrepreneurial interest, and indirectly influences entrepreneurial intention and nascent behavior by total mediation of interests.
H4: Formative support directly influences entrepreneurial intention, and indirectly influences nascent behavior by total mediation of entrepreneurial intention.
H5: Instrumental support directly influences nascent behavior.

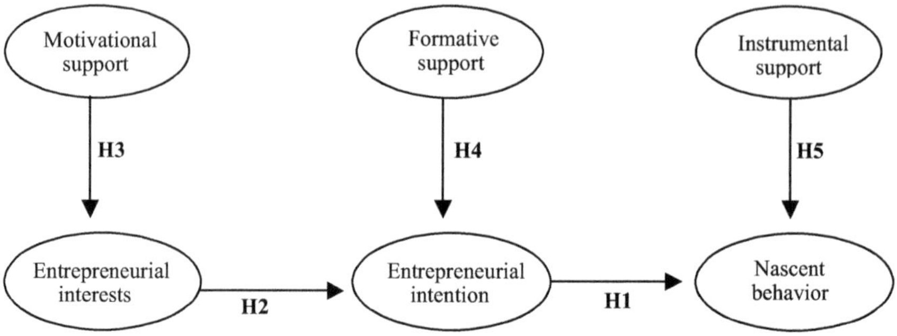

Figure 13.1 Step-based model of entrepreneurship support in the university.

Finally, in our step-based model of entrepreneurship support in the university, we consider the influence of the academic area. In this respect, there is a growing awareness of the need to orientate entrepreneurship education according to the needs of students with different academic background (Enterprise Directorate General 2008b). For instance, within Business schools and Economic studies it is assumed that, since many of the business subjects are offered separately (i.e., marketing, management, etc.), entrepreneurship programs should have a very narrow focus, stressing the start-up phase and the growth of small enterprises. In the case of non-business disciplines, students are usually good in technical aspects and frequently have very strong business ideas, their weaknesses concentrating in the development of specific business knowledge and abilities. Accordingly, experts agree that entrepreneurship education within Science and Technology studies should be concerned with exploiting intellectual property, creating spin-off companies and ventures, selling technology-based ideas, etc., whereas for students in Humanities and Social sciences the focus could be on social entrepreneurship as an emerging area of growth that provides the opportunity to make a difference in community contexts. From this general framework, we assume in this chapter the existence of differential effects of institutional programs on the entrepreneurial orientation of students within different academic areas, according to the following hypothesis:

H6: The effect of motivational, formative and instrumental support on entrepreneurial interest, entrepreneurial intention, and nascent behavior among students differs between academic areas.

4 Research methodology

4.1 Sampling method

We used a survey approach to test the hypotheses of the model. In order to make possible the generalization of results to different institutional contexts, the study sample consisted of undergraduate students at two Spanish universities with different tradition, size, and international prestige: the Complutesian University of Madrid and the University of León.

Established in the sixteenth century, and with more than 87,000 students, the Complutesian University of Madrid is one of the universities of reference in Spain, occupying favorable positions in many international classifications based on research quality criteria, formative capacity and demand, availability of physical and human resources, international presence, and the like. Meanwhile, the University of León belongs to the majority group of Spanish universities with a shorter tradition and more moderate dimensions, comprising nearly 13,000 students after thirty years of history.

Despite the discrepancies described, both universities fit the requirements of the research, since they are engaged in several initiatives aimed at fostering entrepreneurship among students, in the form of business chairs and specific institutional foundations and programs. This ensured that students in the study were exposed to some degree of entrepreneurship education and support.

Once the choice of the two universities had been justified, data collection was performed from February to June 2010. The sample comprised a total of 800 university students (400 from each university), ensuring a criterion of representativeness of 95 percent (being e = ±5%; $p = q = 0.50$). Participants were selected through a procedure of stratified sampling, in accordance with the real distribution of students by field of study in each university. Based on this procedure, 53.1 percent of respondents indicated a main academic background in Social Sciences and Law, 14.6 percent in Technical Subjects, 12.3 percent in Health Sciences, 11.3 percent in Experimental Sciences, and 8.8 percent in Humanities.

Among the total participants, 530 were females (66.3 percent) and 270 males (33.8 percent), aged 18 to 48 years old, the mean age being 23.16 ($SD = 3.14$).

4.2 Measures

In gathering data for the study, we developed a self-reporting questionnaire following a careful procedure to ensure the sufficient validity of scales. Specifically, we used a deductive approach based on the operational definition of the theoretical constructs arising from an in-depth review of the specialist literature on the topic and other similar tools for measurement, intended to identify specific indicators upon which to build the scales. Similarly, the final version of the questionnaire was refined by three experts in the field regarding the suitability of the indicators proposed for measuring the variables of interest.

Data collection was based on a procedure of collective voluntary self-administration of the final questionnaire to groups of students. This was done in the context of scheduled university classes randomly selected for each knowledge area, after obtaining approval from the academic member of staff responsible in all cases and in the presence of a researcher trained for this end.

The questionnaire administered comprised various scales for measuring the variables included in the model: experienced institutional support to entrepreneurship—motivational, formative and instrumental—entrepreneurial interests, entrepreneurial intention and nascent behavior.

Experienced entrepreneurship support was assessed through nine items about the perceived implication of each university in related actions. From a previous revision of the most frequent initiatives implemented in the universities surveyed, we used three items for each of the motivational (e.g., "campaigns to raise the entrepreneurial spirit"), formative ("entrepreneurship training courses") and institutional (e.g., "assistance to entrepreneurial projects") facets of support. For each item, participants had to answer on a Likert-type scale ranging from 0 (*not implicated at all*) to 10 (*highly implicated*).

To measure entrepreneurial interest, we asked students about their attraction towards ten typical entrepreneurial activities (e.g., "to learn about the economy"). For each activity, responses were ranged on a Likert-type scale from 0 (*not interested at all*) to 10 (*very interested*).

Entrepreneurial intention was measured with three items about undergraduates' preferences for self-employment and the likelihood of starting a business versus working as wage employee at the end of higher education (Kolvereid 1996). Accordance responses were ranged on a Likert scale from 0 (*strongly disagree*) to 10 (*strongly agree*).

Finally, to assess entrepreneurial nascent behavior participants were asked to report their degree of accordance with four statements about their involvement in specific actions oriented at new firm creation (e.g. "gathering resources"). Again, they had to use a Likert scale from 0 (*strongly disagree*) to 10 (*strongly agree*).

4.3 Data analysis

The data collected were analyzed through SPSS 17.0 for correlational and descriptive purposes. Next, the statistical treatment of the information recovered was devoted to test the reliability and validity of the measures included in the questionnaire and the fulfillment of the research hypotheses. In the context of these aims, structural equation modeling (SEM) has become one of the most valuable developments of multivariate analysis (Fornell 1982), allowing combination of statistical principles derived from factor analysis and multiple regression for the simultaneous estimation of latent constructs from directly observed measures, and predictive links between exogenous and endogenous latent variables (Barclay *et al.* 1995; Chin 1998a).

Particularly in this case, we chose the partial least squares (PLS) approach to SEM. Unlike covariance-based methods centered on the estimation of model parameters and overall fit measures, PLS aims to maximize the variance explained by indicators and latent variables, through the estimation of ordinary least squares (OLS) and principal components analysis. In this context, data treatment responds to a kind of *flexible modeling* (Wold 1980) based on the creation of optimal linear predictive paths with minimal demands with respect to measurement scales, residual distributions, and sample sizes (Chin 1998b; Fornell and Bookstein 1982; Wold 1985). Therefore, compared to maximum likelihood methods, the PLS approach fits better the requirements of exploratory and theory-building applications (Barclay *et al.* 1995; Chin *et al.* 2003), as in the case of the present research.

In more precise terms, data analysis was performed with the software package SmartPLS 2.0 (Ringle *et al.* 2005). Based on this procedure, the estimation of the model is carried out in two stages. For reflective measures of constructs (Gerbins and Anderson 1998), the first stage involves the evaluation of the strength of the measurement model (also known as *outer model*), by looking at individual item reliability, internal consistency, and construct validity. Once the measurement model's reliability and validity has been verified, the second stage focuses on the estimation of fit parameters for the structural model (or *inner model*), thus providing information about the fulfillment of the research hypotheses through standardized path coefficients and $R2$ indexes. Following the purposes of the study, this stage of structural analysis was first run for the total sample to examine the overall fit of the model (Hypotheses 1–5), and then repeated separately for the five subsamples of university students considered by academic area, in order to perform an explorative group comparison (Chin 2000) and test the robustness of the model across segments of students with different academic backgrounds (Hypothesis 6).

5 RESULTS

5.1 Correlational and descriptive analysis

Table 13.2 presents the means, standard deviations, and correlation coefficients among the study variables. As expected, most variables were positively correlated, thus showing a significant relationship between experiences of entrepreneurship education and support in the University and results in terms of entrepreneurial interests, intentions, and behavior.

Table 13.2 Means, standard deviations (SD), and correlations among variables

	1	*2*	*3*	*4*	*5*	*6*
1. Motivational support	—					
2. Formative support	.75*	—				
3. Instrumental support	.76*	.84*	—			
4. Entrepreneurial interests	.14*	.16*	.15*	—		
5. Entrepreneurial intention	.14*	.13*	.14*	.32*	—	
6. Entrepreneurial behavior	.23*	.30*	.26*	.27*	.52*	—
M	2.96	2.40	2.31	5.95	3.36	1.46
SD	2.10	2.14	2.07	2.21	2.03	1.94

Note: * $p < .01$.

An inspection of mean scores discloses a poor assessment of the motivational, formative and instrumental supports provided by the universities surveyed to satisfy the entrepreneurial aspirations of undergraduates, with average scores of 2.96, 2.40, and 2.31, under the intermediate value of 5. In the same line, participants reported a generalized low preference for entrepreneurship as a career choice, the mean values being of 3.36 and 1.46 in the entrepreneurial intention and behavior scales. Only entrepreneurial interests were slightly higher, with an average score of 5.95.

5.2 Evaluation of the measurement model

As previously mentioned, the first stage of PLS modeling involves assurance that the measures used reflect the underlying theoretical constructs, by looking at items' reliability and scales' internal consistency and construct validity. In this respect, Table 13.3 shows the indexes obtained for the six latent variables of the model and their reflective measures, according to the SmartPLS program.

Table 13.3 Indicators for measurement model evaluation

Item/Scale	*Loadings (λ)*	λ^2	*t*	α	ρc	*AVE*
Motivational support				.867	.918	.789
Cultivating the entrepreneurial spirit	.875	.766	4.11*			
Receptivity to students' interest	.899	.808	4.78*			
Encouragement to students	.890	.792	4.51*			
Formative support				.910	.943	.847
Entrepreneurship information	.916	.839	3.61*			
Entrepreneurship counselling	.950	.902	3.37*			
Entrepreneurship training courses	.895	.801	3.76*			
Instrumental support				.902	.939	.836
Project assistance .920	.846	32.75*				
Access to resources	.934	.872	34.77*			
Monitoring and follow-up	.888	.788	20.04*			

Item/Scale	*Loadings (λ)*	λ^2	*t*	α	ρc	*AVE*
Entrepreneurial interests				.926	.938	.582
Learn about the economy	.675	.455	9.13*			
Invent ways to cover needs of consumers	.727	.528	10.56*			
Plan business systems	.570	.325	2.89*			
Develop products and services	.774	.599	13.21*			
Sell products and services	.793	.629	16.85*			
Develop marketing strategies	.766	.587	12.97*			
Maintain relationships with suppliers	.855	.731	25.69*			
Maintain relationships with investors	.841	.707	21.23*			
Learn about business management	.797	.635	13.08*			
Select and coordinate human resources	.791	.626	12.25*			
Manage a business on your own	.758	.574	11.10*			
Entrepreneurial intention				.659	.808	.592
Preference for entrepreneurship	.786	.618	12.15*			
Probability of entrepreneurship	.918	.843	47.25*			
Probability of wage employment (*Reversed*)	.563	.317	3.87*			
Entrepreneurial behavior				.867	.910	.717
Attend courses and conferences	.773	.597	13.32*			
Develop a business idea	.863	.745	18.48*			
Gather resources	.896	.803	27.01*			
Start-up activities	.851	.724	17.25*			

Note: * $p < .001$.

Concerning items' reliability, all item loadings (λ) were above the minimum level 0.5 for acceptability (Chin 1998a, 1998b; Barclay *et al.* 1995). The significance of loadings was further verified through a bootstrap procedure with 200 subsamples, for obtaining significant *t*-statistic values at a 0.95 level (based on *t*(199), two-tailed test). Likewise, communalities (λ2) were well above the minimum 0.25 (Bollen 1989), the latent constructs explaining between 56.3 percent and 95 percent in their respective observed indicators.

Regarding the internal consistency of scales, the examination of Cronbach's alpha (α) and the composite reliability (ρc) reveals values exceeding the minimum threshold of 0.7 (Nunnally 1978; Nunnally and Bernstein 1994; Barclay *et al.* 1995; Hair *et al.* 1998), which proves that the occurrence of random error in measures was minimized.

Convergent validity was tested by the index Average Variance Extracted (AVE). In all cases, values were above the minimum benchmark of 0.5 (Fornell and Larcker 1981), meaning that 50 percent or more variance of the indicators was accounted for. Next, we checked the correlation matrix of latent variables, with the square roots of AVE values as diagonal elements, and no problem was detected, thus suggesting adequate discriminant validity.

5.3 Evaluation of the structural model

Having analyzed the reliability and validity of the measures, a second stage of the PLS technique was used to assess the explanatory power of the hypothesized relationships among constructs.

The statistical significance of path coefficients was tested by performing bootstrapping with 200 subsamples, to obtain the *t*-statistic values associated with these predictive links (Chin 1998a). Table 13.4 shows the direct and indirect effects obtained by this procedure.

According to Hypothesis 1, it was confirmed that there is a direct effect of entrepreneurial intention on entrepreneurial behavior, with a path coefficient of 0.524. Hypotheses 2 was also supported, as results confirmed a direct effect of interests on intention, and an indirect effect on nascent behavior by total mediation of entrepreneurial intention. These effects explained 10 percent of variance in intention and 4.23 percent of variance in behavior. According to Hypothesis 3, there was a positive direct effect of motivational support on entrepreneurial interests, with a path coefficient of 0.151. Hypothesis 4 was not supported, since the effect of formative support on entrepreneurial intention was not significant. Finally, instrumental support explained a significant 4.34 percent of variance in nascent behavior, thus supporting Hypothesis 5. Coefficients for all these hypotheses are summarized in Figure 13.2.

Table 13.4 Direct and total effects

	Direct effects			Total effects		
	Path	*Variance*	*t*	*Path*	*Variance*	*t*
Intention → Behavior	.524	28.87%	7.97**			
Interests → Intention	.307	10.00%	3.30**			
Interests → Behavior				.161	4.23%	2.90**
Motivational support → Interests	.151	2.28%	1.65*			
Motivational support → Intention				.046	0.79%	1.34
Motivational support → Behavior				.024	0.57%	1.28
Formative support → Intention	.114	1.86%	1.07			
Formative support → Behavior				.060	1.82%	1.03
Instrumental support → Behavior	.171	4.34%	2.05*			

Note: * $p < .05$; ** $p < .01$;

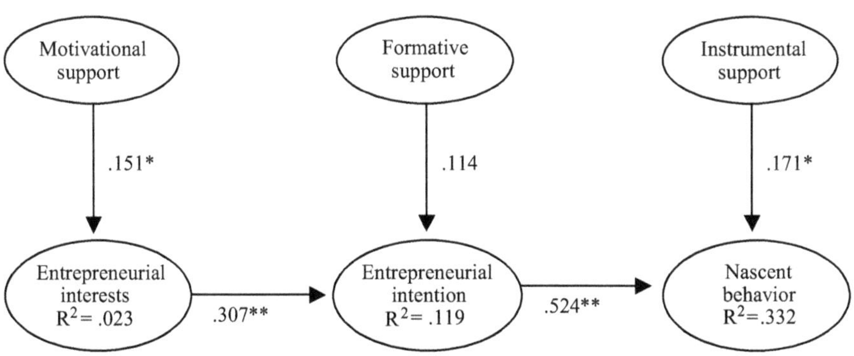

Figure 13.2 Results for the structural model.

Note: * $p < .05$; ** $p < .01$.

Beyond the analysis of the magnitude and significance of path coefficients, R2 indexes for entrepreneurial intention and behavior were above the minimum threshold of 0.10 (Falk and Miller 1992), the overall model explaining 11.87 percent of variance in intention and 33.16 percent in behavior. The R2 index was lower for entrepreneurial interests, the overall model explaining a reduced 2.28 percent of variance.

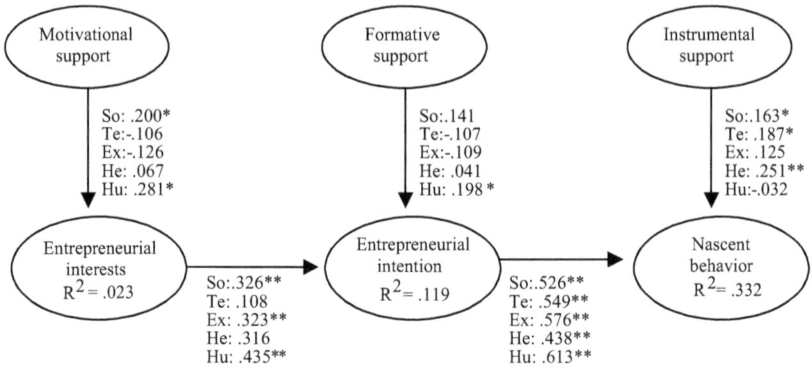

Figure 13.3 Results for the structural model by academic area.

Note: * $p < .05$; ** $p < .001$
So = Social; Te = Technical; Ex = Experimental; He = Health; Hu = Humanities.

Finally, the jackknife Q2 statistic for predictive relevance was 0.212 for nascent entrepreneurial behavior and under the minimum 0.20 for entrepreneurial interest and intention, concluding a relatively low overall fit of the model.

Considering the potential effect of academic area in the model (Hypothesis 6), Figure 13.3 presents comparative results for each subsample. In general, the paths expected between intention and behavior (Hypothesis 1) and between interests and intention (Hypothesis 2) reached statistical significance for most subjects, with the only exception of the no significant effect of interests on intention in the Technical sample.

Institutional support was found to play a differential role in the model depending on the academic field of students. Regarding Hypothesis 3, motivational support predicted the entrepreneurial interests of students within Social sciences and Humanities, but not in the Technical, Experimental and Health samples. Similarly, while the expected impact of formative support on entrepreneurial intention (Hypothesis 4) was rejected for most samples, the path coefficient was significant for students within Humanities. Finally, Hypothesis 5, concerning a direct path between instrumental support and nascent behavior, was only confirmed for the Social, Technical, and Health samples.

Overall, the model explained between 26.5 percent and 36.3 percent of variance in nascent behavior and between 10.5 percent and 25.6 percent of variance in the entrepreneurial intentions of students in all academic areas, whereas the R2 indexes were lower for entrepreneurial interests. Finally, the jackknife Q2 statistic for predictive relevance was above the minimum 0.20 for nascent entrepreneurial behavior in the Social, Technical and Experimental areas, while the overall fit of the model was low for students within Health sciences and Humanities.

6 Discussion

The main purpose of this chapter was to analyze the influence of university programs in the Spanish area on the entrepreneurial career choice of students. From this framework, we first hypothesized a sequential relationship between entrepreneurial interests, intentions, and nascent behavior to explain the development of entrepreneurial vocations of university students.

In congruence with general career models (Shapero 1982; Ajzen 1991; Lent *et al.* 1994), a direct positive effect was found from entrepreneurial intention upon the degree of involvement shown by students in early activities relating to putting into practice a business idea. Therefore, even if the number of enterprises created by recent graduates from higher education in the countries of our continent has often been small, thus making it difficult to test this prediction among groups of university students, the present study shows that the commitment felt by individuals toward self-employment constitutes a fundamental determinant for the implementation of specific actions aimed at starting a business in a phase prior to the establishment of a new venture.

Furthermore, one of the most novel contributions of the theoretical approach adopted in this work with respect to previous literature relates to the consideration of entrepreneurial interests as antecedents in the process of career choice, this view being closer to the work reality faced by university students. On these lines, the results supported a direct positive effect of interest on entrepreneurial intention, and an indirect effect on the nascent behavior totally mediated by it. Hence, we can conclude that the early attraction of students toward entrepreneurial activity plays an important part in leading their immediate professional aims toward the creation of enterprises.

On the other hand, we used the expected sequence of interests–intention–behavior to test the effect of entrepreneurship university support programs on the career behavior of students. Findings supported a positive link between sources of motivational support and the entrepreneurial interests of future graduates, thus confirming the effectiveness of encouragement actions and awareness campaigns in stimulating the entrepreneurial spirit of students. Also, the effect of instrumental support on nascent behavior was significant, pointing to the importance of assisting the first steps of undergraduates' nascent initiatives with financial and non-financial resources. However, the hypothesis concerning the direct link between formative support and entrepreneurial intention was not supported, the effectiveness of training programs and courses to raise preferences for self-employment being of no significance.

In the context of these findings, a second objective of the study was to test the validity of the model in subsamples of university students with different academic backgrounds. While the sequence between entrepreneurial interests, intention, and behavior was common, in general terms, for undergraduates within all fields—thus supporting the generalization of results in this sense—the effect of institutional support was different by academic area. Taken together, results point to the general conclusion that institutional supports to entrepreneurship in the universities surveyed fit mainly the needs of students within Social Sciences and Humanities, as opposed to students in other areas. In particular, current programs seem not to be fulfilling the needs of students within experimental fields, since the effects of motivational, formative, and instrumental support were non-significant in this subsample.

To summarize, our model proved to be useful as a general framework for analyzing the behavioral impact of entrepreneurship programs in university settings, according to the academic and professional circumstances of young people finishing higher education. Furthermore, results support the importance of using academic background as a segmentation variable to better understand the effectiveness of programs. From this angle, the implications of this study for entrepreneurship practitioners in Spanish universities call for attention to the step-based relationships between interests, intention, and behavior when designing appropriate actions to foster entrepreneurship among students, as well as the need to adapt educational and institutional support to the future professional circumstances of students within various academic areas.

Finally, the reach of the results obtained in the study carried out should be interpreted in the light of certain methodological limitations, which leave the door open for further work in this field. First, it should be noted that the majority of scales used for measuring the variables in the model were drawn up ad hoc for the purposes of this research. Hence, they will require future validation to check their usefulness for the purposes assigned to them in this work. Furthermore, despite the usefulness of the comparative cross-sectional design used in understanding the aspirations and work projects of future university graduates, further longitudinal analyses are needed to give an account of the development of initial processes of selection of entrepreneurial careers into the tangible form of new successful enterprises in the marketplace. This would require following up the entrepreneurial sequence as it develops over time among the same group of students. Likewise, while the fact that the empirical study was carried out in two different Spanish universities demonstrates that the conclusions drawn from it are sufficiently solid, further studies are required to allow generalization of the results to other Spanish or European institutions.

Acknowledgment

This research was supported by the FPU Program of the Ministry of Education in Spain.

References

Ajzen, I. (1991) "The theory of planned behavior," *Organizational Behavior and Human Decision Processes*, 50, 179–211.

Autio, E., Keeley, R. H., Klofsten, M. and Ulfstedt, T. (1997) "Entrepreneurial intent among students: Testing an intent model in Asia, Scandinavia and the USA," *Frontiers of Entrepreneurship Research, Babson Conference Proceedings*. Available online at www.babson.edu/entrep/fer (accessed 14 September 2011).

Barclay, D., Higgins, C. and Thompson, R. (1995) "The Partial Least Squares (PLS) approach to causal modeling: Personal computer adoption and use as an illustration," *Technology Studies*, 2, 285–309.

Bird, B. (1988) "Implementing entrepreneurial ideas: The case for intentions," *Academy of Management Review*, 13, 442–454.

Block, Z. and Stumpf, S. A. (1992) "Entrepreneurship education research: Experience and challenge," in D. L. Sexton and J. D. Kasarda (eds), *The Stage-of-the-Art of Entrepreneurship*. Boston, MA: PWS-Kent Publishing Company, pp. 17–42.

Bollen, K. A. (1989) *Structural Equations with Latent Variables*. New York: Wiley.

Bonnett, C. and Furnham, A. (1991) "Who wants to be an entrepreneur? A study of adolescents interested in a Young Enterprise scheme," *Journal of Economic Psychology*, 12, 465–478.

Chin, W. W. (1998a) "The partial least squares approach to structural equation modeling," in George, A. (ed.), *Modern Methods for Business Research*. New Jersey: Laurence Erlbaum Associates, pp. 295–336.

Chin, W. W. (1998b) "Issues and opinion on structural equation modeling," *MIS Quarterly*, 22(1), 7–1.

Chin, W. W. (2000) "Frequently asked questions: Partial least squares & PLS-graphs." Available online at http://disc-nt.cba.uh.edu/chin/plsfaq.htm (accessed 14 September 2011).

Chin, W. W., Marcolin, B. L. and Newsted, P. R. (2003) "A partial least squares latent variable modeling approach for measuring interaction effects: Results from a Monte Carlo simulation study and an electronic mail emotion/adoption study," *Information Systems Research*, 14, 189–217.

Communication 2010/2020 from the Commission. Europe 2020. "A strategy for smart, sustainable and inclusive growth, 2010." Available online at http://eur-lex.europa.eu/LexUriServ/LexUriServ.do?uri=COM:2010:2020:FIN:EN:PDF (accessed 14 September 2011).

Corduras, A., Urbano, D., Rojas, A. and Martínez, S. (2008) "The relationship between University support to entrepreneurship with entrepreneurial activity in Spain," *International Advances in Economic Research*, 16, 395–406.

Corduras, A., Levie, J., Kelley, D. J., Saemundsson, R. and Schott, T. (2010) *Global Entrepreneurship Monitor Special Report: A global perspective on entrepreneurship education and training*, Babson Park–London: Babson College–London Business School.

Council of the European Union (2000) *Conclusions of the Presidence: Lisbon strategy for growth. Towards a green and innovative economy*, Brussels: European Commission.

Directorate General of Small and Medium Enterprise Policy (2006) *Iniciativas emprendedoras en la universidad española*. Madrid: Centro de Publicaciones del Ministerio de Industria, Turismo y Comercio.

Enterprise Directorate General (2008a) *Survey of Entrepreneurship in Higher Education in Europe*. Main report, Brussels: European Commission.

Enterprise Directorate General (2008b) *Entrepreneurship in Higher Education, Especially within Non-Business Studies*. Final report of the expert group. Brussels: European Commission.

Falk, R. F. and Miller, N. B. (1992) *A Primer for Soft Modeling, Akron*. Ohio, IL: The University of Akron.

Fayolle, A. (2005) "Evaluation of entrepreneurship education: Behavior performing or intention increasing?" *International Journal Entrepreneurship and Small Business*, 2, 89–98.

Fornell, C. (1982) "A second generation of multivariate analysis: An overview," in C. Fornell (ed.), *A Second Generation of Multivariate Analysis* (Vol. 1), New York: Praeger Publishers, pp. 1–21.

Fornell, C. and Bookstein, F. L. (1982) "A comparative analysis of two structural equation models: Lisrel and PLS applied to market data," in C. Fornell (ed.), *A Second Generation of Multivariate Analysis* (Vol. 1). New York: Praeger Publishers, pp. 289–324.

Fornell, C. and Larcker, D. F. (1981) "Evaluating structural equation models with unobservable variables and measurement error," *Journal of Marketing Research*, 18, 39–50.

Galloway, L. and Brown, W. (2002) "Entrepreneurship education at university: A driver in the creation of high growth firms?" *Education + Training*, 44, 398–405.

Galloway, L., Anderson, M., Brown, W. and Wilson, L. (2005) "Enterprise skills for the economy," *Education + Training*, 47, 7–17.

Garavan, T. N. and O'Cinneide, B. (1994) "Entrepreneurship education and training programmes: A review and evaluation. Part I," *Journal of European Industrial Training*, 18, 3–12.

Gerbins, D. and Anderson, J. C. (1998) "An updated paradigm for scale development incorporating unidimensionality and its assessment," *Journal of Marketing Research*, 25, 186–192.

Gibb, A. A. (1987) "Enterprise culture: Its meaning and implications for education and training," *Journal of European Industrial Training*, 11(2), 3–38.

González, J. and Wagenaar, R. (eds) (2003) *Tuning Educational Structures in Europe*. Bilbao: University of Deusto.

Gorman, G., Hanlon, D. and King, W. (1997) "Some research perspectives on entrepreneurship education, enterprise education and education for small business management: A ten year literature review," *International Small Business Journal*, 15(3), 56–78.

Hair, J. F. J., Anderson, R. E., Tatham, R. L. and Black, W. C. (1998) *Multivariate Data Analysis*, 5th Edn. London: Prentice Hall International.

Hartshorn, C. and Hannon, P. D. (2005) "Paradoxes in entrepreneurship education: Chalk and talk or chalk and cheese? A case approach," *Education + Training*, 47, 616–627.

Hegarty, C. (2006) "It's not an exact science: Teaching entrepreneurship in Northern Ireland," *Education + Training*, 48, 322–335.

Heinonen, J. and Poikkijoki, S. (2006) "An entrepreneurial-directed approach to entrepreneurship education: Mission impossible?" *Journal of Management*, 25, 80–94.

Henry, C., Hill, F. and Leitch, C. (2005) 'Entrepreneurship education and training: Can entrepreneurship be taught? Part II', *Education + Training*, 47, 158–169.

Izquierdo, E. and Buyens, D. (2008) "Impact assessment of an entrepreneurship course on students' entrepreneurial competencies: A constructivist perspective," *International Advances in Economic Research*, 14, 395–406.

Jack, S. L. and Anderson, A. R. "Entrepreneurship education within the condition of entreprenology," paper presented at Conference on Enterprise and Learning, Aberdeen, United Kingdom, 1998.

Johannisson, B. (1991) "University training for entrepreneurship: Swedish approaches," *Entrepreneurship and Regional Development*, 3, 67–82.

Katz, J. and Gartner, W. (1988) "Properties of emerging organizations," *Academy of Management Review*, 13, 429–441.

Kolvereid, L. (1996) "Prediction of employment status choice intentions," *Entrepreneurship Theory and Practice*, 20(3), 47–57.

Kolvereid, L. and Moen, Ø. (1997) "Entrepreneurship among business graduates: Does a major in entrepreneurship make a difference?," *Journal of European Industrial Training*, 21, 154–160.

Krueger, N. F., Reilly, M. D. and Carsrud, A. L. (2000) "Competing models of entrepreneurial intentions," *Journal of Business Venturing*, 15, 411–432.

Lent, R. W., Brown, S. D. and Hackett, G. (1994) "Toward a unifying social cognitive theory of career and academic interest, choice, and performance," *Journal of Vocational Behavior*, 45, 79–122.

McGee, J. E., Mueller, S. L. and Sequeira, J. M. (2009) "Self-efficacy: Refining the measure," *Entrepreneurship Theory and Practice*, 33, 965–988.

McMullan, C. A., Chrisman, J. J. and Vesper, K. (2001) "Some problems in using subjective measures of effectiveness to evaluate entrepreneurship assistance programs," *Entrepreneurship Theory and Practice*, 26, 37–54.

Matlay, H. (2008) "The impact of entrepreneurship education on entrepreneurial outcomes," *Journal of Small Business and Enterprise Development*, 15, 382–396.

National Agency for Quality Assessment and Accreditation (2007) *Mapa de las actividades de empleo realizadas por las universidades*. Madrid: ANECA.

Nunnally, J. C. (1978) *Psychometric Theory*. New York: McGraw-Hill.

Nunnally, J. C. and Bernstein, I. H. (1994) *Psychometric Theory*, 3rd Edn. New York: McGraw-Hill.

Okudan, G. E. and Rzasa, S. E. (2006) "A project-based approach to entrepreneurial leadership education," *Technovation*, 26, 195–210.

Oosterbeek, H., van Praag, M. and Ijsselstein, A. (2010) "The impact of entrepreneurship education on entrepreneurship skills and motivation," *European Economic Review*, 54, 442–454.

Placer, J. L. and Vázquez, J. L. (2005) *Actividades y resultados del Plan Legio en 2002–05 y proyección para 2006*. León: Secretariado de Publicaciones y Secretariado de Promoción Empresarial de la Universidad de León.

Pollitis, D. (2005) "The process of entrepreneurial learning: A conceptual framework," *Entrepreneurship Theory and Practice*, 29, **399–424.**

Rasmussen, E. A. and Sørheim, R. (2006) "Action-based entrepreneurship education," *Technovation*, 26, 185–194.

Recommendation 2006/962/EC of the European Parliament and of the Council of 18 December 2006 on key competences for lifelong learning. *Official Journal of the European Union* L 394, 10–18.

Ringle, C. M., Wende, S., and Will, A. (2005) *SmartPLS 2.0 (beta)*. Hamburg: University of Hamburg.

Robinson, P. B., Huefner, J. C. and Hunt, H. K. (1991) "Entrepreneurial research on student subjects does not generalize to real world entrepreneurs," *Journal of Small Business Management*, 29(2), 42–50.

Sandberg, W. R. and Hofer, C. W. (1987) "Improving new venture performance: The role of strategy, industry structure and the entrepreneur," *Journal of Business Venturing*, 2, 5–28.

Schröder, E. and Schmitt-Rodermund, E. (2006) "Crystallizing enterprising interests among adolescents through a career development program: The role of personality and family background," *Journal of Vocational Behavior*, 69, 494–509.

Shapero, A. (1982) "The social dimensions of entrepreneurship," in C. Kent, D. Sexton and K. Vesper (eds), *The Encyclopedia of Entrepreneurship*. Englewood Cliffs, NJ: Prentice-Hall, pp. 72–90.

Shaver, K. and Scott, L. (1991) "Person, process, and choice: The psychology of new venture creation," *Entrepreneurship Theory and Practice*, 16(2), 23–45.

Sierra, J. (2010) *Creación de empresas y emprendedores en la Región de Murcia*. Murcia: Consejo Económico y Social de la Región de Murcia.

Soutaris, V., Zerbinati, S. and Al-laham, A. (2007) "Do entrepreneurship programmes raise entrepreneurial intention of science and engineering students? The effect of learning, inspiration and resources," *Journal of Business Venturing*, 22, 566–591.

Stone, M. (1974) "Cross-validatory choice and assessment of statistical predictions," *Journal of the Royal Statistical Society*, 36, 111–133.

Storey, D. J. (2000) "Six steps to heaven: Evaluating the impact of public policies to support small business in development economies," in D. Sexton and H. Landstrom (eds). *The Blackwell Handbook of Entrepreneurship*. Oxford: Blackwell, pp. 176–193.

Varela, R. and Jimenez, J. E. (2001) "The effect of entrepreneurship education in the universities of Cali," *Frontiers of Entrepreneurship Research, Babson Conference Proceedings*. Available online at www.babson.edu/entrep/fer (accessed 14 September 2011).

Vázquez, J. L., Naghiu, A., Gutiérrez, P., Lanero, A. and García. M. P. (2009) "Entrepreneurial potential in the University: Intentions and attitudes towards new venture creation," *Bulletin UASVM*, 66(2), 507–512.

Vázquez, J. L., Georgiev, I., Gutiérrez, P., Lanero, A. and García, M. P. (2010a) "Promoting regional development from university. Students' perceptions on entrepreneurship motivation and training programs," *Trakia Journal of Sciences*, 8(1), 1–7.

Vázquez, J. L., Lanero, A., Gutiérrez, P., García, M. P., Alves, H. and Georgiev, I. (2010b) "Entrepreneurship education in the university: Does it make a difference?" *Trakia Journal of Sciences*, 8(3), 64–70.

Vesper, K.H. and Gartner, W.B. (1997) "Measuring progress in entrepreneurship education," *Journal of Business Venturing*, 12, 403–421.

Westhead, P., Storey, D. J., and Martin, F. (2001) "Outcomes reported by students who participated in the 1994 Shell technology enterprise programme," *Entrepreneurship and Regional Development*, 13, 163–185.

Wold, H. (1980) "Soft modeling: Intermediate between traditional model building and data analysis," *Mathematical Statistics*, 6, 333–346.

Wold, H. (1985) "Systems analysis by Partial Least Squares," in P. Nijkamp, H. Leitner and N. Wrigley (eds), *Measuring the Unmeasurable*. Dordrecht: Martinus Nijhoff Publishers, pp. 221–251.

14 The role of university medical centres in an innovation network

Dennis Häckl, Hans Wiesmeth and Oliver Fiala

1 Introduction

1.1 Background: the relevance of universities and the medical sector for Germany

Studies on the economic impact of universities or their facilities can be traced back to the times of the foundation of the first universities in Europe. Supporters of the University of Basel, for example, which was founded in 1460, showed in an analysis for the city council the regional economic benefit of the university.

Interestingly enough, they did not only focus on the additional consumer spending in the city of around 20 guilder per student, but also on the effect 'that the sons of those who have to take care of decisions would find easier the way into positions for which a study course was a precondition' (translated analogously by the author, cf. Boockmann 1999: 93). This shows that the direct financial effects of universities have been discussed for at least 550 years and that effects, which are far beyond direct monetary issues, were also in the field of interest. These latter effects are those realized in the long run and would nowadays be regarded as 'network effects'.

The idea of a city benefiting from a university is of course accepted nowadays without any doubt. However, as we will argue in our paper, some effects remain unexposed and – more problematically – some potential that a university could offer to the city remains unused. Indeed, we see missing networks between scientists and their facilities in a region, namely, insufficient cooperation.

From the point of view of a German federal state, which is usually responsible for the financing of universities, the problems are quite different. On the one hand, the benefits of the university are (or at least should be) accepted without any doubt, but on the other hand there are high costs associated with running the universities – money, that could be invested in other projects such as culture or infrastructure. From this point of view, costs and benefits have to be well balanced. However, in financially difficult times the benefits might be outweighed by the costs. In this context, especially those benefits that will be realized in the long run such as network effects, will rarely find attention. Additionally, those effects might be discounted at a very high rate or cannot be quantified clearly. This might lead to a situation in which long-term effects are underestimated or completely neglected.

As we will argue, the German system of medical schools and teaching hospitals can be found in such a situation. History shows that the institutions were not seen as an economic factor but rather as a cost factor. Thus, spending cuts of more than 10 per cent within two years for a medical school or complex privatization of university hospitals occurred and harmed the long-term effects. This has led to a situation in which world-renowned German medical schools and teaching hospitals (in the remainder of the text termed university medical centre) are not used adequately for the benefit of the region.

Indeed, university medical centres should be regarded as an economic factor and treated as such. In order to achieve this, sound and stable financing is needed for the university, but this requires the courage to change the central framework conditions. These changes, in some cases, do not require any additional expenditure.

1.2 The objective and structure of the chapter

In our chapter we discuss the short- and long-term regional effects of medical schools and teaching hospitals. We will examine the current framework conditions that have a significant impact on the extent to which the potential can be realized. We will further highlight the relationship between university medical centres and regional innovation networks.

Our chapter is structured as follows: we start with a short description of innovations in the medical sector and their relevance for the German export sector. Then, we discuss the different roles of university medical centres with regard to their economic impact. The chapter distinguishes between the short- and long-term effects of the medical institutions in question. After that, we draw a conclusion in which we address the long-term effects of university medical centres by showing the case of a large medical school in Germany. The remainder of the chapter is dedicated to some recommendations for changing the framework conditions of the university medical centre.

2 Innovations in the health care sector

An innovation is not only an invention or a new idea. Instead, it first requires a successful application as well as a specific diffusion into the market. In the medical sector we primarily observe pharmaceutical and medical–technological innovations. Nevertheless, there are also innovative services, such as new treatment options, and hybrid innovations combining process and product innovations.

Owing to the German history of successful engineering with a lot of patents and visible products, innovations in Germany are frequently reduced to the technical field, neglecting innovations in the health care sector (cf. EFI 2011: 77). This view is supported by the fact that process innovations make up most of the medical–technological progress, which does not generate 'classical' marketable products, but optimizes, for instance, treatment techniques. Innovative services bringing about better health care provision or more efficient processes could be a sound basis for an interesting business idea from which national and international companies in the field of medical technology could benefit.

To sum up, we regard innovations as product innovations (e.g. pharmaceuticals, medical technology, etc.), innovative services (e.g. treatment or surgical techniques, national and international consulting services, etc.) and hybrid innovations (as in the field of health services research). Furthermore, the non-linearity of the innovation process in knowledge-based sectors, including the medical sector, has to be taken into account. Thus, an innovation process involves different agents who interact with each other so that new knowledge is created, which can then be translated into new products or processes (Fritsch *et al.* 2008: 7). This 'production process' can be understood as being organized in the form of a network, and highlights mutual dependencies between the actors involved. In our opinion, a central institution coordinating the process and advocating for discipline is a necessary establishment (Ramlogan *et al.* 2007).

The most important countries for the export of German medical and pharmaceutical products are the USA and several European countries. Large emerging countries with a fast growing economy such as China, India or Brazil, however, are of less relevance to German exporters (cf. Institut Arbeit und Technik 2009: 7f.). Furthermore, if the number of clinical studies is taken as a proxy for the innovativeness of a country in the medical sector, there might be a shift to those emerging countries, as recent statistics on clinical studies show (Institut Arbeit und Technik 2009: 11).

In sum, there will be many challenges for the export-driven areas in the health care sector with regard to the development of new markets; the export sector of the German health care sector has thus to defend current markets or re-enter them. To tackle these emerging problems, innovations are needed which will only be 'generated' within a strong innovation system.

3 The economic impact of medical schools and teaching hospitals

3.1 Short- and long-term effects

The economic effects of a university medical centre and its roles in a regional innovation system or a regional network are various. We observe monetary and non-monetary effects with different regional and temporal impacts, which can be divided into direct and indirect effects. Bauer (1997) examined one university with regard to its economic impact and identified employment, income and information effects. This typology is also applied to the analysis of a university medical centre in this paper and will further be enriched.

Employment effects consider the function of a university as an employer, that is, creating new jobs in a region, and as a provider of human capital, that is, increasing the supply of qualified graduates on the job market. Income effects cover additional income that is created by the existence of a university and thus contain possibilities for consumption in a specific region. Information effects describe effects that result from the knowledge exchange between different actors and from the application of specific knowledge. Information effects contain, for example, scientific know-how, which can be transferred to cooperating companies and consequently supports their innovativeness (Bauer 1997).

Studies about scientific institutions usually attribute two different roles to academic institutions: that of a consumer of inputs and generator of outputs (cf. Niermann 1996; Bauer 1997). The amount of input consumed by one academic institution consists of its various expenses for construction, material, personnel, etc. In many studies student expenditure is likewise regarded as an input. This input induces further – short-term – demand effects. These monetary demand effects create income and employment effects. Obviously, information effects cannot be integrated into such an analysis.

When observing the generation of output in an academic institution, we examine the effects of its research, teaching and (executive) education. This induces – in the long term – so-called supply-side effects, for example, the provision of human capital in the form of graduates and the supply of information through research activities. This supply finally leads to innovations and creates locational advantages.

3.2 Demand side effects

The foundations for such impact models can be found in the methodology of studies that consider the economic impact of universities (cf. Bauer 1997 for Munich; Leusing 2007 for Flensburg; Niermann 1996 for Bielefeld; Pfähler *et al.* 1997 for Hamburg).

Some of these studies include teaching hospitals in their analysis. The authors highlight the special task of the hospital in health care provision; furthermore, they identify higher expenses for construction and a lower regional demand for materials.

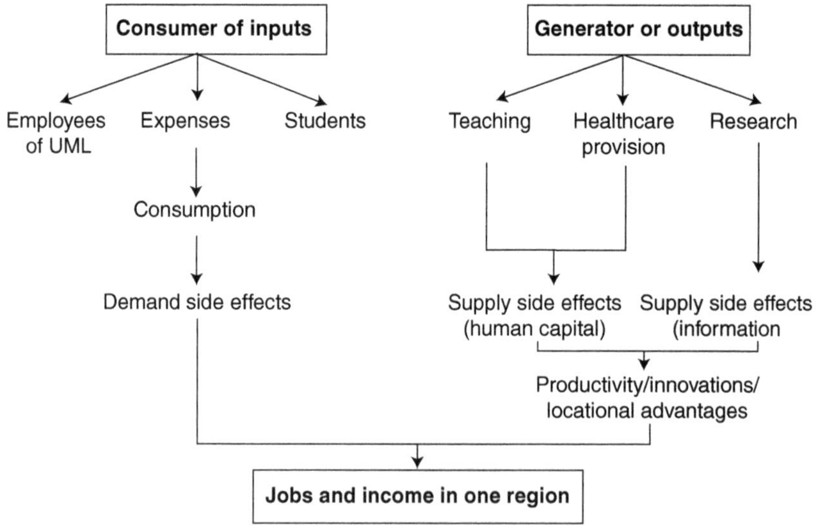

Figure 14.1 The economic impact of a university medical centre.

Source: adapted from Leusing (2007).

Drawing comparisons between different studies is very difficult, as a wide variety of methods can be applied. Moreover, differences may exist owing to the size of the region (different impact on regional effects) and the institutions that were included (some studies include libraries, etc.). Nevertheless, some authors compare the conclusions of such studies (e.g. Leusing 2007) to find patterns in regional economic impact. For our own study, the following findings are of interest:

1 Expenses of personnel and students tend to have the highest regional economic impact.
2 The existence of universities attracts more extra-university (research) institutes to the region.

As highlighted above, such an impact model can neither explain the impact of the university medical centre on the economic structure of a region nor the positive effects of knowledge transfer on the innovation capacity.

The central problem of such analyses is the correct attribution of indirect effects to the university medical centre. There are many factors affecting the decision as to whether and where to locate a research institution or a company. In such analyses very strict assumptions have to be made; for example, the university medical centre is the only reason for the location of such institutions, which might not hold in reality (Bessette 2003).

There are different methods to estimate such short-term effects. Easy impact models are frequently used calculating direct effects (primary income, consumption and employment effects, which are directly related to the university), indirect effects (generated by expenses of the institution with respect to construction and materials; third-party funded positions, effects of research institutes, effects caused by students and visitors) and induced effects ('induced' by the consumer spending of all employees). The ratio of direct effects to the sum of all effects can be interpreted as multiplier. Another way is the application of a Keynesian-like multiplier using a formula. This income multiplier is calculated by using the consumption rate, the tax rate and the share of the amount of goods and services consumed that remains in the region. However, this method neither allows the consideration of different household types, that is, varying income situations and thus different consumption rates, nor a sector-specific analysis. A third alternative is the input–output analysis, which allows a sector-specific analysis. A precondition for such a detailed analysis is a regionalized input–output matrix, containing relationships of intermediate consumption for each sector. A disadvantage of this method is obviously missing data for calculating regionalized input–output matrices.

For policy makers, it would be very easy to justify governmental budget division on the basis of a multiplier analysis for different projects. The money is best invested in projects with higher multipliers. However, concrete multipliers depend on the correct attribution of indirect effects and on the methods applied. For Germany, multipliers of different academic institutions vary between 1.0 and 3.41 (Spehl *et al.* 2005).

3.3 Supply side effects

As shown in Figure 14.1, the university medical centre impacts on the regional economy not only by its spending, but also by generating outputs. Regarding teaching, there are no specialties of the university medical centre against other academic institutions. The output of teaching and (executive) education can be seen in the alumni of the institution, generating human capital for a region. What distinguishes a university medical centre from other institutions is the provision of health care. The teaching hospital uses high-class medical technology and is able to treat severe cases, which other hospitals cannot treat. Thus, the hospital also contributes to the human capital in one region since – in a health economic sense – it reduces the number of days lost owing to illness and gets people back to their work earlier: thus the productivity losses of an economy are reduced.

In our paper, we do not want to focus on these human capital effects in more depth. Instead, we want to contribute to the understanding of the positive external effects of research and the interaction with other actors in the context of generating innovations. To do so, we will discuss the role of the university medical centre as a central node of a network.

3.3.1 Vision: university medical centres as a central node of an innovation network

As highlighted in the introductory chapter, it is hard to define innovations in the health care sector. Innovations can range from pharmaceutical (e.g. new agents) to medical–technological, or include process innovations offering, for example, new treatment options (cf. Häckl 2010: 59).

Many of these innovations cannot be transferred into 'marketable' products, which accounts for the difficulty in defining their monetary value. In this regard the economic impact cannot be discovered as in the case of short-term effects, for example, estimated via a multiplier or by fiscal effects. This leads to a situation in which the effects of the innovation process cannot be quantified correctly. Solely focusing on 'marketable' products would systematically undervalue the innovative aspects within the health care sector.

But how can these factors of the innovation process be operationalized? Which data should be gathered and to what extent?

It is equally important to consider the surroundings of innovations, namely, the framework conditions under which innovations are to be generated. The proverb 'necessity begets ingenuity' does not apply to most areas, least of all to top-class medicine. Demanding the establishment of close ties between doctors and the industry, one has to take into account the time-consuming character of such a process. The invested time will be missing elsewhere and thus co-workers are required to compensate for it.

According to the theory of regional innovation systems, an innovation process is 'an event based on the division of labor, during whose course the cooperation of different actors creates new knowledge that is translated into new products or manufacturing processes' (Fritsch *et al.* 2008: 7). Of major importance is the transfer of knowledge between the involved parties, which is influenced to a high extent by the regional surroundings. This is why a positive correlation between the geographical concentration of industry and science and innovation capacities can be detected. It is to be assumed that innovative motors, namely, central network nodes, will establish within the regional innovation system instead of all involved parties contributing equally to the innovation process.

The innovation network's success depends decisively on its individual actors, who obviously weigh the accruing costs against emerging utilization effects before and throughout their participation in the network. The following institutions, as well as the included persons, constitute the participants of such a network: public and private universities, non-university research institutes, private companies, demanders (particularly in systems with so-called Open Innovation) or state actors (e.g. ministries), as well as organizations (associations etc.) (cf. Fritsch *et al.* 2008: 7).

As mentioned earlier, the division of labour between the involved parties characterizes the innovation process. While previous papers assumed a linear process, that is, the results gained in (mostly university) foundational research are used for application-oriented research, which ultimately creates marketable products in interaction with (private) companies, nowadays a process is assumed that is characterized by frequent feedback between individual actors. It is no longer possible to separate the areas of foundational and application-oriented research and the involved actors in the corresponding areas – particularly in interdisciplinary research projects – as the borders become blurred (see Häckl and Wiesmeth 2009). This clarifies the mutual dependencies among actors and shows that the innovation process needs to be coordinated by an institution, that is, requires a central node within the network.

Therefore, individual actors have different roles within the network: in particular, public institutions such as a university medical centre, consisting of a teaching hospital and a medical faculty, combine several functions and are thus of significant relevance:

- provision of medical assistance, that is, the function of medical service provider;
- generator of new knowledge, that is, the function of research centre requiring the corresponding personnel and technical resources, particularly in the field of top-class research;
- preservation of foreign as well as own knowledge, a so-called reservoir and antennae function for explicit as well as implicit (personalized) knowledge. This includes the cooperation of doctors/researchers with colleagues from research institutes belonging to separate national or international networks;
- transmitter of knowledge, that is, the function of teaching centre. Newly generated as well as externally acquired knowledge is transmitted in a personalized way, creating benefits for the society in the area of medical assistance as well as in the form of employment effects through the transfer of knowledge and technology.

Of particular importance is the way in which individual actors cooperate, that is, the network's strength. This can be influenced to a great measure by external circumstances (the existence of different actors in the region, the actor's spatial proximity and platforms for knowledge exchange). A central, coordinating node – just like the university medical centre with its numerous functions – is of high relevance for the possibility of such a network developing. Further, this node can provide common panels for exchange in the form of (inter-) national congresses or smaller discussion groups.

In considering the cooperation of individual actors, several surveys (see Fritsch *et al.* 2008: 24) assign a significant role to informal knowledge exchange and commissioned research. Cooperative ties are preferably established with big companies while the lack of spatial proximity is not considered detrimental. This is particularly interesting for smaller states with hardly any big medical engineering or pharmaceutical companies. However, spatial proximity can influence the building of trust between actors, which is essential for cooperation.

3.3.2 Healthcare provision as an economic factor

In the metropolitan area in and around Munich, the capital of the Free State of Bavaria, the growth and employment potential of the health care 'industry' has been identified for at least twenty-five years and has been realized through governmental support. Ideas were born to establish a 'Bavarian Silicon-Valley' around the university hospital Großhadern in Munich (cf. Universitätsklinikum Großhadern der LMU München 1999). The City of Munich established a Council for health care and – in cooperation with actors from politics and academia – set up a master plan for the health economy in Munich (cf. GesundheitsBeirat der Landeshauptstadt München 2008). One of the targets of the master plan was – as part of the provision of high-class health care – to promote the importance of the health economy for Munich. Munich's position as a centre for world-class education in medicine, life sciences and health sciences with two university hospitals was highlighted.

The most interesting aspect concerning these activities is that the health care sector was no longer regarded as a cost factor but as a sector with high economic impact. Several analyses were conducted in order to improve the understanding of how the university medical centre's economic potential could be realized.

Several areas for action were identified; among them the optimization of health care provision, the attraction of international health care demand and the support of life sciences industry (biotechnology, pharmaceutical industry and medical technology). The university medical centre is related to all such areas and can function as a catalyst for development.

The relevant question in this context is to what extent the university medical centre can contribute to the generation of innovations, an element that mainly depends on the attitudes of the involved actors. Are the medical staff and the researchers, both in the university hospital and the medical faculty, willing to cooperate with the industry? The medical staff has primarily to provide medical care while the researchers have to fulfill their teaching requirements and do research and publish the results in order to improve their reputation. If there is time left, then there will be an interest in cooperative action.

However, apart from the time restrictions and the interest in scientific research, there are further obstacles to practice-oriented research generating innovations as output. In many cases, researchers report a lack of support from their universities when trying to commercialize their research findings. They mention difficulties finding adequate business partners and the time-consuming process before receiving a patent.

In compliance with the findings of interviews conducted with leading personnel in hospitals, it is very hard to set up a direct cooperation between researchers and companies. They argue that an intermediary institution is often needed to translate research results into marketable products.

Based on this result, we will now argue that the state has to reconsider its impact on the framework conditions. Adequate framework conditions and mediators are necessary to introduce innovations – generated in the university medical centre – to the market.

4 Policy implications

4.1 Tapping the potential of the innovation network

The innovation network, which is going to be established by the university medical centre, leads via product and process innovation to new, interesting positions in a region. Policy makers do not consider this positive 'external effect' entirely, which leads to underfinancing. Furthermore, structural parameters that might lead to more flexible framework conditions should be reconsidered.

According to basic economic theory, adequate framework conditions are necessary in order to tap the full potential of an institution or economic agent. Some of these possibilities for internalizing this external effect will be discussed in the remaining section.

As shown earlier, researchers need to be motivated to take part in this innovation network more intensively. To encourage them, financial incentives offering a bonus for transfer activities should be provided. Moreover, shifting additional tasks to other members of staff (e.g. administrative work) could bring relief to the researchers and improve their research activities. The German Council of Science and Humanities, for instance, recommends this measure (Wissenschaftsrat 2007: 54). From an economic point of view, however, forms of monetary compensation will have a higher influence on cooperativeness than non-monetary incentives. To intensify research, which is supported by third-party funds, achievement-orientated compensation and support related to administrative issues are recommended (Fritsch *et al.* 2008: 26).

4.2 Establishing intermediary institutions

As the case of the university hospital Großhadern in Munich exemplified, an institutionalized department for knowledge and technology transfer can contribute to the innovativeness of a region, under the condition that there is an institution, like the centre of the Helmholtz Association, fulfilling an intermediary function between the university medical centre and private companies.

Such an institution could be a specialist point of contact for both actors and influence the cost–benefit considerations about cooperation positively. It could take up research ideas at the university at an early stage and provide ideas for the commercialization of those research projects.

Although the university medical centre can function as a catalyst or central node of the regional innovation network, the eventual result depends on the external circumstances and the behaviour of the individual actors. If the network comprises only small and medium-sized enterprises, research projects requiring a lot of resources might not be undertaken on a regional basis. Furthermore, not every innovation 'generated' by the university medical centre can be adopted by such companies. In many cases the private (business) partners in such networks lack absorptive capacity.

5 Conclusion and future research

In our paper we focused on a university medical centre as a consumer of inputs and a generator of outputs. While the regional economic effects of the first role can be measured relatively easily, the effects of the latter role can hardly be measured by applying traditional methods. Usually, patent statistics give an idea about regional innovation systems and the various activities of the respective partners; however, patents are of little relevance in the health care sector (with the exception of biotechnology or the pharmaceutical industry). Future research will therefore have to develop different measures, such as the analysis of joint publications, to identify the position of the university medical centre in the regional innovation network.

Apart from these methodological issues we wanted to contribute to the understanding of the positive external effects of research and the interaction with other actors in the context of generating innovations. The state has to be aware of its impact on the framework conditions, which directly impact the success of a university medical centre. Adequate framework conditions and mediators are necessary in order to introduce innovations – generated with or within the university medical centre – into the market. As a result, many institutions in the medical sector will no longer be regarded as cost factors but more as economic factors. The aim of our paper was to start a discussion on the role of a university medical centre in a regional innovation network. Future research will have to resume this discussion and show concrete case studies on this role and generate empirical evidence.

References

Bauer, E.-M. (1997) *Die Hochschule als Wirtschaftsfaktor: Eine systemorientierte und empirische Analyse universitätsbedingter Beschäftigungs-, Einkommens- und Informationseffekte – dargestellt am Beispiel der Ludwig-Maximilians-Universität München.* Kallmünz/Regensburg, Münchner Studien zur Sozial- und Wirtschaftsgeographie, Vol. 41.

Bessette, R. W. (2003) 'Measuring the economic impact of university-based research', *Journal of Technology Transfer*, 28, 355–361.

Boockmann, H. (1999) *Wissen und Widerstand – Geschichte der deutschen Universität*. Berlin: Siedler Verlag.

EFI Expertenkommission Forschung und Innovation (2011) 'Research, innovation and technological performance in Germany, Report 2011'. EFI, Berlin. Available online at www.e-fi.de/fileadmin/Gutachten/EFI_2011_en_final.pdf (accessed July 2011).

Fritsch, M., Henning, T., Slavtchev, V. and Steigenberger, N. (2008) 'Hochschulen als regionaler Innovationsmotor?' Hans Böckler Stiftung – Arbeitspapiere Nr. 158.

GesundheitsBeirat der Landeshauptstadt München (2008) 'Masterplan Gesundheitswirtschaft München, Bericht und Entwicklungsmöglichkeiten'. Available online at www.gesundheitsbeirat-muenchen.de/html/pdf/geswirtsch_0708.pdf (accessed July 2011).

Häckl, D. (2010) *Neue Technologien im Gesundheitswesen. Rahmenbedingungen und Akteure*. Wiesbaden: Gabler Verlag.

Häckl, D. and Wiesmeth, H. (2009) 'Kosten-Nutzen-Betrachtung der Biotechnologie-Offensive des Freistaats Sachsen', Report to the Saxonian State Ministry for Science and the Arts.

Institut Arbeit und Technik (2009) 'Die deutsche Gesundheitswirtschaft – Defizite ihrer außenwirtschaftlichen Aktivitäten und wirtschaftliche Handlungsoptionen', Report on behalf of the Federal Ministry of Economics and Technology. Available online at www.iat.eu/aktuell/veroeff/2009/bandemer01.pdf (accessed July 2011).

Leusing, B. (2007) 'Hochschulen als Standortfaktor – Eine empirische Analyse der regionalökonomischen Effekte der Universität Flensburg', Universität Flensburg Discussion Paper Nr. 15.

Niermann, U. (1996) 'Wirtschaftsfaktor Universität – Eine Input-Output-orientierte Analyse am Beispiel der Universität Bielefeld', in Joachim Frohn (ed.) *Empirische Wirtschaftsordnung und Ökonometrie*, Bd. 2. Münster: LIT Verlag.

Pfähler, W., Clermont, C., Gabriel, C. and Hofmann, U. (1997) 'Bildung und Wissenschaft als Wirtschafts- und Standortfaktor. Die regionalwirtschaftliche Bedeutung der Hamburger Hochschulbildungs- und Wissenschaftseinrichtungen'. Baden-Baden: Nomos Verlagsgesellschaft.

Ramlogan, R., Mina1, A., Tampubolon1, G. and Metcalfe, J. S. (2007) 'Networks of knowledge: The distributed nature of medical innovation', *Scientometrics*, (70)2, 459–489.

Spehl, H., Feser, H.-D. and Schulze, P. (2005) 'Regionalwirtschaftliche Wirkungen der Hochschulen und Forschungseinrichtungen in Rheinland-Pfalz'. Available online at www.mbwjk.rlp.de/fileadmin/Dateien/Downloads/Wissenschaft/wirkung_forschung_lang.pdf (accessed April 2011).

Universitätsklinikum Großhadern der LMU München (1999) *Festschrift anläßlich des Jubiläums 25 Jahre*. Munich: Universitätsklinikum Großhadern.

Wissenschaftsrat (2007) 'Empfehlungen zu Public Private Partnerships (PPP) in der universitätsmedizinischen Forschung'. Available online at www.wissenschaftsrat.de/download/archiv/7695-07.pdf (accessed March 2011).

Part III

Academic footstep in the city

15 The adaptive university

Responding to new expectations and coping with established tasks

Jan Sadlak

Introduction

Historically, there is much evidence that multiple relations existed between universities and their host cities and regions through following their traditional activities such as generation of scientific knowledge and providing education and training. In a particular way universities were also improving the physical characteristics of the city or region. It is worth pointing out that students and other members of the academic community bring positive economic and other types of activities to urban settings, providing energy, determination, curiosity and creativity, which add to the social and cultural atmosphere of cities. Historically, the cities which has recognized, among other things, the importance of support for universities and research institutions has been able to assure a lasting economic growth and generation of wealth (Michalski, 2011).

The long history of the European university provides evidence that seeking a state of 'equilibrium' between the university and the city is an important feature of the broader context –that of relations between the university and the state (Ruegg 1992). What characterizes the present situation is that the universities, particularly those that demonstrate intellectual vigour and research excellence, are looked upon as the engines of economic growth and technological advancement. Consequently, they are increasingly being spurred towards urban and local development plans aiming at 're-imagining the future' of a particular city or urban setting.

The main cause of this accelerated engagement is that 'knowledge', 'innovations' and 'creativity' are recognized as 'the fuel' of the twenty-first-century economic engine. Accordingly, it gives, on the one hand, a new dimension to what universities have always been doing, and, on the other hand, higher education has become drawn into national, regional and local efforts to enhance competitiveness and innovativeness, which should result in capitalising on opportunities in the so-called 'knowledge-based economy'.

This chapter looks at recent initiatives of engaging universities in meeting such developmental expectations in the context of urban transformations and the consequences of such engagement on the university and its mission.

Shifting of the urban world and its universities

There is much evidence that scientific and higher learning activities are unevenly distributed not only between nations but also within them, and concentrated in certain regions and cities. As the study of current scientific landscape carried out by the Royal Society points out: 'Moscow accounts for 50% of Russian research articles; Tehran, Prague, Budapest and Buenos Aires each top 40% of their national outputs, and London, Beijing, Paris and Sao Paolo are each responsible for over 20%' (Royal Society 2011).

The source of this power comes from several areas. First, as in the past, urban areas benefit from agglomeration – or economies of scale – which enables industries and service sectors to have higher productivity than they do in a rural setting. It is also much less expensive to provide goods and services in concentrated population centres. But what needs to be pointed out is that cities of a certain size usually have a critical number of educational and research institutions to attract talent and inward investment. They can also facilitate knowledge exchange, benefiting from network effects between various institutions, including universities, research organizations and cultural institutions that spur economic growth and productivity. As these institutions are the 'fuel' of economic development, it is evident that urban areas that have highly performing universities and research organizations are better placed to attract talent and financial resources.

The above-presented urban centres, as well as some others such as New York, Shanghai, Hong Kong, Tokyo, San Francisco, Berlin, Barcelona, Dublin and Singapore, are in a relatively privileged position for the attracting and retention of talent and creativity. In this context the universities are recognized as people and talent rich communities.

The economic role of cities varies widely among regions today – as do their future growth patterns. It is clear that there is no 'one size fits all' approach to tapping into the advantages deriving from knowledge, innovation and creativity. There is, however, one common element: that the active role of universities is no longer an alternative but a necessity; although the role of universities is generally not restricted to a given urban area, region or country, nevertheless, it is of paramount importance for universities that they are able to function in a sound local infrastructure and a favourable urban environment attractive for both the student population and the academic community at large.

In this context, a strategic global transformation in the urban world should be pointed out. A recently published report (McKinsey Global Institute 2011) confirms the influence of major cities. Half of global GDP in 2007 came from 380 cities in developed regions, with more than 20 per cent of global GDP coming from 190 North American cities alone. The 220 largest cities in developing regions contributed another 10 per cent. But by 2025, one third of these developed market cities will no longer make the top 600. By 2025, 136 new cities are expected to enter the top 600, all of them from the developing world.

Interaction between 'town and gown'

The increased interaction and cooperation between the university and the urban authorities undoubtedly serves the mutual interests of both parties. A survey evaluating urban policies which was conducted in September 2010 by *The Economist Intelligence Unit* has asked the following question (*Economist* 2011):

> 'What should mayors do to make their cities more competitive for business?' The findings were the following:
>
> - 61 per cent: Improve public transportation and road systems;
> - 33 per cent: Improve schooling and higher education;
> - 11 per cent: Create conditions to attract major sporting/cultural/corporate events.

It is not surprising that cities are looking at collaboration with higher education and research with renewed interest. A good example of an active stand by a city to enhance their competitiveness by involving the university is an initiative by the mayor of New York City, Michael Bloomberg, to regain its standing as an 'innovation capital'. The city, which has more than 600,000 students and a number of top-flight universities, is offering real estate – at virtually no cost – plus up to $100 million in infrastructure upgrades to a university willing to build a world-class engineering or applied sciences campus in Brooklyn's Navy Yard or on Roosevelt Island. It is expected that such establishment will also attract a 'critical mass of technology entrepreneurs', capable of creating new companies, construction jobs as well as permanent jobs (Bloomberg 2011).

The winning proposal for this environmentally-friendly and energy-efficient university of applied sciences and engineering was presented by Cornell University and Technion-Israel Institute of Technology. It will be organized around three interdisciplinary hubs in the field of connective media, health and environment. The officially called 'NYCTech Campus' will host some 2,500 students and 280 academic staff. It is expected to be completed by 2027. It should generate more than $US 23 billion in economic activity and $US 1.4 billion in tax with spin-off effects – around 20,000 construction jobs, 8,000 permanent jobs and 600 new businesses that will create an additional 30,000 jobs (Travers 2012).

From a policy point of view it is relevant to look at the key criteria that were set up for the selection of a 'winner'. The following have been indicated as being of particular importance:

- develop research that will lead to the formation and expansion of companies in and attraction of companies to New York City in industries that demonstrate the most potential for growth;
- create construction and permanent jobs and generate tax revenue;
- develop a financially self-sustaining campus;
- contribute to the diversification of New York City's economy by expanding its applied sciences sector.

There are also European examples of policies to lift a city's economy by promoting the city as a 'creative hub' in supporting existing institutions or creating new 'out of paradigm' type of learning and research organizations. An example of the latter is the Skolkovo Institute of Science and Technology, or SkTech – a new elitist graduate research establishment which is a joint undertaking of the Skolkovo Foundation and the Massachusetts Institute of Technology (MIT). This institution, located near Moscow, aims to create a new model for graduate education and research in science and technology, but also to facilitate interaction between researchers, entrepreneurs and investors in five 'clusters': information technology, biomedical, energy efficiency, space and nuclear technologies, all of which have deep roots in the country's science. SkTech, which as foreseen should have around 1,800 students, is conceived, together with 40 corporate research and development (R&D) centers and a 'Technopark', housing up to 1,000 start-ups, as important element of making this Moscow's suburb 'one of the world's biggest high-tech cities in a few years' (*The Economist*, 2012).

An even more pioneering arrangement is proposed in the context of the project called CREATE – Campus for Research Excellence And Technological Enterprise – which was initiated in 2010 by the Government of Singapore. It is based on the idea of bringing together in one location a selected number of top-level research universities and institutions representing diverse academic disciplines to interact readily in order to generate innovation. It aims to host a couple of research centres from each regions of the world. The specially constructed environmentally sustainable building complex located in the proximity of the National University of Singapore expects to host some 1,000 researchers and to be a magnet for attracting scientific talent from all over the world.

It is nowadays quite common that forward-looking cities are making efforts to attract investment by promoting their particular advantages in which reference to local universities and science and technological parks is a norm. Even countries and regions that are coping with present crises see investments in innovation as part of a strategy for economic recovery. An example of such political planning is the Scientific and Technological Park of Terrassa – *Orbital 40* – which has been recently created within the metropolitan area of Barcelona. Its functioning is concentrated on so-called 'value industries', which includes aeronautics, optics and photonics. One of the key partners of this park is the Polytechnic University of Catalunya (UPC).

If high tech, health and environment represent the dominant tendency of the economic growth of cities, there are also other options in support linking universities to the cities by emphasizing fashion, art and music, hoping that the magnetic effect of such policies would rub off on the economy. A good example of such policy is in Berlin, which supports various start-ups and markets itself as 'a creative capital'.

An even more ambitious plan of interconnection with academe is in Paris, where universities are going to benefit from the opportunities offered by highly competitive funding of 35 billion euros to modernize the French higher education and research sector, commonly referred to as the 'Big Loan' (*Grand Emprunt)* (Juppé, Rocard *et al.* 2009). Out of the total amount, 16 billion euro is dedicated to creating poles of excellence – *l'initiative d'excellence (Idex)*. In the first round, one of three selected by an international jury was a project – *Paris science et lettres*. At the beginning of 2012, five new winning projects were announced, three of which are located in Paris or its vicinity – *Sorbonne-Universités*, *Sorbonne-Paris Cité* and *Saclay*. Each project will receive in the course of four years funds between 700 and 750 million euros for the upgrade of infrastructure and living conditions of students, as well as the creation of centres of excellence in research and learning.

Two observations seem relevant in the context of this policy initiative. It is expected that the received funds will reinforce the 'international standing' of Paris as an academic city, as well as leading to better integration among the higher education and research institutions. The government expects that it will gradually lead to a more integrated institutional structure of French higher education. For the latter reason, the applications must be presented by consortia – so-called PRES (*pôle de recherche et d'enseignement supérieur*). Several such consortia have been developed contradicting a traditional view about the silo mentality academic community. It is still to be seen how this initiative will actually be implemented and how the process of adaptation will be carried out at the ground level. There is already evidence that effective collaboration between universities, research organizations and the state has resulted in the 'boom in the big cities', not only in Paris and its vicinity but also in other dominant, highly urbanized regions such as Rhône-Alpes, Provence-Alpes-Côte d'Azur, Strasbourg, Toulouse and Bordeaux. The downside of this development has been that disparities have grown between these metropolises and other regions where the number of academics and researchers was insufficient to attain critical mass to develop viable research proposals (Marshall 2011).

It should also be pointed out that the examples of 'investing in universities' as engines for development of urban economies are not limited to America or Europe. Actually, policies of boosting research, innovation and enterprise talent are a global phenomenon. Thanks to the new era in communication, no country, region or city has a lock on research, innovation and technology anymore. However, it has always been dependent on creative people and entrepreneurship. In this regard cities with an adequate intellectual base and cultural diversity can attract creative people and be well positioned to draw more talent for which nowadays there is intense and global competition.

It is expected that a final result of various initiatives involving universities and other research organizations will result in turning cities into entities of sustainable and inclusive economy, delivering high levels of employment, productivity and social cohesion. Leaders of the cities and universities growingly recognize that in the era of 'global urbanization' their collective future depends on investment, human capital and innovation, which, on one hand, raises new expectations, but, on the other, brings new challenges.

What all this means for the 'university'

The above-presented policies and initiatives give rise to questions about their impact on the university, keeping in mind that the active role of universities in transferring knowledge and innovation to industry, the service sector and public institutions is relatively new. Traditionally, universities have offered education and professional training at the highest level, whilst undertaking research as support activity for educational provision. In this model (often referred to as the Humboldtian model of the university), knowledge transfer was an individual and strongly regulated or institutionalized process, based foremost on graduates transferring the state-of-the-art knowledge attained during their studies to their respective professional occupation. This mechanism still today constitutes one of the main forms in which knowledge and innovation is disseminated throughout the community, be it local, regional, national or international. It is, however, a relatively slow and unstructured way of passing on new knowledge to industry and the community. As has already been presented, policy makers have actively sought more direct, faster and more effective means of knowledge transfer. The case for a greater role of the university is to some extent influenced by policies aimed at reviving manufacturing based on technological innovation.

It will come as no surprise that one of the side effects of this more active role assumed by the universities is not only that they have become important but that this has increased interest in and attention to what the universities are doing, commonly referred to as the need for 'accountability', evidence of 'productivity', greater 'transparency' and more accessible information. In other words, the university has become too important to be left to the academe.

This is where university rankings come in. University rankings were originally conceived as 'a study guide' (most likely inspired by various consumer guides) providing information to future students and their parents. A great number of rankings still have this objective in mind and present themselves as a 'good university guide'. If informing students is still the prevailing reason for producing rankings, experience shows that their use is much wider and includes:

- providing information to policy-making bodies, foundations, funding agencies, research councils, corporate sectors and employers, and international organizations;
- fostering competition among higher education institutions;
- providing a basis for awarding quality labels;
- facilitating marketing of institutions and study programmes;
- encouraging creation of centres of excellence; and
- providing supplementary rationale for allocation of funds.

Let us mention that 'university rankings' have entered into many comparative studies, including those of cities, by taking into consideration how many universities in the given city are in leading national or international rankings.

In order to deal with those challenges, universities need to be better primed for a highly competitive environment for talent, funding and resources. Universities often use the shield of 'public service' and exceptionality to perform what in the business world is called 'mature enterprise', tending to be increasingly risk-averse, at times self-satisfied, and in some countries such as the USA, unduly expensive.

The process of adaptation will also imply modification concerning the design of studies in order to stimulate creativity and entrepreneurship, which are important for dealing with the challenges of an increasingly interconnected and complex world. Indeed, just as the advantages the developed countries enjoyed in the past were based in large part on scientific and engineering advances, today it is cognitive flexibility, inventiveness and design thinking that have become equally important. The study, conducted in 2010 through IBM's Institute for Business Value, reflecting the consolidated view of more than 3,600 students in more than forty countries, demonstrates the gap in this domain, as only four out of ten students believe their education has prepared them well to address these issues.

No less important is the adoption, as an integral part of the university mission, of its role in 'continuous learning' (also referred to as 'life-long learning' – LLL). The growing need for a professional workforce to keep upgrading and adapting their skills is recognized as a necessity. Universities will have to pay attention to equipping their graduates with propensity and awareness for such learning. Universities should open up and become not only a cluster of academic staff and students, but also a natural place for interaction with talented, open-minded people. This 'hub function' is particularly important in a world where more and more people may work from home and the dangers of isolation, burn-out or de-professionalization become an issue.

And some final observation. No matter how much we argue about the importance of universities, it needs to be pointed out that universities cannot play a direct role in resolving present and future economic and social problems. However, no country, region or city can count on a lasting lead in scientific know-how, education and innovation. This is why it is so important to assure a supportive political climate stimulating research and innovation. It is a very difficult task taking into consideration that education in general and higher education in particular is seen as one of the most important issues of society – until something else is. During election campaigns, education is always floated about as an important issue. But in reality, the economy, health and welfare reign supreme in the electorate's consciousness. It is therefore important that universities learn how to argue about their place in society.

The analyses presented also confirmed the increased gravity in the relationship between cities and their universities and other research organizations. It derives from better understanding that the growth and prosperity of cities critically depend on identifying and supporting prospective engines, of which the primary and main source is research, innovation and creativity.

References

Bloomberg, Mayor Michael R. (2011) 'Future of New York City', keynote speech on 19 July at the Crain's New York Business conference. Available online at www.nyc.gov/portal/site/nycgov.

Juppé, Alain, Rocard, Michel *et al.* (2009) *Investir pour l'avenir: Priorités stratégiques d'investissement et emprunt national.*

Economist (2011) 'Liveconomics: By the numbers', *The Economist*, 19 February.

Economist (2012) 'Can Russia create a new Silicon Valley?' *The Economist*, 14 July.

McKinsey Global Institute (2011) *Urban World: Mapping the Economic Power of Cities.* Available online at www.mckinsey.com/global.

Marshall, Jane (2011) 'France: City universities "booming" says evaluation', *University World News*, 23 January. Available online at www.universityworldnews.com.

Michalski, Wolfgang (2011) *Capitalising on Change in a Globalising World: A View from Hamburg.* Hamburg: Murmann-Verlag.

Royal Society (2011) *Knowledge, Networks and Nations: Global scientific collaboration in the 21st century.* London: RS Policy document 03/11.

Ruegg, Walter (1992) 'Themes: The social role of the European university', in H. de Ridder-Symoens (ed.), *A History of the University of Europe.* Cambridge/New York: Cambridge University Press, pp. xix–xxviii.

Travers, Eileen (2012) 'New York campus bid to be innovation capital', *University World News*, 15 January. Available online at www.universityworldnews.com.

16 Attracting power of the academic city

Measurement in the context of economic growth

Mikołaj Herbst

Introduction

Human capital stock affects economic growth by increasing labour productivity in a given area or by enhancing the ability of the regional economy to create and absorb innovations. In the case of an academic city, this process can be reinforced by attracting students and researchers to study and work at the local universities. To do this successfully, the city needs not only high quality academic institutions but also a wider labour market for educated individuals and, more generally, an ability to attract the creative class to settle. This chapter provides a comparative analysis of the capacity of the eighteen largest Polish cities[1] (regional capitals) to attract and absorb human capital. The research is based on a unique dataset from the nk.pl website (which allows users to contact their former classmates).

The chapter is organized as follows: section 1 introduces some basic theoretical concepts and empirical results on the relationship between human capital and local development. Section 2 overviews data used in the research. Section 3 develops a conceptual framework for the education-driven and post-educational mobility of individuals between cities. Section 4 is devoted to defining the empirical indicators used in the research. Section 5 discusses the results. Section 6 develops a typology of academic cities based on the results from the previous sections. Section 7 concludes.

1 Human capital and regional development – theoretical foundations

Human capital is considered to be an important factor in regional and local development. The theoretical foundations for this view are provided both by development models outlined in works in the field of Regional Studies (Florida 2000; Porter 2001), and by economic growth models developed by Lucas (1988) and Nelson-Phelps (1966). Although these particular economic theories were formulated mainly in reference to the development of national economies, the mechanisms they unveil can be observed – sometimes even more easily – at lower levels of aggregation.

The impact of human capital (mainly knowledge potential) on economic development is well documented both nationally and regionally. In their major work published in the *American Economic Review*, Mankiw *et al.* (1992) argue that the level of education can largely explain differences in per capita GDP between various countries. The model proposed by Mankiw *et al.*, known as the 'augmented' Solow model (as, unlike the original, it takes into account the human capital factor), has considerably enhanced the quality of GDP per capita forecasts in relation to the original model which only employed such factors as labour resources and physical capital to explain growth. The results obtained by Mankiw *et al.* inspired many researchers who strived to verify the influence of human capital on economic growth in the subsequent years. The key works in this respect were published by Benhabib and Spiegel (1994), Barro (1999), Bils and Klenow (2000) and Bernanke and Gurkaynak (2001).

The role of human capital in development has also been corroborated by many empirical studies conducted regionally. Using a panel dataset for the period from the 1960s to the 1990s, Di Liberto and Symons (2001) prove that, in the past, the level of education significantly influenced the regional rates of growth in Italy, even though this positive effect was primarily owing to the popularization of primary education in the poorly developed south of the country. In contrast, Badinger and Tondl (2002) argue that, contemporarily, in the whole of Europe (the study was carried out using data for 128 regions in ten EU Member States), economic growth is strongly correlated with education resources mostly at the tertiary level. Furthermore, the significance of human capital for the pace of regional development is also corroborated in a paper by de la Fuente (2002) focusing on Spain. In this study, de la Fuente proves that the equalizing of education levels as well as the import of technological solutions have been two main factors responsible for the economic convergence of the Spanish regions. Another study, by Persson and Malmberg (1996), which analyses factors of economic growth in US regions in the period1920–90, shows that, in the past, the level of human capital (measured by the average number of years of schooling) had a substantial, positive impact on regional rates of growth.

Similarly, in post-1989 Poland, regions with high human capital potential reported faster growth. Studies at the subregional level carried out for the years1995–2003 (Herbst 2007) (extended till 2006 in later, draft versions) suggest a strong statistical correlation between the level of the residents' education, the regional per capita GDP and the rate of economic growth.

When comparing the pace of growth of regional economies, what we in fact compare is the potential of the regions' largest cities. Decades of accelerated metropolization, an increasing share of services (including financial) in the GDP, and the growing role of tertiary education in the labour market, are factors that make regional metropolises the key players in determining the development prospects of their regions.

According to the aforementioned theories formulated by Lucas and Nelson-Phelps, human capital affects regional development by improving labour productivity in a given area or by enhancing the capacity of a given economy to generate and absorb innovations in various fields. Seen from the perspective of an academic city, the necessary precondition if these two mechanisms are to succeed is the ability to attract students and academic teachers to the local universities and colleges. To absorb human capital, a city needs not only to have quality higher education institutions but also to offer a ready labour market for its graduates and an attractive place to live for well-educated people. If we limit our discussion to students and graduates (without taking academic teachers into account), we can say that the success of a city in building human capital potential through the strengthening of its academic function is in fact limited to two issues:

- Can a city (its higher education institutions) attract students?
- Can a city retain the graduates in the local labour market?

Finding answers to these questions would make it possible to evaluate the city's efficiency in attracting and absorbing human capital which, according to the theory and results of empirical studies, considerably determines the capacity for rapid development – of both a given city and the surrounding region.

2 Data

Regrettably, the data available in the public statistics system do not allow any in-depth analyses. Although the collected data include information on the number of students and their chosen higher education institutions, fields and modes of study, the official publications say nothing about the past of the students studying in various cities – what secondary schools they attended or what distance they are willing to travel in order to study in the city and the university or college of their choice. Such data can be found in the USOS software now being widely introduced by Polish universities, but access is given rather reluctantly owing to fears of infringing upon the personal data protection regulations and because of technical difficulties. However, even if such a set of data were compiled for one higher education institution, it would not be sufficient to evaluate the 'academic attractiveness' of the whole city or to make comparisons between individual universities or cities. Therefore, studies on the capacity to attract students published in Poland to date have been either monographs that analyse the situation at individual higher education institutions (Wasielewski 2004) or have relied on estimates based on very general data published by GUS – the Central Statistical Office (Chojnicki and Czyż 1997).

Data concerning the situation of university graduates are even scantier. Research of this type is conducted almost exclusively at the level of individual higher education institutions, by the university career bureaus, bureaus of promotion or graduates' associations, such as the studies by the Socio-Economic Association 'Absolwent' (The Graduate) of the University of Information Technology and Management in Rzeszów (*Losy absolwentów – raport* 2008). These studies cover samples that are dissimilar in terms of numbers and methodologies used and therefore their results cannot be compared. Moreover, in many cases, the questionnaires used in such studies in no way tackle the issue of the spatial mobility of graduates but focus on their career achievements instead.

The study discussed in this chapter relies on a source of data outside the sphere of public statistics, and does not require statistics to be collected at higher education institutions. The source is nk.pl, a social networking website where users can voluntarily register in the list of their current or past class or student group in order to maintain or renew social contacts. The advantage of this mechanism in terms of its use for scientific research is a high probability that the information provided is reliable. In order to find classmates, the potential user needs to virtually register in real schools and classes that he or she attended, and cases of fabrication (as a joke or for commercial purposes) are easy to uncover and subsequently can be omitted in research.[2]

For those users who 'register' in their current or past student groups and to one of their classes at an earlier stage of education, information on the distribution of these educational establishments in Poland becomes available.

Moreover, what distinguishes the nk.pl website as a scientific tool is its mass popularity. At the end of 2008, the website had over 11 million registered users. If, from the total population of Poland, we deduct preschool children and the elderly, the majority of whom lack sufficient computer skills, we will see that over one third of Polish citizens are registered in nk.pl. This rate will be much higher if we take into account only the population below a given age threshold (set e.g. at 50 years of age) because young people are considerably over-represented among the website's users.

The problem of the over-representation of data (or rather a lack thereof) is frequently mentioned in connection with attempts to use information from social networking websites for scientific purposes. It should be noted that any evaluation of the usefulness of data depends on the research context. In a situation where – as in the study at hand – the goal is to assess the current capacity of cities to attract students and retain graduates in the local labour market, even a considerable over-representation of young people in the research sample should not be regarded as an adverse feature. This is because, in terms of the objective of the study, the current situation and most recent changes in the phenomenon under analysis are more important than the situation and trends observable in the more distant past. The same argument applies to another feature of samples originating from social networking websites: the absence of people without Internet access. Even though this problem still affects a considerable number of Polish households (according to the Polish Central Statistical Office in 2010 about 39 per cent of households did not use the Internet), it can be assumed that this is of no significance as far as the population of students and university graduates is concerned.

The research is based on the sample of individuals who were registered users of nk.pl in January 2009, and at this time were virtually enrolled at any tertiary school as either a student or former student. In the case of graduates, the sample was restricted to individuals who graduated in 1990 or later. The resulting total number of observations was 2.0 million, of which 1.3 million were graduates and 0.7 million were students at the time the data were collected from nk.pl.

The average age in this user group (taking into account only those users who revealed their age) was twenty-eight years. Women accounted for as much as 62 per cent of the sample. About 730,000 users (35 per cent) were students, and the remaining 65 per cent were university graduates.

3 Conceptual framework – mobility of human capital in the context of benefits for academic cities

It is useful to consider, following Hoare and Corver (2010), a simple two-city model of educational and post-educational mobility. Figure 16.1 illustrates possible strategies taken by an individual at two stages of life – when leaving home (H) to enrol in a tertiary school, and when graduating from university (U) and deciding where to live and work (L). At both stages young adults choose between an Eastern and a Western city.[3]

Considering both the educational and post-educational mobility of individuals, Hoare and Corver identified four possible types of behaviour:

- local path, chosen by those who study and work locally (in their home city);
- return path, chosen by those who migrate to study, but return to work;
- settlement path, chosen by those who migrate to study, and stay in the new location to work;
- outsider path, chosen by those who study locally, but migrate to work after graduation.

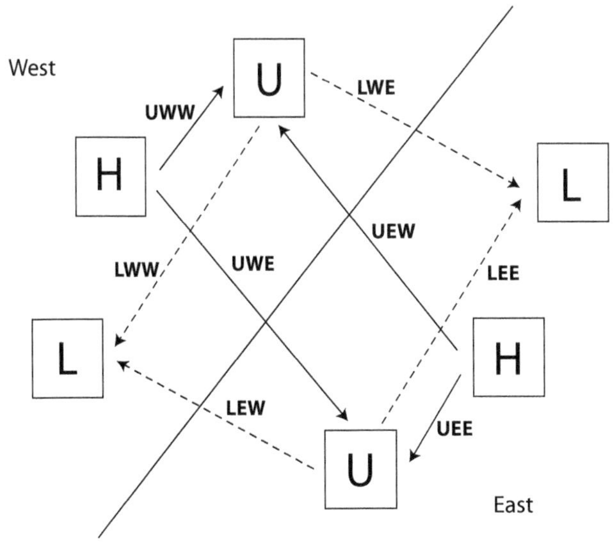

Figure 16.1 Educational and post-educational mobility in a two-city model.

Source: adapted from Hoare and Corver (2010).

In the case where a city benefits from its academic function, the more human capital is attracted and absorbed, the better it serves the local economy. Thus in the two-city model shown in Figure 16.1, West wants to maximize:

that is, to hold as many graduates from local secondary schools at the local university as possible, and also to attract the highest possible number of students from East. Besides this, West maximizes:

which means it wants local students to stay on the Western labour market after graduation, and to persuade graduates from East to work in West.

In the real world students and graduates choose their place of studies and work from among a large number of available destinations, and not just two cities. Moreover, before enrolling at university, students grow up in different places, some in big cities, others in smaller settlements spread over the territory of a country (if one considers mobility on a national scale). When migrating for education or work, individuals will take into account not just the attractiveness of different destinations (the quality of the university, expected wages on the local labour market), but also the cost of migration. The net benefits of migrating to a given city is therefore the difference between the expected benefits and the costs.

The cost of migration is difficult to estimate, because it includes both pecuniary and mental elements. Monetary cost can be measured as the difference in prices of goods and services between the current place of residence (A) and the potential destination (B). The cost of travelling from A to B may also be taken into account, especially if one plans to come back to the place of origin on a regular basis. The mental cost of migration may in turn be considered as the negative consequences of moving away from the family home or, broader – from the hometown. Migrating limits support from family, and places one outside the social network he or she has been part of. It also forces one to establish new relationships in the destination place.[4]

Many studies use distance between place of origin and destination as a proxy measure for the cost of migration (see for example Ghatak *et al.*). This is a useful approach, as one can assume that at least two costs related to migration – the cost of travelling and the mental cost, are proportional (and positively correlated) with the distance between the original place of residence and the destination. In turn the data on the third cost component – the difference in prices of goods and services (cost of living) is frequently unavailable at the level of individual cities.

If the cost of migration is proportional to the distance, then an individual considering two destination cities offering similar benefits will always choose the one that is located closer to his or her current place of residence. Based on this assumption, we can recognize the attractiveness of a city as a migration destination for human capital by the fact that it is able to attract educated people even from distant localities. Alternatively, with the eighteen Polish cities examined in this chapter being regional capitals, we can measure the attracting power of a city by verifying how good it is in attracting migrants from outside its own region.

4 Indicators

In the introductory part of this chapter I formulated two questions related to the city's potential benefits from performing an academic function. The first question was whether a city, or its higher education institutions, could attract students. The other was whether a city was able to retain the local university graduates on the local labour market. Let us now, based on the above discussion, split these questions into five more detailed issues:

- How far are students willing to travel to study in the considered city?
- How many students does the city attract from outside its own region?
- What proportion of local university graduates settles in the city?
- What proportion of graduates originating from outside the city settles there after graduation?
- How far from the city do graduates go after they complete higher education?

Based on the above questions let us propose a set of indicators measuring the performance of an academic city with respect to the attraction and absorption of human capital. The formulas and their explanations are shown in Table 16.1.

In addition to measuring separately the performance of cities with respect to their ability to attract and absorb human capital, I propose two composite indicators, taking into account both these aspects.

Table 16.1 Measures of city performance in human capital attraction and absorption

Measures of human capital attraction (N=number of students)	
$AT_{1i} = d_{i\ sec\_tert}$	Mean distance from secondary school to tertiary school in i
$AT_{2i} = \frac{N_{i\ stud\_outreg}}{N_i}$	Share of tertiary students from outside of i's region in the total student population of i
Measures of human capital absorption (N= number of graduates)	
$AB_{1i} = d_{i\ tert\_resid}$	Mean distance from tertiary school in i to graduate's place of residence
$AB_{2i} = \frac{N_{i\ settlers}}{N_{i\ grad} - N_{i\ local}}$	Share of tertiary students settled in i among those who came from outside i's region
$AB_{3i} = \frac{N_{i\ settlers} + N_{i\ local}}{N_i}$	Share of i's university graduates settling in i.

The first composite indicator is the ratio of indicators AT1 and AB1, that is the mean distance between secondary school and tertiary school for all students enrolled in the university in the city *i* and the mean distance between the city *i* and the current place of residence for all graduates of *i*'s universities.

$$CI_1i = \frac{d_i\,sec\_tert}{d_i tert\_resid}$$

The other composite indicator is based on the concept of migration effectiveness . It is computed as the difference between the number of students originating from outside the local region, but studying in *i* and settling there after graduation (settlement path), and the number of students that have grown up in *i*, studied there but left the region after graduation (outsider path). The result of this subtraction is then divided by the sum of its components.

$$CI_2i = \frac{N_i\,sttlers - N_i\,outsiders}{N_i\,sttlers + N_i\,outsiders}$$

5 Results

A map of Poland indicating the eighteen Polish academic cities, including population information, is shown in Figure 16.2.

Figure 16.2 Regional capitals in Poland.

Source: based on GUS data.

In turn, the values of AT_1 and AT_2 indicators for the eighteen academic cities in Poland are shown in Table 16.2. They reveal substantial discrepancy in the ability of different academic cities to attract students from areas outside the local region. In the low performing cities, such as Białystok or Rzeszów, incoming students account for less than 20 per cent of the total student population.[5] Meanwhile, in the highest ranked cities, more than 40 per cent of students finished secondary school outside the region.

Surprisingly, the cities with the highest value of AT_2 include Zielona Góra and Opole, which are average size cities located in western Poland. Unlike other top-scoring cities, according to the AT_2 indicator (such as Warsaw, Kraków or Wrocław), they are not known for having a well-developed academic function. The possible explanation of this puzzle is that the value of the AT_2 indicator is influenced by the size of region in which a city is located. Cities in smaller regions (including the two discussed here) have to widen their academic catchment areas beyond the region's borders, as their borders are relatively close to the capital city, making it easy for students from the other side of the border to access higher education institutions in the region. This suggests that measuring student attraction performance with AT_2 favours centrally located cities in large regions.

Table 16.2 Measures of attracting power of eighteen academic cities

	AT_1	AT_2
City	*Mean distance from secondary school to tertiary school (km)*	*Share of tertiary students from outside the region (%)*
Białystok	53.9	18.5%
Bydgoszcz	56.7	23.8%
Gdańsk	73.6	31.6%
Gorzów Wielkopolski	54.4	29.0%
Katowice	38.3	15.5%
Kielce	49.8	29.2%
Kraków	79.7	42.3%
Lublin	69.0	26.5%
Łódź	46.6	20.5%
Olsztyn	83.3	38.8%
Opole	57.7	36.3%
Poznań	85.5	32.7%
Rzeszów	47.9	18.4%
Szczecin	87.0	23.4%
Toruń	79.6	37.7%
Warszawa	92.4	38.0%
Wrocław	82.0	35.4%
Zielona Góra	67.2	42.5%

Source: own calculation based on data from nk.pl.

When measured with the AT_2 indicator (average distance from secondary to tertiary school), the attracting power of the cities is expressed in kilometres, and it is independent of the city's location with respective to regional borders and regional size. According to this measure, Warsaw is the top-performing city in attracting students from distant localities (92.4 km on average). The catchment area of Poland's other largest metropolises – Poznań, Wrocław and Kraków – is also large, with a value of AT_2 ranging from 79.7 km to 85.5 km, although two medium-sized cities – Szczecin and Olsztyn – perform at a similar level (87 km and 83.3 km, respectively).

A striking exception from the rule that the largest cities are most successful in attracting students is the city of Łódź – the third largest metropolis in Poland. The city has one of the lowest shares of students from outside the region and the radius of its catchment area is below 50 km – the second lowest. One possible explanation of this phenomenon is that Łódź has never been a recognized academic centre. The city developed largely in the nineteenth century as an important location for the textile industry (manufacturing) and it remained an industrial centre till the 1990s. During transformation of the Polish economy towards a free market system, most state-owned enterprises collapsed, and Łódź entered a period of high structural unemployment. Soon, many of these textile companies that survived the first years of transformation closed down, being unable to compete with Chinese producers.

Another reason for the weak performance of Łódź in attracting students is the proximity of Warsaw. Located 142 kilometres from Łódź, and about 1.30 hr by train (see Figure 16.2), the capital of Poland constantly drains the city of valuable human resources. With the exception of some specific faculties (e.g. the famous Łódź film school), Warsaw universities are higher ranked and thus more attractive for students.

A specific case in Table 16.2 is the city of Katowice. The capital of the Upper Silesia region is the centre of Poland's largest industrial district in which mining and the steel industry were developing extensively in the 1960s and 1970s. Similarly to Łódź, the Upper Silesia region experienced a severe decline during the transformation period of the 1990s as many of its companies, including mines, were closed. However, the low performance of Katowice in attracting students from distant locations, observed in Table 16.2 (AT_2=38 kilometres), is not caused by the weakness of the city's higher education institutions (they are developing dynamically), but rather by the fact that Katowice is a part of a densely populated multi-city agglomeration. Upper Silesia is not a typical region consisting of a central city and surrounding small towns and villages. The core of the region is a 1,000 km^2 of urban sprawl (twenty-five linked cities) populated by approximately 2 million people. The agglomeration is a natural catchment area for Katowice's tertiary schools. Although the surface of this area is smaller than in the case of most large academic cities in Poland, this should not be interpreted as sign of underdevelopment in Katowice's academic function.

Table 16.3 gives measures of cities' performance in absorbing the graduates of higher education institutions. It shows that city size is a decisive factor in determining the attractiveness of cities as a place of residence for the educated population. Poland's largest metropolises – Warsaw, Kraków, Poznań, Wrocław, Gdańsk and Łódź – all rank top, regardless of which of the three measures we consider. Comparing Tables 16.3 and 16.2 we notice that the preference for large cities is more prevalent among university graduates (when choosing a place to live and work) than among students (choosing a place of study). Note, however, that the scores of Łódź are again influenced by the proximity of Warsaw. The short distance between the two cities allows graduates to live in Łódź while working in more developed Warsaw and thus minimize living costs. Thus, the proximity of Warsaw improves the perceived performance of Łódź regarding its ability to absorb graduates.

Table 16.3 Measures of cities' absorbing power

	AB_1	AB_2	AB_3
City	*average distance between the city of studies and graduates' current place of residence*	*share of graduates settling down in the city among those from outside the region*	*share of local graduates who settle down in the city*
Białystok	64.4	15.7%	44.2%
Bydgoszcz	58.9	17.0%	38.7%
Gdańsk	57.6	26.0%	37.2%
GorzówWlkp.	78.1	10.3%	35.4%
Katowice	43.7	9.5%	14.9%
Kielce	64.7	7.8%	30.1%
Kraków	63.4	28.3%	42.5%
Lublin	80.1	15.0%	35.4%
Łódź	45.7	20.3%	47.8%
Olsztyn	80.5	16.3%	30.1%
Opole	61.1	11.8%	24.7%
Poznań	71.5	24.8%	39.6%
Rzeszów	60.9	16.0%	29.4%
Szczecin	86.0	21.2%	42.7%
Toruń	78.3	17.2%	29.3%
Warszawa	58.2	35.4%	52.4%
Wrocław	66.1	27.7%	42.5%
Zielona Góra	77.3	8.7%	24.5%

Source: own calculation based on data from *nk.pl*.

The results in Table 16.3 also confirm that the case of Katowice should be considered separately and it is not really comparable to other cities. Katowice retains only 14.9 per cent of local school graduates within the city borders (the lowest score among the eighteen cities), but the average distance between the city and the graduates' places of residence is also very low – 43.7 kilometres. Thus many graduates leave the city but stay in the agglomeration.

High values of performance indicators referring to both attracting students and absorbing graduates may be interpreted jointly as a sign of a city's ability to benefit from its academic function (by creating and accumulating human capital and thus stimulating economic growth). Meanwhile, high performance in just one of the considered areas (student attraction or graduate absorption) indicates limited benefits from academia. A city with a low ability to attract students and a low rate of graduate absorption suggests that hosting higher education institutions has little influence on the city's development potential.

The two composite indicators, the values for which are shown in Table 16.4, aim at capturing both aspects of human capital accumulation in the academic cities, and thus better reflect the benefits gained from performing an academic function.

Table 16.4 The values of composite indicators for eighteen Polish cities

City	CI_1	CI_2
Warszawa	1.59	0.85
Gdańsk	1.28	0.70
Kraków	1.26	0.79
Wrocław	1.24	0.67
Poznań	1.20	0.67
Olsztyn	1.03	0.52
Łódź	1.02	0.10
Toruń	1.02	0.49
Szczecin	1.01	0.41
Bydgoszcz	0.96	0.31
Opole	0.94	0.35
Katowice	0.88	0.33
Zielona Góra	0.87	0.24
Lublin	0.86	0.08
Białystok	0.84	−0.10
Rzeszów	0.79	0.25
Kielce	0.77	−0.09
Gorzów	0.70	−0.20

Source: own calculation based on data from nk.pl.

Applying this integrated approach we end up with five large metropolises strongly outperforming other cities (see Table 16.4 according to both used indicators). Not surprisingly Warsaw is at the top of the ranking, followed by Gdańsk, Kraków and Wrocław. Again, Łódź performs worse than other cities of similar population number. The bottom of the list is occupied by the relatively small cities of Gorzów Wielkopolski and Kielce.

The CI_2 scores for particular cities are also presented on a map (see Figure 16.3), with the positive values marked with upward pointing arrows, and negative – with downward pointing arrows. The map reveals substantial differences between cities with respect to their ability to accumulate human capital. The large metropolises (Warsaw, Kraków, Wrocław, Poznań and Gdańsk) form a pentagon characterized by high values of CI_2. The scores of cities in the western and particularly eastern periphery are much lower. The weakest performance is observed in Białystok, Lublin, Rzeszów and Kielce, all located east of Warsaw. Despite its central location (in the middle of the pentagon) Łódź also achieves a very low value of CI_2 indicator (0.10). Negative values of CI_2 for some cities suggest that the outflow of locally 'raised' graduates to other regions exceeds the inflow of those who grew up in other regions, but graduated in local universities and settled down in the city after graduation. In other words, these cities should be considered as net losers of human capital.

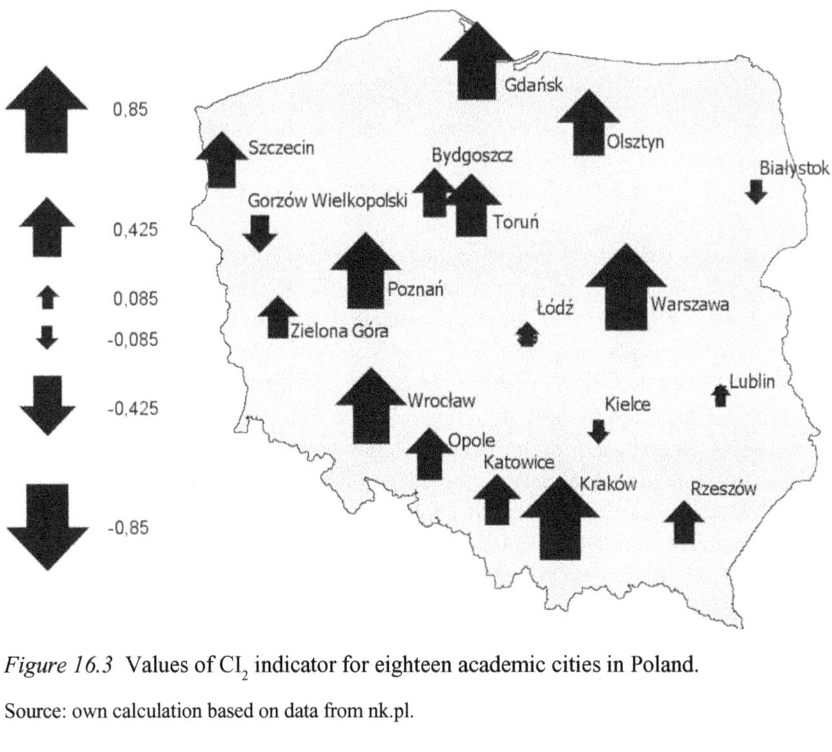

Figure 16.3 Values of CI_2 indicator for eighteen academic cities in Poland.

Source: own calculation based on data from nk.pl.

6 Typology of academic cities

Performing an academic function (hosting a higher education institution) may help cities to accumulate human capital which, according to economic theory, should positively affect regional growth rates in the future. However, the attractiveness of particular cities as destinations for educated individuals is uneven. Moreover, good performance in attracting students does not necessarily mean that the given city will be equally good at absorbing graduates (although performances in these two areas are undoubtedly correlated). We can thus formulate a simple typology of academic cities in terms of the potential benefits they experience from having universities on their territories.

- *Winning academic cities* are these that attract many students, even from distant localities and successfully absorb graduates on the labour market.
- *Supra-regional academic centres* are characterized by a well-developed educational function (they perform well in attracting students), but suffer from an outflow of tertiary school graduates.
- *Post-university migration destinations* are the cities that do not play an important role in the tertiary education market, but perform well in attracting and absorbing the educated workforce.
- *Local academic centres* are characterized both by low ability to attract students (especially from outside the local region) and low attractiveness as a place of residence for graduates.

The scatter plot shown in Figure 16.4 helps to assign particular cities to different categories from the above typology. Values on the *x*-axis represent (for each academic city) the mean distance from students' secondary schools to the city (i.e. to the tertiary school of student choice). This is one of the proposed measures of city performance in attracting students (see Table 16.2). The values on the *y* axis represent the share of tertiary school graduates who settled down in the given city, among graduates who came to the city from outside the local region (they attended secondary school in a region other than the region of the university where they eventually enrolled).

The cities located in the upper right corner of Figure 16.4 are those achieving high performance in both attracting and absorbing human capital. We assign them to the category of winning academic cities (marked with a solid black oval). Within this category, Warsaw can be distinguished as highest ranking city with respect to both measured aspects of human capital accumulation. In addition, the cluster includes Kraków, Wrocław, Poznań and Gdańsk – all large metropolises with population numbers exceeding 500,000.

A dotted black line surrounds the group of cities whose tertiary school catchment areas go beyond the immediate regional neighbourhood. Meanwhile, these cities fail to retain a substantial portion of inflowing students as they graduate from universities. According to the typology proposed above, the cities located close to the right bottom corner can be classified as supra-regional academic centres.

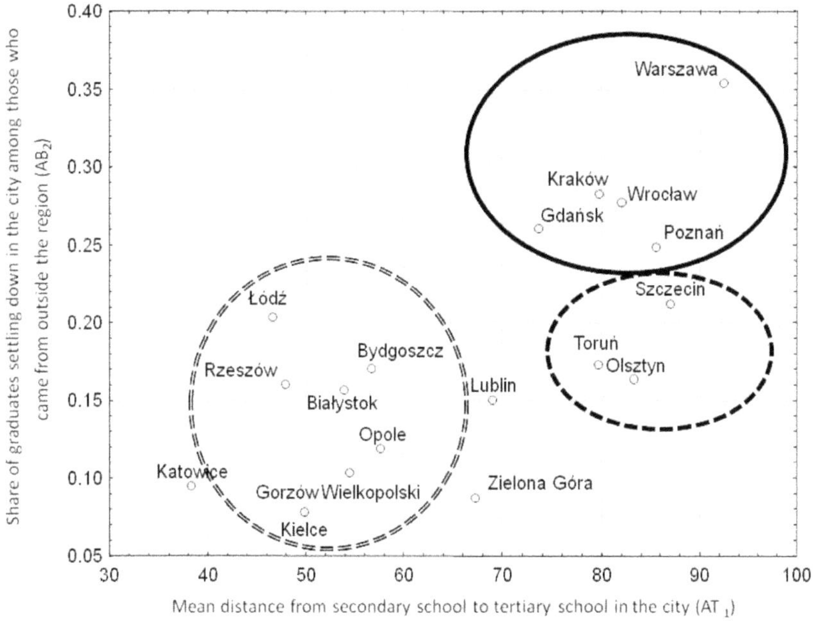

Figure 16.4 Typology of academic cities by their capacity to attract and absorb human capital.

Source: own calculation.

Although their wide catchment area must be considered as an asset, in the end these cities act as providers of formed human capital to other, more attractive destinations. This category includes Szczecin, Toruń and Olsztyn.

Local academic centres are in turn located close to the bottom left corner in Figure 16.4. They are characterized by both a limited catchment area of their universities and low mobility of graduates. This category includes the largest number of Polish cities. Depending on whether we count some border cases (such as Lublin and Zielona Góra), and two specific cities discussed earlier (Katowice and Łódź), the label of local academic centre fits from eight to eleven cities.

7 Conclusions

Economic theory suggests that the accumulation of human capital has positive effects on an economy's future rate of growth and the level of income. If so, then cities engaged in the formation of human capital (hosting higher education institutions) have a natural advantage over those without an academic function. However, the ability of a city to benefit from its higher education sector depends on its performance in human capital accumulation, which presumably differs from one city to another.

In this chapter I proposed two indicators of city performance in attracting students and three indicators of achievement in absorbing graduates. In addition

I defined two composite indicators, capturing both aspects of human capital accumulation.

I calculated the values of all indicators for eighteen Polish cities, using a unique dataset on educational and post-educational mobility in Poland between 1990 and 2008. The source of data was the web page designed to enable contact between former classmates after graduation. The dataset used in the research included information on the location of secondary and tertiary schools as well as the current place of residence of approximately 2 million registered users.

The research led to the following major conclusions:

- Large metropolises perform best in terms of both attraction and absorption of students and graduates, which suggests they strongly benefit from their academic function. Within the group of metropolises Warsaw visibly outperforms other agglomerations. Nonetheless, I labelled five cities (Warsaw, Wrocław, Poznan, Kraków and Gdańsk) as winning academic centres.
- Some Polish cities may have well-developed universities (which attract students even from distant locations), but they usually fail to prevent graduates from migrating to more attractive destinations. In this group, labelled 'supra-regional academic centres' I included Szczecin, Toruń, Olsztyn and perhaps Lublin.
- Mobility of human capital is harmful in eastern regions of Poland. Cities in this part of the country attract students only from the closest neighbourhood, and lose a significant percentage of graduates to other agglomerations. This category of 'local academic centres' which could be also called 'losing academic cities' includes Rzeszów, Białystok, Kielce (all located in the East), as well as Gorzów Wielkopolski, Bydgoszcz and Opole.
- The city of Łódź clearly fails to exploit its privileged, central location in the country, and significant population potential (the third largest city in Poland) to benefit from its academic function. The city's ability to attract students is among the lowest within the reference group, while its relatively good performance in absorbing graduates does not reflect the attractiveness of the Łódź labour market for educated individuals, but rather the fact that many of them choose to live in Łódź and to work in nearby Warsaw.

The research also showed that unconventional data sources may provide a valuable alternative to public statistics for the purpose of analysing human capital mobility. In particular, in countries such as Poland, in which the public statistics programmes provide very limited information on the educational and post-educational mobility of citizens, retrieving data from social media may be the only way to conduct empirical studies in this area. However, given the limitations of such experimental datasets, academic research cannot rely on it in the longer perspective, and the issue of human capital mobility needs to be introduced on a larger scale in the official statistics.

Notes

1 Poland has sixteen administrative regions, but two of them have their capital functions (institutions and services) divided between two cities. We include all four cities in the analysis.
2 It should be added that the nasza-klasa.pl website has already been used for spatial analyses of higher education (although on a smaller scale), for example in the PhD dissertation by A. Bajerski submitted at the Adam Mickiewicz University in Poznań (Bajerski 2009).
3 The original drawing by Hoare and Corver referred to North and South, but this was changed as division and competition between the East and West better reflects the reality of Polish regional development.
4 Dasgupta (2003) claims that one's position within a social network is transferable into monetary benefits.
5 Although Katowice also seems to achieve low values of AT1 and AT2, its case is rather special. Katowice can hardly be discussed in isolation from the rest of the Upper Silesian agglomeration – a highly urbanized and densely populated area inhabited by approximately 3 million people. The agglomeration is a natural catchment area for Katowice's tertiary schools which depresses the values of student attraction indicators, but does not necessarily mean the attractiveness of the city for students is low.

References

Badinger, H. and Tondl, G. (2002) *Trade, Human Capital and Innovation: The engines of European regional growth in the 1990s*. Research Institute for European Affairs, University of Economics and Business Administration.

Bajerski, A. (2009) 'Przekształcenia struktury przestrzennej szkolnictwa wyższego w Polsce po 1989 r.', unpublished PhD dissertation. Poznań: Instytut Geografii Społeczno-Ekonomicznej i Gospodarki Przestrzennej Uniwersytetu im. Adama Mickiewicza.

Barro, R. J. (1999) 'Human capital and growth in cross-country regressions', *Swedish Economic Policy Review*, 6(2), 237–277.

Benhabib, J. and Spiegel, M. (1994) 'The role of human capital in economic development', *Journal of Monetary Economics*, 34(2), 143–173.

Bernanke, B. S. and Gurkaynak, R. S. (2001) *Is Growth Exogenous? Taking Mankiw, Romer and Weil seriously*. National Bureau of Economic Research, NBER Working Papers, No. 8365.

Bils, M. and Klenow, P. J. (2000) 'Does schooling cause growth?' *American Economic Review*, 90(5), 1160–1183.

Chojnicki, Z. and Czyż, T. (1997) *Struktura przestrzenna nauki w Polsce*. Poznań: Bogucki WydawnictwoNaukowe.

Dasgupta, P. (2003) 'Social capital and economic performance: Analytics', in E. Ostrom and T. K. Ahn (eds) *Foundations of Social Capital*. Cheltenham: Edward Elgar.

de la Fuente, A. (2002) 'On the sources of convergence: A close look at the Spanish regions', *European Economic Review*, 46, 569–599.

Di Liberto, A. and Symons, J. (2001) *Education and Italian Regional Development*. London: London School of Economics and Political Science, Centre for Economic Performance.

Florida, R. (2000) 'The learning region', in Z. Acs (ed.), *Regional Innovation, Knowledge and Global Change*. New York: Pinter.

Ghatak, S., A. Mulhern, *et al.* (2008). 'Inter-regional Migration in Transition Economies: The Case of Poland', *Review of Development Economics*, 12(1): 209–222.

Herbst, M. (2007) 'Wpływ kapitału ludzkiego i społecznego na (krótkookresowy) wzrost gospodarczy w polskich podregionach', in M. Herbst (ed.), *Kapitał ludzki i kapitał społeczny a rozwój regionalny*. Warsaw: EUROREG, Uniwersytet Warszawski, pp. 166–203.

Losy absolwentów – raport (2008) Available online at www.absolwent.com.pl/strona.aspx?id=134,8.

Hoare, A. and Corver, M. (2010) 'The regional geography of new young graduate labour in the UK', *Regional Studies*, 44(4), 477–494.

Lucas, R. E. (1988) 'On the Mechanics of Economic Development', *Journal of Monetary Economics*, 22, 3–42.

Mankiw, N. G., Romer, D. and Weil, D. (1992) 'A contribution to the empirics of economic growth', *American Economic Review*, 107(2), 407–437.

Nelson, R. and Phelps, E. (1966) 'Investment in humans, technological diffusion, and economic growth', *American Economic Review: Papers and Proceedings*, 51(2), 69–75.

Persson, J. and Malmberg, B. (1996) *Human Capital, Demographics and Growth across the U.S. States 1920–1990*. Stockholm University, Institute for International Economic Studies, Seminar Papers, No. 619.

Porter, M. (2001) *Porter o konkurencji*, transl. by A. Ehrlich. Warsaw: Polskie Wydawnictwo Ekonomiczne.

Wasielewski, K. (2004) 'Młodzież wiejska na UMK w Toruniu', *Wieś i Rolnictwo*, 1, 126–138.

17 The innovative potential of Kyiv

Overestimated and underused

Olga Mrinska

Introduction

In the modern economy, a country's competitiveness is heavily influenced by its ability to innovate; in industry, the service sector, management and administration. A country with high intellectual and human capital, R&D-intensive businesses, efficient financial and institutional infrastructure for technology transfer and the commercialisation of science has higher chances of being closely integrated into the world economy and thus reap the benefits of ever-expanding knowledge-producing global networks. It has significant impact on the profitability of businesses, productivity of the economy and overall welfare of the state.

Capital cities play a crucial role in this model since they are usually centres of academic and research excellence, national economic centres and a country's gateways to the world economy. This is particularly true for post-Soviet economies such as Ukraine.

Kyiv, as the capital of Ukraine, is the biggest economic, political, cultural and scientific centre of the country; it produces a fifth of the country's GDP, attracts 40 per cent of national FDI, and houses a third of the country's research potential (institutions and experts). Its universities, research and academic institutions, few technological transfer organisations and venture capital facilities are among the most advanced in the country. There are of course other R&D strongholds, albeit with a less diversified knowledge base: the cities of Kharkiv, Lviv, Odessa and Donetsk.

This chapter analyses the structure and quality of the research and educational potential of Ukraine in general and Kyiv in particular, key features of its business climate and the structure of the city economy. This analysis will contribute towards reconstructing the image of Kyiv as the potential centre of an innovative economy.

1 Key drivers of an innovative economy: human capital, R&D, business environment

The Soviet legacy brings several positive features into Ukraine's potential: a well-educated population and a solid scientific and technological base. However, this legacy is not being transformed into strengths that are important in the modern economic model. Ukraine remains one of the least competitive and innovative countries in Europe and the world, despite its educational and scientific potential. For example, in the 2011/12 Global Competitiveness Ranking by the World Economic Forum, Ukraine was only eighty-second while in the 2010 Network Readiness Index Ranking (also by WEF) it was ninetieth (WEF 2012).

Clearly the baggage of the past is not being adequately transformed into the assets of the future. Ukraine's GDP per capita in 2010 was just 6,698 USD purchase power parity (PPP), which is significantly lower than that of its European neighbours (EU average GDP per capita in 2010 was 30,455 USD PPP) (IMF 2011).

The reasons for such failures are diverse, but the most significant are:

1 Historically, a considerable share of Ukraine's R&D potential was concentrated in the defence sector (for example it employed three quarters of all research cadre), and many technologies were inaccessible for the needs of the wider economy – only a fraction of Soviet innovations worked towards creating conventional products/know-how and consumer goods.
2 The process of building up Ukraine's basic state institutions and infrastructure after it gained independence in 1991 took lots of time and effort; thus little or no attention was paid to sustaining and enhancing its science, higher education and R&D sector.
3 The Ukrainian private sector chronically lacks the capital and access to borrowings for developing and acquiring new technologies, processes or know-how, while the state has almost no instruments for stimulating this type of investment.[1] The brain-drain is playing a crucial part in diminishing Ukraine's innovative potential as it is the most talented and entrepreneurial members of society that are leaving for countries that offer better environments for productive research, protecting intellectual property and commercialising inventions.

It is crucial to have at least a brief overview of the core components that underpin the innovative model of Ukraine in general and Kyiv in particular: education; science and research; as well as the business climate. We believe it is a country's human capital, the strengths of its research sector, the availability of a market infrastructure, and beneficiary regulatory environment that determine the dynamics of sustainable economic growth on the basis of innovations, and the expansion of production and export of knowledge-intensive products and services.

1.1 Education

It is important to outline a general state of affairs in the education system in Ukraine, which includes higher education. Owing to the dramatic transformational processes happening in the national economy and society in general, the whole system is going through significant shifts – qualitative (changing demands of the modern economy, growth of private initiative, decline of paternalistic state, lifelong jobs and social security) and quantitative (depopulation and change of age structure of population). These transform the basis of a country's human capital and its ability to innovate presently and in the future. Often these shifts are not underpinned by adequate reform of the educational system, which is still very much rooted in the Soviet standards and approaches.

Ukraine scores highly in many indicators that characterise its human capital. For example, it remains one of the most educated nations according to the UN Human Development Index (HDI); in 2010 it was twenty-seventh in the global education ranking (UN 2010) while its overall HDI ranking was sixty-ninth among 169 nations. Ukraine's literacy rates are close to 100 per cent and on average Ukrainians have 11.3 years of education. Participation in higher education in Ukraine has increased dramatically over the last twenty years and currently there are 2.7 million students in Ukrainian universities and institutes.[2] Their number is continuously growing; the number of students per 10,000 population has leaped from 310 in 1992 to 578 in 2010 (KPU 2010). The key reason is that a degree is the major precondition for getting even a moderately qualified and relatively well paid job.

The formal characteristics of the Ukrainian education system also look good compared to the developed economies. Ukraine spends a healthy 5.3 per cent of its GDP on education, which is comparable with France, the United Kingdom (both 5.6 per cent), Germany (4.4 per cent), or the US (5.5 per cent). However, around 90 per cent of the education budget is being spent on the payroll and supporting the existing infrastructure (Verkhovna Rada 2011) – there is very little investment going into the advancement and modernisation of schools' and universities' infrastructure, and enhancing the quality of the curriculum (Likarchuk 2011). Ukraine has an extensive network of secondary educational institutions where the number of teaching staff is significantly greater than in other European countries. Moreover, the entire administrative and assessment system in education is based on quantitative indicators rather than qualitative.

At the same time, the productivity of the Ukrainian education sphere is low. According to expert assessments in 2009, the Gross Value Added (GVA) per one employee in the education sector was 3,625 USD which is three times less than overall economic productivity (10,735 USD) (Vitrenko 2011). This is especially surprising since 70 per cent of employees in this sector have a higher education, compared to 55 per cent for the economy in general. This demonstrates that the higher level of skills, or at least formal attributes of such skills, is not being converted into higher labour productivity.

The problems of the Ukrainian education sector are better illustrated not so much by the formal participation rates, years in education and skills proliferation, but by the level of satisfaction with the education services. According to a Gallup World Survey, only 38 per cent of Ukrainians responded positively to the question 'Are you satisfied with the quality of education and schools in your city/locality?' This is a very low indicator compared to the 70 per cent satisfaction rate in the USA and the UK, 59 in Germany, 42 in Russia, or 66 in Poland. So despite high spending and participation rates, Ukrainians are unsatisfied with their country's education system (Vitrenko 2011).

It is more difficult to get statistics on satisfaction rates in higher education but it is evident that the qualitative characteristics of the sector are lagging behind the quantitative. Various surveys of employers (see for example Compass-2011 Report by SCM 2011) demonstrate that graduates persistently lack practical knowledge and qualifications that are required in practice. Thus employers need to invest heavily in their young cadre in order to get them up to speed with their business skills.

Many experts believe that the lack of proper investment in higher education and the dramatic rise in the number of private (or rather commercial) higher education institutions (HEIs) with limited financial and intellectual recourses are to blame. For example, the total number of universities in Ukraine increased from nine in Soviet times to more than eighty at the end of the first decade of the 2000s. In total, Ukraine has 358 universities, academies and institutes plus 593 colleges, of which roughly two thirds are state HEIs (KPU 2010).

Ukrainian HEIs are relatively small and have far fewer students and professors than their international counterparts. For example, universities from the Top-100 of the Shanghai Academic Ranking of World Universities have on average 40–80 thousand students and 3–5 thousand research and teaching staff. In Ukrainian universities the average number of students is around 6.5 thousand, and the number and quality of teaching staff is often inadequate, with lots of part-time professors and teachers (Vitrenko 2011). Only the biggest Ukrainian universities have a more significant student base and teaching potential.

Less than a half of all HEIs have some sort of research facilities (mostly state HEIs) and over the last twenty years the number of researchers based at universities has shrunk threefold to 9.6 thousand. The share of HEIs' input into the total volume of fundamental research and scientific work is only 10 per cent compared to 85 per cent in other academic institutions. Their share in applied research work volume is slightly higher – 15 per cent (the academic share is 44 per cent) (SSSU 2011a). So the relative significance of Ukrainian HEIs in R&D is rather limited and thus the quality of their graduates is relatively poor when compared to their counterparts abroad.

The Government of Ukraine has undertaken some steps in the direction of enhancing the research potential of several key universities. In 2009–10 it granted fourteen universities across Ukraine special autonomous research status, which ensures greater freedom in selecting staff, planning and implementing research activities, obtaining funding from non-budget sources, as well as initiating and taking part in international research activities (Cabinet of Ministers 2010). Six of these universities are located in Kyiv. The status brings new opportunities for developing university research and teaching potential, and for taking an active part in international research projects and networks. At the same time, this status gives more formal rather than practical opportunities and means that in order to become truly research-intensive these universities need to go through significant staff changes and attract more highly qualified researchers, as well as change their administrative systems so that they are focused on quality rather than quantity indicators. Also, the most effective way forward for creating true research universities would be to merge existing HEIs with the research institutes of the National Academy of Sciences of Ukraine (see below).

It is worth noting that currently Ukrainian HEIs are facing a double squeeze. On the one hand, the Government needs to cut public spending, including on education, owing to the prolonged crisis and growing public dept. On the other hand, the role of the private and third sector in funding higher education is still marginal. State HEIs have very little autonomy in the financial sphere and thus cannot really tap into alternative sources of funding (as there are strict rules on how their money, including from private or third sector sources, should be appropriated). This creates substantial financial limitations for introducing innovations into teaching processes and research, since universities cannot attract significant private investments and, when they can, they are not able to spend money at their discretion. At the same time there is hardly any private university that has a private endowment, and most of them function on the basis of fees from education services, which makes them more commercial in nature than private (Shulga 2009).

1.2 Science and R&D

If education in Ukraine has rather respectable formal indicators of development despite suffering from qualitative deficiencies, the sphere of science and research cannot be praised even for the formal indicators. Unlike education, science and research does not have the same level of state financial support. The overall spending on R&D in Ukraine in 2010 was only 0.9 per cent of GDP compared to 1.92 per cent in the EU27 (2009) or 2.34 per cent in the OECD (2008). The total spending on R&D in 2009 was around 2.5 billion USD PPP (SSSU 2011a). The state's share in total R&D expenditure is marginal; it increased from 0.4 per cent in 2000 to 1.1 per cent in 2010.

The private sector is the key investor in R&D, though only about 14 per cent of all Ukrainian companies are engaged in innovative activities (which also include purchase of external technologies and advanced equipment produced elsewhere). The relatively high share of private investment in R&D could be explained not so much by vivid investment activity of the private stakeholders, but by rather meagre resources available from the state budget. There are some R&D activities that are wholly or partially funded from state budget, but they are mainly concentrated in strategic sectors of the national economy, such as energy, extracting industry, metallurgy and chemistry, and funds are being allocated according to targeted state programmes for supporting innovations approved annually. Among big R&D investors are large formerly and presently state-owned companies in sectors such as aircraft construction, the space and rocket construction industry, transport and machinery. Statistics for SMEs is unavailable but secondary indicators illustrate that this sector has less capacity to invest in R&D.

The R&D funding relative to GDP has been decreasing over the past decades (from 1.19 per cent of GDP in 1996 to 0.9 per cent in 2010, SSSU 2011a) and there are no indicators that the Government is planning to prioritise this spending despite formal policy commitments to create a more competitive and innovative model of economy (see for Example Cabinet of Ministers 2011a). Budget allocations continue to reflect short-termism in decision making and a lack of strategic vision for the role of R&D in the national economy.

The situation regarding human research potential is even worse and is constantly deteriorating. Between 1992 and 2010 the number of researchers has decreased threefold – from 90 thousand to 30 thousand. At the same time the number of people with the degree of Doctor of Science has increased twofold.[3] In Ukraine, scientific degrees – Candidates and Doctors of Science – are very popular among civil servants and high level managers and that is why today the total number of such 'status' scholars is 10 per cent higher than the total number of career researchers (!). And if in 1995 the proportion between Doctors of Science and researchers was 1:36, in 2010 this proportion was 1:6 (SSSU 2011a).

However, the number of people with scientific degrees does not indicate the quality of the research cadre, as they often work in other sectors and do not contribute directly to R&D activities. It is clear that such inflation of the category of scholar/ researcher does more harm to the system than poor financing and administrative faults. The interest towards research, science and innovative activities in society is diminishing, with the growth of 'status' scholars and the exodus of true researchers and inventors abroad. As a result, the popularity of science as a profession among youth is consistently decreasing while Ukraine produces record numbers of lawyers, economists and political scientists.

There are 1,340 research institutions in Ukraine and 750 of those get state budget funding. Traditionally, the allocation is distributed quite evenly and each institution receives between 120 and 250 thousand USD a year, which is substantially below the level needed for productive activities. There are also targeted budget research programmes that support specific projects implemented in one or several institutions, but they are traditionally underfunded; between 1 and 20 per cent of the required amount is usually allocated a year, and thus the results match the input (Semynozhenko 2011).

The challenges of science and research in Ukraine could also be explained by its disintegrated institutional structure – as in many post-socialist countries, Ukraine has a dual system where education is provided by HEIs and research is conducted mainly by institutes of the National Academy of Sciences of Ukraine (academia). On the one hand, this allows researchers to concentrate on their research without the distraction of teaching tasks. But on the other hand, there is a dramatic gap between teaching processes and research, as well as between present and potential future generations of researchers in the areas that have always been Ukraine's strengths – mathematics, physics, chemistry, biosciences, electronics, engineering and new materials. The popularity of these professions is falling dramatically and the cadre of scientists and researchers is rapidly ageing (see Neprytamannyi 2012). With natural substitution rates being so low, the prospects of Ukraine's strong research areas are looking gloomy.

For many years experts have proposed the merger of the two systems and the creation of truly strong research universities with the teaching infrastructure of current HEIs and the rigour of academic research institutions. However, there is no political will to take such a dramatic step and the two systems continue to live parallel lives without really enriching each other.

The existing institutional system also lacks an effective technology transfer infrastructure, which would link ideas and new technologies produced in R&D institutions with the businesses that are ready to invest in new products and services. There is a network of regional technology transfer centres on the basis of the State Investment Agency of Ukraine. There is also a wider Ukrainian Technology Transfer Network (UTTN) under the auspices of the Ministry of Education and Science, which brings together state and private institutions involved in R&D activities. This network was created using the model of the European Innovation Relay Centres Network (currently the European Enterprise Network). However, it would be fair to say that these institutions are very much separate from the core business support institutions (which are also not that numerous) and there is very limited knowledge about their activities among the business community (Kozachenko and Georgiadi 2010). So it would be premature to consider these developments as consistent steps towards creating an effective and comprehensive system of technology transfer that is an integral part of the business support infrastructure.

It is worth mentioning that, over the last several years, Ukraine has been taking a more active part in various international research programmes. Traditionally this was the prerogative of the big research-intensive companies that produce core knowledge-intensive products in Ukraine (e.g. aircrafts or rocket equipment for space programmes). But today, more universities and academic research institutions are taking part in the Seventh Framework Programme (FP7) and other European and International programmes. For example in FP7 (up to April 2010), there were 742 teams of Ukrainian researchers that applied for funding, with 583 proposals in ninety-nine calls. The success rate for Ukrainian applications was 22 per cent and the total amount of allocated funding was 10.3 million euros (Supel 2010: 20).

1.3 Business climate

Ukraine's overall business climate is difficult to characterise as beneficial. For many years Ukraine has had rather low positions in global rankings which illustrates the quality of the regulatory and business environment. In its 'Doing business 2012' ranking, the World Bank placed Ukraine at a staggeringly low 152 out of 183 countries (WB 2011). Ukraine's score regarding the tax system is even more embarrassing – 181 out of 183.

Since the overall business environment in Ukraine is so poor, it is difficult to expect better conditions for innovation-oriented and knowledge-intensive businesses. It is necessary to underline that the necessary legislative and regulatory framework, though imperfect, is in place, and is often based on best international practices. However, the level of enforcement and especially funding is very unsatisfactory (see Chapter 1.2 above).

One of the key instruments for commercialising the scientific and research findings and for supporting innovation-driven production is the technological park. Since 1999 the Government of Ukraine has created special economic conditions for several technological parks that are based on well-established scientific institutions. The Law of Ukraine on the Special Regime of the Innovation

Activities of Technological Parks stipulates the terms and conditions for the functioning of sixteen technological parks (four of which are still inactive), seven of which are located in Kyiv. However, analysis of the effectiveness of this instrument demonstrates a very limited effect on the economy owing to the small scale of operations.

Technological parks were established from the start as sort of free economic zones, with significant tax benefits and exemptions (VRU 1999). When in 2005 the Government performed a major tax policy review and cancelled most preferences for projects being implemented in technological parks, it significantly affected their activities and in some cases brought them to a halt. Most of the parks have still not recovered. This proves that streamlining the commercialisation of science purely via tax incentives does not have a long-term effect and progress stops as soon as the benefits are reconsidered. Meanwhile, Ukrainian technological parks lack the key elements that would ensure more sustainable and successful collaboration of research and business – simplifying the terms of business operations, stimulating a technology transfer infrastructure and enabling a financial infrastructure for high-risk innovative projects.

Brief analysis of the technological parks' activities over the last ten years based on reports from the Ministry of Education and Science (see MES 2010) demonstrate their relative 'rigidity' as many of them have between one and three commercial projects. These projects are mostly funded by private sector partners, though the state can provide funding in some cases, where projects are in line with the priorities of state innovation policy and targeted innovation programmes funded from the state budget. Technological parks are often used as an opportunity to create a 'commercial' arm for a state research institution (academic or industrial), rather than create an incubator for mass commercialisation of technologies by local researchers or invite entrepreneurs to join innovative projects launched on the basis of the institute's inventions.

The negative effect of such an approach is that there is no 'creative destruction' – even if the company (or innovation project) is unsuccessful there are no easy means for its liquidation and the launch of a new idea and new enterprise. Thus, quite often, unsustainable projects are supported while other more competitive initiatives are ignored. It should be mentioned here that the activities of technological parks and their projects need to comply with the list of priorities of scientific and research activities established by separate legal acts.

Venture funding, which is crucial for supporting innovative projects, is at nascent form in Ukraine. Owing to challenges of protecting and legally enforcing intellectual property rights, ownership rights are not totally guaranteed and thus the risk of such investments are increased. With support from the European Commission, the Government of Ukraine is in the process of establishing a State Venture Fund, which will provide about 100 million USD of funding for innovative and commercially viable projects. There is only a handful of Ukrainian private venture funds/companies that invest in innovative companies or start-ups (many others use the name 'venture fund' but perform conventional commercial operations with investments in property or industry). There is also the Association of Private Investors in Ukraine that brings together Ukrainian business angels (it is a member of the European Business Angel Network). The number of international venture companies that are opening offices in Ukraine is gradually growing. The most attractive sectors for venture funding in Ukraine are traditionally software development and IT and Internet-based business solutions, as Ukraine remains one of the leading 'producers' of licensed software engineers.

2 The economic and human potential of Kyiv

Kyiv always has been, and remains, the unchallenged leader of the national economy. It provides one fifth of the GDP and attracts a hundred times more FDI than the region with the lowest level of FDI. Traditionally, Kyiv has lower levels of unemployment, the most well-qualified and flexible workforce (with constant inflows of fresh labour from other regions, but rarely from abroad), the highest income levels, better social standards, and a relatively well developed and maintained infrastructure. For the same reasons, however, Kyiv also has the highest cost of living, an overcrowded transport and social infrastructure, and exorbitant property prices.

Kyiv's economy is heavily dependent on the services sector, which is quite diversified. There is no other region that could match this scale – services account for 85 per cent of its Gross Value Added (GVA) (see Figure 17.1) and for almost half of its exports (dominated by transport and telecommunications, as well as property and business services). It is crucial to highlight the insignificant share of industry in the city economy – only about 9 per cent of GVA is produced by industrial companies.

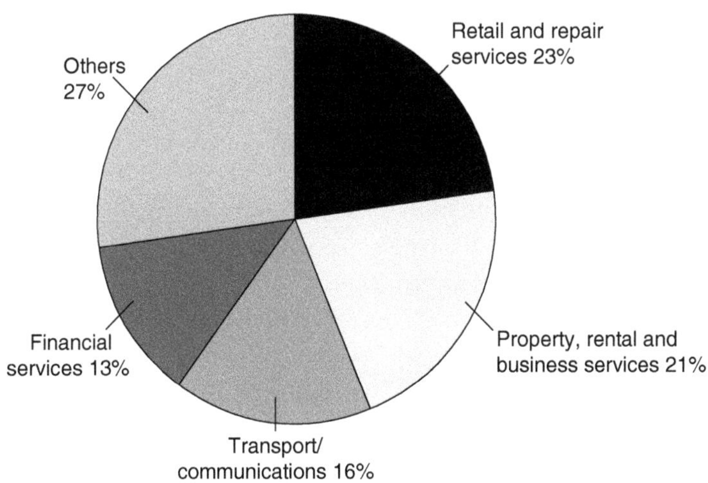

Figure 17.1 Nominal GVA structure in Kyiv, 2010 (est.).

Source: SSSU (2011b).

Over the past ten years, Kyiv has enjoyed an unprecedented economic boom owing to lucrative investments in the property and construction markets, retail, banking and other financial services. Kyiv also benefits from being Ukraine's administrative capital, which brings a substantial concentration of public sector jobs and investments.

It is the structure of the Kyiv economy, however, and its openness to international capital markets, especially over the several years prior to the crisis that left it very vulnerable to the global economic crisis and led to deep recession. Overall, the Kyiv economy contracted by 18.3 per cent in 2009 relative to 2008, compared to a 14.8 per cent decline of the national economy (see Figure 17.2). Financial services, retail, property and construction were sectors that affected the decline most of all. The economy started to pick up in 2010, but the growth rates are still sluggish compared to the pre-crisis trend.

Kyiv is the key player in foreign trade and in 2010 it was responsible for 24 per cent of all commodity export and 35 per cent of all imports (though these impressive figures reflect the administrative structures of companies rather than the real origins and destinations of trade, which is usually performed by head offices located in the capital). The total value of commodity export in 2010 reached 7 billion USD and it was dominated by low value-added commodities such as grain, sunflower seeds and oils (half of the total), chemical commodities (8 per cent) and machinery products (7 per cent). Kyiv is also responsible for 34 per cent of Ukraine's export in services (three quarters of these are transport and communication services, as well as property and business services) and 36 per cent of import. The total value of services export in 2010 was equal to 2.7 billion USD (Kyiv City Statistics Office 2011). So despite the fact that GVA produced in the city is dominated by services, it is agriculture commodities and chemical commodities that dominate city exports.

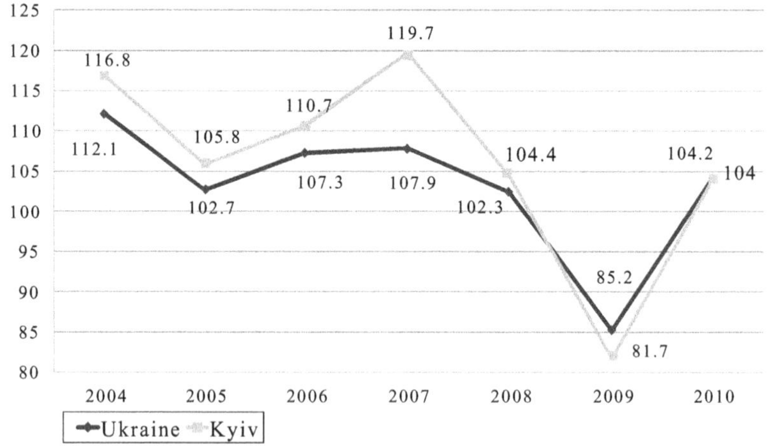

Figure 17.2 Dynamics of Ukrainian GDP and Kyiv GRP, 2004–10 (per cent).

Source: SSSU (2011b).

This mismatch between services and industry is also noticeable in the structure of foreign investments. FDI flows in Kyiv demonstrate the high attractiveness of such sectors as financial services, retail, property and business services, while industry and construction sectors attract 5 per cent and 8 per cent of FDI respectively (Figure 17.3).

The employment structure of the capital city is different from the structure of output and trade as it demonstrates a greater role of public services (including health, social care, education and communal services) that provide almost one third of all jobs in Kyiv. A further 20 per cent of jobs are created in industry and construction while another 55 per cent of jobs are in services (Kyiv City Statistics Office 2011).

The mismatch between the structure of the Kyiv economy and its labour market is even further aggravated by the 'specialisation' of the key educational institutions of Kyiv. It would be true to say that Kyiv's research and educational institutions are more advanced in natural and material science than in social science and humanities. That is why most truly significant new inventions and know-how, which could gain international exposure, are generated in these sectors (Strikha 2012). On the other hand, even in those institutions specialised in preparing highly technical and applied cadres there is a growing trend of domination by 'general' professions such as economists, lawyers and managers, over the 'core' professions of the given institution (though it is necessary to highlight the role of Kyiv as an educational centre for the entire country).

Higher education represents a significant share of the Kyiv economy; there are sixty-nine universities, academies and institutes, and the total student population of the city reached half a million in 2010 (the total registered population of Kyiv is about 3 million people) (Kyiv City Statistics Office 2011). A total of 140 thousand people graduated from Kyiv universities in 2010. Five out of ten top universities in Ukraine (according to UNESCO Chair at KPU) and six out of fourteen autonomous research universities are located in Kyiv (See Table 17.1).

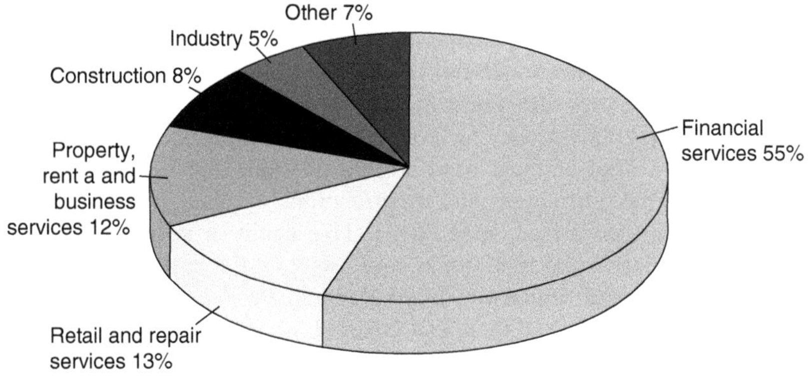

Figure 17.3 Cumulative FDI structure in Kyiv, October 2010.

Source: Kyiv City Statistics Office, 2011b.

Table 17.1 The five biggest universities of Kyiv

Name	*Number of faculties/ research institutes*	*Number of students*	*Teaching staff*	*Autonomous Research University status*	*Position in national HEI ranking**
Kyiv National Taras Shevchenko University	14/5	26,000	2,360	Yes	1
National University 'Kyiv Polytechnics'	20/10	36,000	3,290	Yes	2
National University of Bio-resources and Environmental Management	20/12	37,000	3,000	Yes	3
Kyiv National Economic University	9/13	38,000	640**	Yes	23
National Aviation University	21/9	35,000	1,170	Yes	30

*Prepared by UNESCO Chair at Institute of Applied System Analysis, Kyiv Polytechnic University.
**Staff with scientific degrees only.

Sources: www.osvita.com.ua and www.osvita.org.ua/abitur/entrance/ratings/2.html.

Kyiv universities and academies provide plenty of graduates for 'conventional' sectors that represent the core of the city economy – retail, business services, property and construction. For example, Kyiv National University of Construction and Architecture is crucial for supplying cadres for property development, construction and for reshaping the architectural, infrastructure and landscape 'face' of the city. There are also strong educational bases for preparing 'core' cadres such as teachers or doctors. But if we look at the structure of the Kyiv economy (Figure 17.1) we can see that it lacks 'edge' sectors that are built on knowledge and innovations. There is a plethora of more sophisticated educational and research institutes that exist outside the economy and have very little commercial exposure.

Very few Kyiv universities have significant collaborative links with the private sector – most of them have ad hoc relations and hardly interact in order to bring the curriculum and qualifications of their graduates in line with businesses' needs (see SCM 2011). This could be partially explained by a lack of initiative and motivation, and partly by the limited autonomy of Ukrainian universities, which was highlighted above (see section 2.1). There are of course exceptions and there is a growing number of companies that are becoming proactive in engaging with universities, offering them their support in preparing special courses, providing opportunities for internships and sometimes even opportunities for joint research projects with the company's staff. These initiatives are mainly limited to the engineering and 'technical' HEIs with a more distinctive specialisation.

One such example is Kyiv Polytechnic University (KPU), which is the national leader in terms of corporate collaboration, establishing joint labs with companies and stimulating the commercialisation of new ideas produced by their staff and students. Kyiv Polytechnic has one of the most successful technological parks in Ukraine (currently it is implementing nine projects) which exists in the form of a corporation where more than fifty Ukrainian and international companies are listed as partners. Over the past five years, technological park partners have produced 150 innovative products and services. The university has practical collaborative arrangements with such companies as Microsoft, Intel, Motorola, and many others, as well as with a number of big Ukrainian IT, telecommunications and engineering giants that use it as a main recruitment base. Kyiv Polytechnic also produces a significant share of Ukrainian software engineers that are widely employed by international companies via offshore projects.

3 Research and innovations in Kyiv

Around a third of all the research resources of Ukraine are located in the capital city: there are more than 300 research institutions and 35 thousand R&D jobs. The total spending on scientific and technical activities in the city in 2010 was 469 million USD of which 250 million USD was from the state budget. The total volume of implemented scientific and research works in 2010 was equal to 512.5 million USD in current prices (total in Ukraine – 1.2 billion USD), of which 30 per cent was fundamental research and around 40 per cent – new developments (Kyiv City Statistics Office 2011).

The most successful and internationally renowned scientific and research institutions located in Kyiv work in the areas of physics, mathematics, chemistry, biology, electric welding and material sciences. They are responsible for the greatest share of international collaboration projects and occasional citations in international journals (see Strikha 2012). However they are rarely involved in commercial activities and their involvement in the wider economy is very limited. Table 17.2 presents the six technological parks in Kyiv which are mostly created on the basis of the research institutes of the National Academy of Sciences of Ukraine (NASU). The table also presents the number of commercial projects that are being implemented in these technological parks – fifteen out of a total seventeen projects registered in Ukrainian technological parks are based in Kyiv (see more on technological parks above in section 2.3). Sources of funding are mainly commercial with some indirect support from the state.[4]

Table 17.2 Technological parks in Kyiv

No.	*Name*	*Affiliation*	*No. of projects*
1	Kyiv Polytechnic	Kyiv Polytechnic University	9
2	Paton Institute of Electric Welding	NASU	1
3	Semiconductor technologies and materials, optoelectronics and sensor techniques	Institute of Semiconductor Physics, NASU	3
4	Institute of technical thermo-physics	NASU	1
5	Intellectual information technologies	Glushkov Institute of Cybernetics, NASU	1
6	Ukrinfotech	Ukrtelecom (state corporation)	–

Source: Ministry of Education and Science of Ukraine (2010).

There are no statistics on the number of companies that are involved in innovative activities for the economy in general, only for those that are working in the industrial sector. In 2010 there were only 140 Kyiv companies that performed innovative activities and that were involved in the commercialisation of new inventions/processes. Among these companies are the true giants in their sectors (often state-owned enterprises) working in the spheres of aircraft construction, transport and shipbuilding, machinery and electronics. Since industry represents only about 9 per cent of the city economy, the relative number of such enterprises in Kyiv is much higher than the national average – 24 per cent, compared to 13.8 in Ukraine in general. And this figure has shown a strong downturn trend since 2004 (see Figure 17.4), when the share of innovative industrial enterprises in Kyiv was 41.5 per cent. Moreover, the share of innovative products in the overall industrial production of Kyiv is also going down; in 2005 it was 16 per cent while in 2010 it dropped to 7.7 per cent. Overall, Kyiv companies spend 137.5 million USD on innovative activities, of which 90 per cent is funded from their own capital (Kyiv City Statistics Office 2011).

Venture capital firms that are active in Kyiv are happy to invest in start-ups but mainly in IT and Internet segments, where the time frame needed for bringing projects to fruition is much shorter than that needed for developing new materials, new methods of electric welding or new biological substances that can be used in pharmacology or industry. There are a number of business incubators, some of which are more virtual than physical, which are mainly oriented towards more mobile and commercially oriented developers in the IT and Internet segments of innovations.

There are also some peculiar forms of support for innovative ideas that are quite typical for Ukraine and Kyiv in particular. This comes from rich businessmen who have made their fortune in resource-intensive industries (mainly the metallurgy and energy sectors) and who decide to support specific innovative projects that require significant financial support in the mid-term perspective but which have already passed the crucial 'development' stage and have very positive indications of commercial success.

One such example is the collaboration between Igor Taruta, one of Ukraine's richest businessmen, and researchers from Paton Institute of Electric Welding. They have developed a unique method for obtaining nano particles of various metals of extreme purity that could be used in the pharmaceutical industry as well as in animal welfare and agriculture. The initial investment in the joint venture called NanoMedTech is estimated at 5 million USD (there are no official figures) and the first commercial results from this joint venture are expected in three years time. This example is particularly illustrative since NanoMedTech was created in 2010 on the basis of an invention made at Paton Institute which already has its own technological park (see Table 17.2). The logical conclusion would be that big Ukrainian investors prefer stand-alone agreement and direct partnership rather than joining the institutional framework of a technological park, clearly indicating the level of effectiveness of the latter.

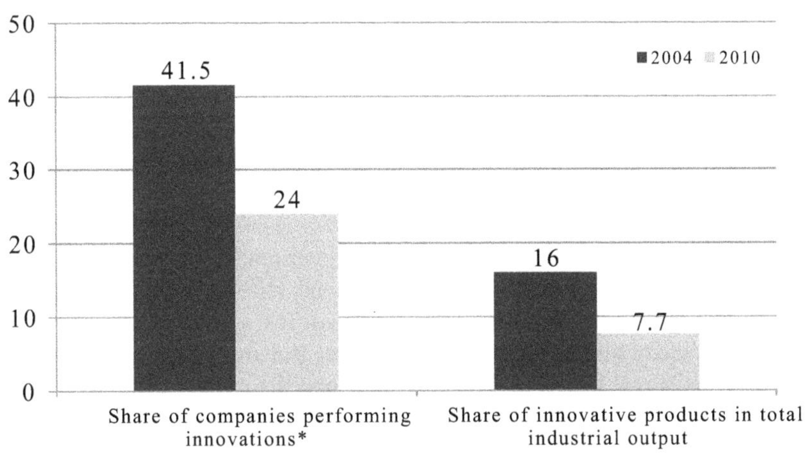

Figure 17.4 Innovations in Kyiv industry (per cent).

* Industrial companies of all sizes and forms of ownership.
Source: Kyiv City Statistics Office (2011).

4 Idle innovation

Innovations and innovative development are buzzwords used by Ukrainian politicians and experts. There are numerous strategies and official decrees that proclaim an innovative model of development as priority for Ukraine. However there is very little substance both in terms of policy decisions/ instruments and analysis of factors and opportunities for innovation-driven development. There are very few studies that concentrate on the realistic assessment of the current educational, scientific and research potential and its impact on the economy. Such lack of insight only aggravates the overall level of ignorance about the significance of innovations for various sectors of Ukraine's economy.

A very illustrative example would be the process of developing the Strategy of Development of Kyiv till 2025. It is an ambitious project, launched by the new head of city administration in 2011, which aimed to engage a wide range of experts, community organisations and businesses in a lively and productive debate over the future of Kyiv. The Strategy mission aims to make Kyiv 'an innovative centre of Central and Eastern Europe which attracts significant investments and highly qualified personnel'. In a SWOT analysis Strategy authors indicated that Kyiv has very high educational, research and intellectual potential (KCSA 2011). However, they completely fail to explain how the city is going to use these valuable assets for building an innovative and competitive economy. There is no thorough 'audit' of existing institutions – either academic, educational or industrial – that can move some sectors of the city economy forward, through technological advancements and exceptional qualifications of graduates. There is no knowledge about local technologies and know-how that already exist and how they can be incorporated into the city's economy and national economy in general. That is why local businesses and public sector institutions often go abroad to obtain new technologies, equipment, know-how and solutions while the local analogues exist, thus undercutting the potential for new jobs and income for the city.

The city does plan to stimulate its innovative economy via several projects aimed at creating an integrated and coherent environment for the functioning of innovative businesses; namely, it is building Bionic Hill – a large innovation park on the outskirts of the city with office premises, research labs, business facilities, accommodation and leisure infrastructure. But the initial project concept indicates that it is oriented towards very few and the most trendy sectors – IT, Biotechnology – and big companies, including internationals, without paying much attention to existing city research capacity and assets. Developers are even planning to create yet another university ('The Specialised University' as it is called in the project concept) (see UDP 2012) instead of offering to relocate the departments and labs of Kyiv specialised technical universities that are already leading in the respective areas (as is often done in other countries, for example in the case of English science city complexes).

The current logic of city leadership illustrates nothing but a sort of ignorance of the fact that city research and educational institutions could become potentially powerful drivers behind new areas of economic growth and sources of innovation for already existing businesses that wish to upgrade their production facilities/services and reach new markets.

Another good illustration of the lack of systemic understanding of the spatial aspects of the socio-economic development of Kyiv and its agglomeration is the recent proposal by one of the members of Government to create a 'Silicon valley' on the outskirts of Kyiv, on virgin agricultural land. Silicon fever has finally reached Ukraine, but instead of identifying already existing hot spots in the city – such as the above-mentioned Kyiv Polytechnic University or the cluster of research institutes in the northern part of the city – the government is keen (at least tentatively) to throw billions of dollars on a project that totally lacks any cost–benefit analysis of this location decision and the impact it will have on the territory, human flows and transport infrastructure.

5 Conclusions

Kyiv is the biggest node of Ukraine's human capital where a significant share of national research and educational facilities are located. At the same time, the structure of the city economy barely reflects the existence of some very prominent but unfortunately commercially incapable institutions whose inventions and developments could bring real economic success. This, in combination with an old-fashioned, centrally planned system of funding allocation in education, and especially in the science and research sector, means that there is no real competition of ideas and projects. Institutes have no stimuli for producing competitive and viable ideas that could serve local and international markets. It is, unfortunately, still common to meet researchers whose mentality is defined by the logic 'we only invented it and it's up to you how to use it in practice'. On the other hand, business is very distant from the high concepts and elaborated language that is often used in R&D and thus cannot even identify potential projects for commercialisation.

This situation is even more aggravated by a lack of vision among state leaders and by a paucity of mechanisms for creating a beneficial business climate, where the regulatory environment, law enforcement and intellectual property rights' protection are strong and fair. This will give businesses, either Ukrainian or international, large or medium, an opportunity to think beyond daily routine and invest time and resources into long-term projects, aimed at developing new knowledge-intensive products and services.

Ukraine has one of the worst regulatory environments in the world, which is confirmed by various international rankings and by assessments of its own entrepreneurs. And it spends a meagre amount of public and private money on R&D. It also lacks the financial infrastructure that could support risky innovative projects with longer life cycles that might bring high profit and significant effects for the local market in the long term.

The third problem that undermines the friendliness of the city towards innovative projects and companies is a lack of strategic vision by state and municipal government officials regarding the role innovations might play in the present and future of Kyiv's economy and society. Most current strategies and plans are built around the extensive use of territory, resources – including human – and infrastructure for satisfying growing consumer needs. At the same time there is a neglect of energy efficiency, sustainability and competitiveness – factors that are implicit in the majority of knowledge-intensive products and services.

As recent experience from preparing Kyiv city strategy illustrates, Kyiv's leadership is keen on promoting products and brands with low value added but higher public appeal (culinary- or history-related) in order to increase the visibility and recognition of the Kyiv brand, while still struggling to comprehend the role of Kyiv's engineering achievements – exceptional inventions in the spheres of environmental protection, energy, new materials, aircraft building and many others. They do not include these in the 'assets' of the city, or at least not in the active assets, and are as yet unable to calculate what effect these sectors could have on the local economy, employment and well-being of the city's population.

This lack of vision and ambition for the future that could be based on knowledge-intensive high value-added products and services produced in the city, as result of close collaboration between business, local HEIs and research institutions, and with due support from the state, is the greatest challenge on the way to opening the intellectual and research potential of Kyiv. And unless radical steps are taken in order to change the current status quo, Kyiv innovations will remain as they are today – overestimated and underused.

Notes

1 In February 2011 the Cabinet of Ministers of Ukraine approved the Programme of Development of Investment and Innovative Activities in Ukraine, which presents a range of instruments and plans for establishing the infrastructure necessary for spurring innovative growth (Cabinet of Ministers 2011a). Later in 2011 the Government announced the creation of the State Innovative Non-banking Financial and Credit Institution 'Fund for Supporting Small Innovative Businesses'. Its activities – interest-free loans for SMEs, compensation of interest payments to commercial banks, state guarantees, insurance etc. – are still to be launched (Cabinet of Ministers 2011b).

2 In Ukraine there are four levels of accreditation for higher educational institutions. The two first stages (I and II) are for professional and technical colleges while the two higher grades (III and IV) are for universities, academies and institutes.

3 Ukraine has a two-tier system of academic degrees; candidates and doctors of science.

4 It might be in the form of a state guarantee for commercial loans, partial or total compensation of interest payments, direct funding from state budget (partial or total), state credits and subventions (Cabinet of Ministers 2011a).

References

Cabinet of Ministers of Ukraine (2010) *Decree on Approval of the Regulation on Research University*. No. 163, 17 February 2010. Available online at http://zakon1.rada.gov.ua/cgi-bin/laws/main.cgi?nreg=163-2010-%EF&myid=4/UMfPEGznhhm3c.ZiOXhkA6Hdl0QsFggkRbI1c (accessed 10 April 2012).

Cabinet of Ministers of Ukraine (2011a) *Decree on Approval of the Programme of Development of Investment and Innovative Activities in Ukraine*. No. 389, 2 February 2011. Available online at http://zakon2.rada.gov.ua/laws/show/389-2011- (accessed 22 April 2012).

Cabinet of Ministers of Ukraine (2011b) *Decree on Establishment of State Innovative Non-banking Financial and Credit institution 'Fund for Supporting Small Innovative Businesses'*. No. 1396, 12 December 2011. Available online at http://zakon2.rada.gov.ua/laws/show/1396-2011- (accessed 25 April 2012).

International Monetary Fund (2011) *World Economic Outlook Base*, September 2011. Available online at www.imf.org/external/pubs/ft/weo/2011/02/weodata/index.aspx (accessed 10 April 2012).

Kozachenko, V. and Georgiadi, N. (2010) 'The current state of technology transfer networks abroad and problems of their development in Ukraine'. Lviv Polytechnic National University Institutional Repository. Available online at http://ena.lp.edu.ua:8080/bitstream/ntb/10011/1/26.pdf (accessed 10 April 2012).

Kyiv City State Administration (2011) 'Strategy of Development of Kyiv till 2025' (Draft). Available online at http://issuu.com/kiev2025/docs/kiev_strategy?mode=window&backgroundColor=%23222222 (accessed 25 April 2012).

Kyiv City Statistics Office (2011) Online database of Kyiv socio-economic indicators. Available online at www.gorstat.kiev.ua/p.php3?c=532&lang=1 (accessed 25 April 2012).

Kyiv Polytechnic University (2010) Top-200 Ukrainian Higher Education Institutions. Available online at www.osvita.org.ua/abitur/entrance/ratings/2.html (accessed 10 April 2012).

Likarchuk, I. (2011) 'Collapse of Ukrainian education', *Weekly Mirror*, 8, 4 March 2011. Available online at http://dt.ua/articles/76811 (accessed 10 April 2012).

Ministry of Education and Science of Ukraine (2010) *Analysis of Activities of Ukraine's Technological Parks in 2009*. Kyiv: Ministry of Education and Science of Ukraine.

Neprytamannyi, S. (2012) 'The same songs about main things', *Weekly Mirror*, 14, 13 April 2012. Available online at http://dt.ua/SCIENCE/ti_zh_sami_pisni_pro_golovne_-100496.html (accessed 10 April 2012).

Semynozhenko, V. (2011) 'Three steps away from target', *Weekly Mirror*, 21 May 2011. Available online at www.dknii.gov.ua/2010-10-05-14-41-28/404-2011-05-23-19-29-14.html (accessed 10 April 2012).

Shulga, N. (2009) 'Ukrainian universities in the global context: External and internal insights'. *Ukrainian Science Club portal*, 3 September 2009. Available online at http://nauka.in.ua/news/archive/article_detail/3475 (accessed 10 April 2012).

State Statistics Service of Ukraine (2011a) 'Special bulletin of R&D statistics, April 2011'. Available online at www.ukrstat.gov.ua (accessed 10 April 2012).

State Statistics Service of Ukraine (2011b) 'Online database of socio-economic indicators'. Available online at www.ukrstat.gov.ua/ (accessed 10 April 2012).

Strikha M (2012) 'Challenges of Ukrainian science', *Weekly Mirror*, 8, 2 March 2012. Available online at http://dt.ua/SCIENCE/ukrayinska_nauka_pered_viklikami_-98148.html (accessed 22 April 2012).

Supel, J. (2010) 'Ukrainian participation in FP7: A statistical approach' (second monitoring report). JSO-ERA Project. Available online at www.jso-era.org/ua/rp7-era-dokumenty.html (accessed 10 April 2012).

System Capital Management (2011) Compass-2011: Opinion of Employers and Alumni. Rating of Ukrainian Higher Educational Institutions by the Level of Satisfaction from Education, SCM, Kyiv. Available online at www.yourcompass.org/PDF%20Tables/Compas2011Report.pdf (accessed 10 April 2012).

UDP (2012) Bionic Hill Innovation Technopark Presentation. Available online at http://bionic-hill.com/BionicHills_booklet_mini.pdf (accessed 25 April 2012).

UN (2010) Human Development Index Data. Available online at http://hdr.undp.org/en/data/explorer/ (accessed 10 April 2012).

Verkhovna Rada of Ukraine (2011) Law of Ukraine on State Budget 2011. Available online at http://zakon1.rada.gov.ua/laws/show/2857-17 (accessed 10 April 2012).

Verkhovna Rada of Ukraine (1999) Law of Ukraine on Special Regime of Innovation Activities in Technological Parks. Available online at http://zakon.rada.gov.ua/cgi-bin/laws/main.cgi?nreg=991-14 (accessed 10 April 2012).

Vitrenko, Y. (2011) 'If we are so educated why are we so poor? Economist's Perspective on the Reform of the Education System', *Weekly Mirror*, 3, 28 January 2011. Available online at http://dt.ua/articles/74249 (accessed 10 April 2012).

World Bank (2011) Doing Business Report 2012. Doing Business in a More Transparent World. Available online at www.doingbusiness.org/reports/global-reports/doing-business-2012 (accessed 22 April 2012).

World Economic Forum (2012) The Global Competitiveness Report 2011–2012. Available online at www3.weforum.org/docs/WEF_GCR_Report_2011-12.pdf (accessed 10 April 2012).

18 University and regional development in the Northern European periphery

The case of the University of Tromsø[1]

Kazimierz Musiał

Introduction

Similar to many other countries in Europe and the world in the second half of the twentieth century, the modernization of Norwegian socio-economic life meant a transition from an industrial to a knowledge-based pattern of development. For the country's peripheral regions, which had enjoyed a relatively special status in the Norwegian settings since the post-Second World War reconstruction, the transition meant a necessity to redefine their position in relation to the central government and to prove their value for the national interests in an increasingly global environment (Eriksen 1996; Østerud 2006). The University of Tromsø (UiT) has played a crucial role in the redefinition of the regional epistemic agenda, while along with this process the institution itself has experienced a significant expansion.

The location of the university on the edge of the Arctic, which is of increasing global importance, implies its current mission that is necessarily defined in relationship with regional and, ever more often, international development plans. Apart from its capacity of providing state-of-the-art scientific and academic knowledge, which normally would be required from an institution in the higher education sector, this northernmost university of the world has managed to become a regionally specific hub of knowledge creation and dissemination. Neither regional politics nor the regional economy can do without an institution that is able to provide them with well-skilled, educated and innovative citizens. Furthermore, the national High North policy or transnational cooperation in the Barents region can hardly be conceived without the participation of UiT or its staff as experts and knowledge-brokers.

In the context of paradigms explaining the relationship between the university and its locality or the linkages between the university and regional development, three OECD publications have provided referential frameworks for research on developments in the past two decades. The publications constructed different perspectives, the first of which paid attention to the response of higher education institutions to regional needs and to the level of interaction with local stakeholders (OECD 1999). The second perspective focused on the role of regions in promoting learning, innovation and raising economic efficiency in the context of strategies aimed at realizing a knowledge-based society and learning economy (OECD 2001). The third perspective dealt with the input from higher education institutions into regional development within the framework of globalization (OECD 2007).

Notwithstanding the topical relevance of these three perspectives, the case of the University of Tromsø escapes an easy pattern of analysis owing to its strong regional and national relevance in the Norwegian setting. To account for this, the topic has to be investigated also from other angles, among which the historical and political specificity of the Norwegian regional development policy constitutes the necessary point of departure for the study. Hence, the specific geographical determinism and historical dependencies that had a bearing on the institutional developments in the Northern European periphery provide a starting point. To begin with, however, the interplay between the university and the North Norwegian peripheral region will be placed within the broader perspective of strategic liaison between social development and higher education in the national peripheries of the Nordic states.

Norwegian regional policy and higher education within the context of the Nordic welfare state

In the whole Nordic area, the establishment of regional universities, often located in peripheral regions, goes back to the 1960s. This phenomenon is related to the expansion of the welfare state in the Nordic countries and the whole system of education, including HEIs, has become an instrument to realize the goals of the welfare politics. The political imperative to pursue welfare policies was based on a consensus forged by Social Democratic parties in Sweden, Norway and Denmark and on agrarian and small reform-bourgeois parties in Finland and Iceland (Christiansen and Åmark 2006). Apart from social security, full employment and decent health-care services, education became an emblematic issue for the social engineers who saw it as a means of providing equality for all in Nordic societies (Husén 1998; Arnesen and Lundahl 2006).

The widely accepted welfare state broadened its underlying principle of providing social equality through education and added the geographical dimension to it. Creating regional colleges and university centres even in peripheral regions may be interpreted as the embodiment of this approach. Educational reforms and policies at all stages were embedded in the machinery of the welfare state and transformed into components of social policy, labour market policy and regional policy. Ever since, educational reforms and changes within the educational sector have been considered inherent to social change and not treated specifically as pedagogical innovation or didactic reform (Rinne 2010: 99–101; Husén 1998; Telhaug *et al.* 2006). In particular, the so called 'alternative' universities established in the late 1960s attempted to distinguish themselves from the traditional higher education institutions through an interdisciplinary educational offer, a thematically rather than disciplinarily motivated administrative structure and, last but not least, through wide use of problem-based learning, as an innovative didactic method (Huisman *et al.* 2002).

Regionalization of higher education was accompanied by its mass expansion, providing an additional argument for creating further HEIs in peripheral regions and, indeed, in some cases was an immediate reason for their establishment. The expanding welfare state often became an employer of new graduates, as it needed new administrative officers and personnel in the growing social and health service sectors. In all Nordic countries in the years 1955–65 the overall number of students doubled, and this figure doubled again between 1965 and 1987 (Kyvik and Ødegård 1990: 12–13). Affiliate colleges of traditional universities were created along with small regional university colleges, in order to guarantee access to higher education for all young people who fulfilled the requirements and wanted to study. The colleges often offered only a few study programmes linked to the needs of the local labour market and the local public sector. Hence, both the ideologically fuelled imperative to provide equal access to education and the expanding welfare state became key driving forces behind the mass expansion and regionalization of the higher education sector in the Nordic countries in the 1960s and 1970s.

The University of Tromsø – a political, economic and educational project

In Norway it was only in 1859 that the University of Oslo, founded in 1811, was given competition within domestic higher education when a post-graduate agricultural college was established. In 1897 it became a scientific university college (*vitenskapelig høgskole*), commonly known as the Agricultural University of Norway (*Norges landbrukshøgskole*). In 1910 the Norwegian Institute of Technology (*Norges Tekniske Høgskole*) was founded in Trondheim, some 500 kilometres away from the capital. In 1936 the Norwegian School of Economics (*Norges Handelshøyskole*) was established in Bergen and in 1946 the University of Bergen started operating. In 1968 the University of Tromsø was established as the fourth university in the country and it became fully operational in 1972.

With respect to the University of Tromsø, three main aspects appear to have been fundamental for its institutionalization, survival and renewal over the years since its foundation. The first aspect concerns the domain of Norwegian politics and spatial planning. From this perspective a regional university, such as UiT, can be seen as a political project resulting from state interventionism in the years following the Second World War. The second aspect is related to the economic development of the area, including sustained profits from the fishing industry and an efficient use of natural resources. While a combination of fishing and farming have traditionally been life-sustaining resources in the region's economy, the discovery of oil and gas in recent decades has turned North Norway – formerly the most inaccessible and the least developed part of the country – into a potential powerhouse for the national economy. The third aspect concerns the external and internal framework that UiT itself has developed since its inception, partly based on an older tradition of local vocational colleges, but mainly adopted – in the spirit of 1968 – as a non-traditional university incorporating the idea of a more democratic and student friendly higher education institution. To a great extent, the identity of an 'alternative' institution, as compared to the older and more established universities in Oslo and Bergen, has allowed for greater student participation in governance as well as (inter)disciplinary flexibility. Broadening the offer of study programmes was a means to attract students in the initial years, after which the number of students stagnated (Fulsås 1993: 172).

As this chapter will demonstrate, an interplay among these three aspects made it possible for UiT to be established, survive and remain a vibrant learning institution in the rather challenging natural environment of the Arctic region.

1 The regional university as a political project at the top of Europe

The university has played a vital role in the development of the northernmost region of Norway, which by national standards has remained thinly populated and the least developed. It is the city of Tromsø, its municipality and, to a certain degree, the vast area and people of all three counties in northern Norway that provide the setting for the university. Today, the closest locality of the university, Tromsø municipality, has approximately 68,600 inhabitants and covers an area of 2,566.3 km^2. It is the capital of Troms county which is said to be the gateway to the Arctic, at 70 degrees north. Apart from the university, with some 9,000 students, three quarters of the vocationally active inhabitants are employed in the service sector, represented by both public and private companies, while fewer and fewer people draw their income from traditional industries, such as fishery and tourism.

Spatial Keynesianism – Norwegian-style

Along with the social changes that came about in the aftermath of the Second World War, and in an attempt to provide for sustainable and more or less equal economic development over the whole territory of the country, there evolved different conceptions of dealing with the underdeveloped and dilapidated northern districts. Similar to many other countries, the Norwegian state started considering a more active regional policy. The roots of this conception reached back to the first half of the twentieth century, when attempts were made to improve the efficiency of state administration in peripheral localities. The strategy fitted well with the policy practised in the second half of the twentieth century, which Neil Brenner (2004) dubbed 'spatial Keynesianism', thereby referring to the interventionist measures implemented by more developed states and relating to the restructuring of their regional space. The policy of spatial Keynesianism aimed at providing relative equality of the administrative units within a given state and, in a narrow sense, referred mainly to the services offered within the public sector. It was carried out through a policy of redistribution and public spending that were accompanied by providing and guaranteeing common living standards and health services as well as welfare provided by the state (Brenner 2004). In a broader sense, spatial Keynesianism may refer to any interventionist measure of the national government geared towards stimulating regional development in both economic and social domains.

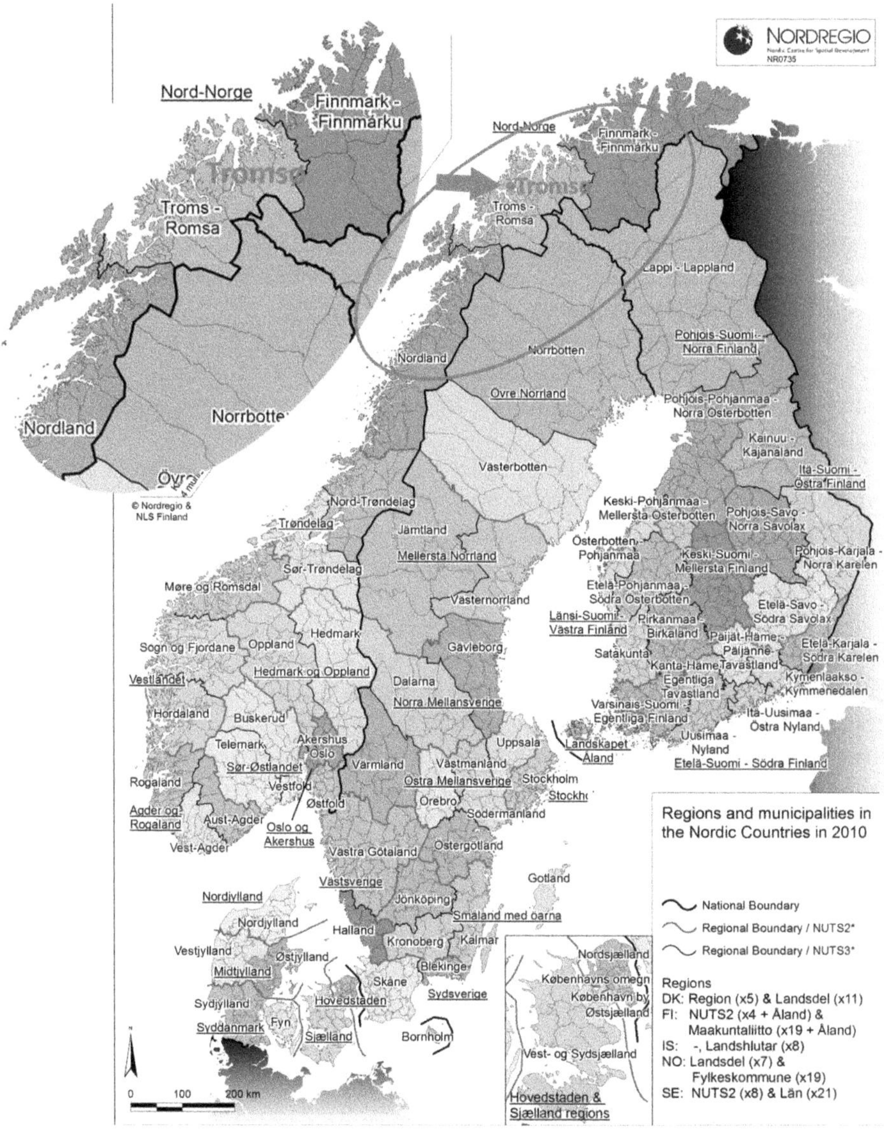

Figure 18.1 Municipalities in Northern Norway, based on Johanna Roto's map of Regions and municipalities in the Nordic Countries (2010).

Source: Nordregio at www.nordregio.se.

With respect to the northern regions of Norway this policy started to take shape immediately after the Second World War. When in 1945 German troops withdrew, leaving behind most of the buildings and infrastructure burnt to the ground, the region became an issue of central government concern and a large-scale modernization scheme under the name of the North Norway Plan was passed by the

Storting (national parliament) in 1952. While the plan was linked to the pre-war plans of the governing Labour Party to modernize North Norway, its aim was to modernize industry, improve living conditions, generate growth and employment, and in the long run, establish patterns of economic redistribution that would be more favourable for Norway's northern districts (Reiersen 1982: 130). Last, but not least, at the moment of its inception, the plan also addressed the increasing fear of communist influence in North Norway, in close vicinity to the Soviet border (Teal 1953; Lloyd 1956). Nevertheless, well into the 1950s, shortages in housing and structural unemployment frustrated the governing Labour Party, that preached full employment and rapid economic growth nationwide. The situation was a breeding ground for radicalism in Northern Norway. As mirrored in the words of Rokkan and Valen, 'the polarized periphery – economically the most backward of all the regions – has a persistent record of radicalism' (Rokkan and Valen 1970: 197).

In 1966 the growing disillusion with the interventionist policy of the national government with respect to North Norway was voiced strongly by the leftist political grouping and the issue became ubiquitous because of the criticism included in Ottar Brox's book *Hva skjer i Nord-Norge?* ('What's happening in North Norway?'). Brox voiced a strong critique of the government's fishery policy and pointed out the increased political peripherality of this part of the country. As noted by David Arter (2001), Brox's book, which itself prompted a widespread debate, coincided with the advent of television into most living rooms, which gave rise to a political movement in the north. In its institutional form it took the shape of the Regional Authority for North Norway (*Landsutvalget for Nord-Norge*), founded in 1974 as a political voice for the three most northerly counties of Finnmark, Troms and Nordland (the whole of North Trøndelag joined in 1995). Hence, in the 1970s, the policies of spatial interventionism promoted by the central government were concurrent to local regional initiatives and civic movements.

The conviction about the advantageous role of higher education institutions in implementing spatial Keynesianism emerged when regional development came to be regarded as a function of aggregation of the social and intellectual capital in a given locality. Modernization of the national economy, based on boosting industrialization in only a few localities and an increasing wealth disparity between rich centres – mainly in the south of the country – and the relatively underdeveloped peripheries, triggered off migration pressures from the countryside to the few larger industrial districts and from the climatically less attractive North to the more comfortable South. Regionalization of higher education was one of the measures to counteract this development and, in the late 1960s, both the politicians in the capital as well as regional decision makers were convinced that regionally anchored but universal higher education institutions were prerequisite for providing competitive and attractive living conditions nationwide. Hence, planned regionalization and proliferation of higher education became an unshakable paradigm when considering the sustainable socio-economic development of the whole territory of Norway (Bie 1981). The developments were particularly favourable for localities in the peripheral regions and North Norway was undoubtedly one of the beneficiaries.

The strategic placement and development of a fully fledged university of Tromsø in the 1960s was the result of an active application of 'spatial Keynesianism', which included an ideal of both social and political equality, and the belief in a strong state that is able to engineer development on its territory. As can be read in the historical documents, it was the political will of a few politicians in the Ministry of Education that was decisive, rather than the seemingly more relevant argument about the lack of doctors and medical education in the northern part of the country (Bertelsen 1991: 200–207). The spatial Keynesianism policy consisted of a set of structural policies that promoted interventionist measures in order to balance out real or conceived territorial disparities in Norway. The policy of spatial interventionism or, as Svein Kyvik (2009: 67) calls it, the 'geographic decentralisation of public goods,' has won increasing support in Norwegian society. In 1957 only 9 per cent of the Norwegian population considered regional policy to be one of the three most important political issues. This figure rose to 27 per cent in 1965 and 59 per cent in 1969 (Valen and Martinussen 1972), which was most likely a reaction against the tendency towards centralization in the first post-Second World War decades.

2 The university and socio-economic development in a peripheral region

As the vast body of relevant literature presently suggests, regional universities arguably have a very important – and sometimes decisive – role to play as dynamos or powerhouses for regional economy and development (OECD 2001; Harding *et al.* 2007). To begin with, many of the examples quoted in topical literature come from regions where the end of the industrial era caused problems with unemployment and the structural dysfunctionality of the obsolete infrastructure, and where regional HEIs were expected to contribute to the recovery of both social and economic structures (Huggins *et al.* 2008). The new social values emerging in the post-industrial era saw learning and knowledge as the basis of a flourishing society. Thus, institutions of higher education were expected to contribute by getting involved with their environment and by becoming more flexible and entrepreneurial (Rinne and Koivula 2005). This also concerned the regional level where, by the end of the 1990s, the triple helix metaphor came to epitomize the most desired, dynamic form of cooperation in a triangle of academia, local business and public administration (Karlsson and Zhang 2001; Etzkowitz 2002). In the specific, geographically determined conditions of the European North and the Arctic, the focus was more on using regional HEIs to boost the economic development of the peripheral regions where otherwise negative migration trends would increase and gradual socio-economic decline could be expected.

For a long time in the Nordic region, like elsewhere, decision makers at all levels have shared the conviction that a university is a regional booster, as in the title of a book by the Dutch economist Raymond Florax (1992). However, the postulated engagement of universities in the economic development of their regions has been reciprocal. The involvement of universities in developing the knowledge economy has been explicitly expected by the polity (European Commission 2003), but at the same time many national systems of higher education used economic arguments to make HEIs utilize regional resources and tap into new sources of income. Local and regional knowledge was to be created by regional HEIs to provide competitive advantage and innovation in the regional economy. When, in the 1990s, the national government increasingly started using management by objectives and introduced development contracts with the universities, many HEIs turned to their closest localities to seek supplementary funding. Through the analytical lens of resource dependency, it is possible to view this development as one of the most important reasons for the willingness of the regional HEIs to get involved with their regions. Arguably, resource dependency may be seen as one of the decisive factors that brought about cooperation between universities and their environment (Gornitzka 1999).

With respect to the University of Tromsø, its importance for the regional economy was recognized fairly early. As noted by Peter Arbo, UiT was lucky to be founded as the fourth university in Norway (after Oslo, Bergen and Trondheim), still in the era when the strong state supported such deeds with funding. Since UiT was regarded as a political project and a 'gift from the Storting' to North Norway, it did not suffer from financial problems in its initial phase (Arbo 2011). Furthermore, the effect of UiT for the regional economy has been very palpable ever since, even if only the injection of basic means from the state budget to the university is taken into account. As an example, UiT received 1.8 billion NOK from the state budget and secured some 0.5 billion NOK from external income in 2011. About 75 per cent of the total amount was used as a means to cover personnel costs, a large proportion of which was used again to purchase goods and services locally. Hence the university's employees, students and visitors feed the local economy as consumers, not to mention their contribution to local tax incomes. From the perspective of the local economy, it is also important that three out of four graduates of the university remain in the region, as their qualifications allow them to find jobs easily, mainly in the public sector (Arbo 2011).

Indeed, the importance of the university as a supplier of a local, well-educated workforce cannot be underestimated. As noted by Hjort (1973, 1976) and Fulsås (1993), one of the most acute issues in Tromsø at the time of UiT's foundation was the lack of doctors and medical personnel in the region, which resulted in the medical sciences becoming one of the more important study fields in the new institution and a strong argument for the university's establishment.[2] Training of medical doctors was related to the development of the welfare state in the national periphery where the ambition was to provide living standards comparable to those in the more developed parts of Norway. Thus, although the public debate was very long and heated, with commissions being appointed and much lobbying in Oslo taking place (Bertelsen 1991), medical education was finally included in a regional development plan that focused on the importance of training specialists for the region at the University of Tromsø. The main argument boiled down to the assumption and expectation that most of the future nurses, medical doctors, school teachers and other graduates should come from the region and most of them would continue to work in the North after graduation. As expressed emphatically in a retrospect of the university's activities, for many university proponents producing graduates was considered to be

> the true purpose of a university, and it was the expected correlation between place of birth, place of study, and region of employment, especially in the case of medical students, which had been of such political significance in the campaign to establish a North-Norwegian university. Now one can conclude that this line of reasoning has been proven correct.
>
> (Fulsås 2002: 29)

The triple helix model – a way to prosper or an empty promise for the regional economy?

The politically motivated conviction about the necessity of spatial interventionism led to the proliferation of the higher education sector in Norway in the 1970s. Since then, regional HEIs have attracted attention as potential engines of regional growth, though without the triple helix metaphor being ever explicitly used in Norway to term an ideal relationship between university, industry and government. By the end of the century it seemed that the model offered almost a universal key to swift regional development, though for many it remained merely a theoretical model, often without options for universal application in the Scandinavian context. In Norway, the relative underdevelopment of the triple helix policy instruments was mainly owing to the specificity of national research and innovation policy that in the 1990s was still being realized through strategic programmes involving large and centrally located companies and non-university research institutes.

The idea of the triple helix started arriving in Norway in the 1990s and one of the key moves facilitating the emergence of this kind of dynamic partnership was readjusting research activities of the universities to make them more applicable and attuned to the needs of the local stakeholders. The previously mentioned resource dependency explains the reason for the triple helix arrangements occurring, triggered off by the changing funding framework of research and higher education. A reform process in this domain has been under way since the early 1990s, when national research councils were reformed to play a broader role in society and to change the organization and normative structure of academic research. In 1993 a unified research council was formed with the explicit task of integrating application- and discovery-oriented research (NOU/1991:24). The new structure has developed a new organizational matrix where research disciplines are combined with societal sectors. New support mechanisms have been established, in the form of targeted programmes with long-term planning horizons and new managerial structures. Furthermore, a more pronounced role for industrial interests has been prioritized (Benner and Sandström 2000: 299). Arguably, this has been a part of a more general social and political change taking place in Norway since the 1980s, when the overall purpose of politics shifted from a mission linked to the building of the welfare state (based on education and public services) to fostering innovation and knowledge-based growth (relying on research and the private sector). Peter Arbo argues that the paradigmatic change from a modernization policy based on industrial development to the knowledge-based development ideal has had a very profound influence on North Norway and its civilizational progress in the long run (Arbo 1996).

Over the years, and especially when the ideas of New Public Management permeated the public administration, seeking for a more economic relationship between the university and the state, the University of Tromsø has renewed itself in order to meet the requirements of present day society and, to a greater extent, of its region (Stensaker and Norgård 2001). However, the triple helix-like patterns of development have evolved in Tromsø very slowly and rather late. This is owing to the fact that the university has always remained heavily dependent on basic funding from the national government and has been subjected to the government's vision of science and technology policy. Well into the 1990s the triple helix ideas did not have much impact on the behaviour of the university, as research and innovation policy was mainly realized through strategic programmes that typically involved large and centrally located companies and non-university research institutes. It is significant that, even by the end of the 1980s, the ensuing reforms in higher education still aimed at strengthening the traditional roles of the universities and colleges in what is known as the Humboldtian spirit, rather than preparing them for a more dynamic interaction with their environment in the spirit of an entrepreneurial institution. Only very slowly did the practice of triple helix cooperation and awareness of its beneficial effects become more prominent, even though cooperation patterns did not need to be built from scratch, as both historical and structural foundations already existed.

When the University of Tromsø was founded, the business sector expected neither to recruit candidates from the university nor to benefit from university research. The distance between the university and industrial sector grew even bigger when UiT turned 'red', that is, when the left-wing students' movement gained hegemony at the university in the 1970s. In the 1980s both the radical politics at the institution and the distrust of industry began to recede, while the discovery of crude oil in the North Sea and goodwill agreements that the Norwegian authorities imposed on foreign oil companies operating there provided seed money for different forms of cooperation. The idea was to make oil extraction contribute to indigenous research and industry development, and as new drillings were expected in the close vicinity of the North Norwegian coast, the communities of North Norway were the primary beneficiaries of the deal (Arbo and Eskelinen 2006). While at the time of tight budgets scientists and researchers at the university were keen to tap into new research money, the quickly expanding oil industry was also more eager to make use of the know-how of the university-based specialists. The most prominent role model was provided by the Institute of Fisheries

Technology, where ambitions to develop research into aquaculture and biochemical processing were matched by means provided by the oil industry in exchange for the assistance of the institute's staff in solving some specific problems.

The concurrent devolution of central government meant that prerogatives concerning regional development issues were passed down to the counties, established in the 1975 county reform, and this gave regional decision makers greater autonomy. Although no further funding was provided from the state budget, the Ministry of Local Government wanted to make sure that the new national strategies for knowledge-based growth would have a regional dimension. These developments stipulated the establishment of FORUT, an institute of applied and contract research linked to the university, which was an equivalent to the system of industrial research institutes. After merging with the Institute of Fisheries Technology in the early 1990s, it created the NORUT Group Ltd., which in 2007 was rebranded as Northern Research Institute. Apart from the Tromsø Science Park that opened in 1994, the Northern Research Institute provides the infrastructure for bridging university research and industry with a special focus on the High North (Arbo and Eskelinen 2006; NORUT 2011).

Owing to the different interests of the changing governments, a pattern of periodic decentralization and then re-centralization of decision making with respect to regional research funding on the part of the Norwegian Research Council could be witnessed. By and large, however, the number of programmes with regional anchoring has been on the constant rise, which was also enhanced by the activities of SIVA, the Industrial Development Corporation of Norway. Thanks to SIVA, science parks and business gardens became central components of the Norwegian innovation support system. As a result of this policy, Tromsø Science Park received the world's northernmost incubator in 2001 and in the following years the science park has greatly expanded (SIVA 2011).

3 An 'alternative' university for the northern region or a forerunner of educational innovation in higher education

Although founding UiT in the northern Norwegian periphery in 1968 was a beacon of the new regional approach in Norway's higher education policy, the policy linking regional development with the establishment of higher education institutions got into full swing only with the proliferation of regional university colleges (*distriktshøyskoler*) after 1969. Especially in the 1970s and 1980s, their task was to provide shorter and more vocationally applicable courses at the university level. As noted by Kyvik and Ødegård (1990:10, 14), the reason for establishing the university colleges, which realized more educational than research related tasks, was to provide vocationally oriented education in peripheral regions, as well as promote regional policy, with the higher education institution as one of its key actors. In recent decades many of the university colleges raised their expectations and ambitions in academic achievement and became unquestionable leaders in their regions, thus skilfully combining the role of an educational and, ever more often, a research institution, while remaining an innovative force in regional development (Kyvik 2009). By this means they also became important players in regional politics.

The University of Tromsø was established as an 'alternative university' with innovative study programmes, including a pre-programmed interdisciplinary approach and some links to the local stakeholders – mainly in the public sector – to realize its third mission, namely, to meet the demands of local politicians, researchers and the community. Being the first real university in the northern part of Norway, UiT was explicitly branded as the University of the Arctic, with medicine, social sciences, marine biology and Arctic research as its important areas of research and education. This brand has been further strengthened by links between the university and the Norwegian Polar Institute residing in the city.

Since the foundation of the university in 1968 the whole of North Norway has remained its referential locale. It is a rather large area of 113,000 km2, consisting of three administrative districts: Nordland, Troms and Finnmark. Some 9.8 per cent of the total population of Norway lives there, which amounts to over 460,000 inhabitants, including Sámi and Finnish minorities. Tromsø is unquestionably the largest city of North Norway, with some 68,600 inhabitants, which makes it the seventh largest city in the whole country. Most of the villages and settlements in the area are very small and their population does not exceed 5,000 inhabitants (Arbo and Fulsås 2002: 47).

The 1968 decision of the Storting was strategic for the Norwegian regional development programme, as it was to provide intellectual and human capital for this large and economically underdeveloped area. The political decision that made an institution of higher education key to the modernization of this part of the country was met with scepticism in traditional universities; they pointed at the weak academic potential of the new institution and seemingly poor perspectives for hiring qualified staff and attracting students (Fulsås 1993). It was probably owing to the scepticism of the academic community towards the new institution that the university decision makers in Tromsø were given more freedom to experiment with the institutional structure, interdisciplinary approach and participatory institutional democracy, with the appreciation of students as valuable stakeholders. For instance, the University of Tromsø was a forerunner of changes in the university management model when it introduced external representatives on its board as early as in 1975 (Fulsås 1993: 208). In many regards, UiT offered an alternative to the established universities in Oslo and Bergen, where democratic participation of the staff and students in the institutional management was not so advanced. Hence, since its inception, the identity of UiT was characterized by innovativeness in academic matters and institutional governance, though in the 1980s a certain return to a structure based on single-discipline academic fields took place (Stensaker 2000).

Despite some problems with recruiting students in the initial years, that is, only 420 students in 1972 and 1,294 students in 1977, the number of students at UiT increased slowly to some 3,158 in 1989, then exploding to some 6,000 in the 1990s, and further increasing to almost 9,000 at present (Arbo and Eskelinen 2006; Frølich 2006; DBH 2011). The growth has been a result of demographic and cultural changes in the region as well as, perhaps more importantly, a result of the economic crisis ensuing in the late 1980s. When unemployment rose in the late 1980s and early 1990s, UiT decided to maintain an open door policy, admitting all students who fulfilled the criteria, without quotas that, owing to financial limitations, started to be introduced in other HEIs in the country. Thanks to a large influx of students, teaching programmes were consolidated, efficiency of the university administration improved and the service mission of the university with regard to its environment became clearer. Apart from its regional mission, the university started to internationalize its educational and research activities to the extent that this may be defined as its fourth official mission, apart from teaching, research and regional services (Musiał 2010). In 2011 there were 724 international students, that is, about 10 per cent of all students at the university, which indicates a constant growth since 2002, when 492 foreigners studied at UiT (DBH 2011). The figures point at an unceasing implementation of the internationalization strategy through academic mobility, where special priority has been attached to attracting students and researchers from Africa and other countries in the south (Frølich 2006). Attracting African students and researchers to the northernmost university of the world may seem a peculiar endeavour, yet the university has been an active participant in the Norwegian Programme for Development, Research and Education (Norad) and in Norad's Programme for Master Studies that aimed at building bridges in education and research with the developing countries.

One of the traditional arguments that spoke for the foundation of a university in Tromsø has been its reputation as a 'gateway to the Arctic' (Arbo and Fulsås 2002: 52). This issue has remained a universal element of the institutional strategy building and has been realized through providing education and research relevant to the Arctic. In practical terms, since the 1990s it has been translated into the establishment of several research institutes and centres, such as the Institute of Marine Research, Northern Research Institute, the Norwegian Institute of Food, Fisheries and Aquaculture Research, Akvaplan-niva AS – Polar Environmental Centre, the Research Centre for High North Operations and others, which cooperate within the so called Tromsø High North Cluster (THiNC 2011). Research related to the Arctic region, its nature, natural resources, sustainable development and climate undoubtedly constitute the university's research specialism but they also provide an opportunity to become an internationally and globally renowned resource and research centre. The current strategy and action plan of the university for 2009–13 includes a vision of a regionally anchored university with global relevance (UiT 2009a; UiT 2009b: 9–11). Even though branding a regional university as an international research institution may be sometimes problematic, the heteroglossia of the institutional identity building process appears to have been a very efficient strategy in the case of Tromsø over the years (Stensaker and Norgård 2001; Wæraas an Solbakk 2009).

The most recent development that will undoubtedly have an impact on strengthening and rebranding the university as a provider of research and competence in local and regional knowledge, was its merger with Tromsø University

College (*Høgskolen i Tromsø*) in 2009. The university college was itself a result of a merger that took place in 1994, and offered education for medical personnel, engineers, economists, artists, musicians and teachers. Its merging with the university strengthened the academic capacity of Tromsø both for the staff and the students, along the lines suggested by the Quality Reform of 2002. It also allowed for the adjusting of the study programmes to the needs and preferences of the students. With the merger of the two institutions the capacity of UiT to provide state-of-the-art knowledge in some specific disciplines has been strengthened. The enlarged institution has also raised its competitiveness when seeking funds in the Norwegian Research Council's programmes.

Even before the merger, the university was quite successful with two large project applications, both of which also demonstrate an increasingly global and international approach at UiT. Thanks to the projects, attempts have been made to attract the best researchers in some disciplines, in which local competence has already been established and is strong. Two new research centres have been established, namely CASTL – the Center for Advanced Study in Theoretical Linguistics – and CTCC – the Centre for Theoretical and Computational Chemistry. Both have the status of a centre of excellence and, despite their localization in Tromsø, they are both sites of cooperation for international teams of researchers conducting research in their relevant disciplines. At the same time, local specificity and local affairs are included as research issues. In CASTL, the Sami languages, spoken by the indigenous minority in the North Nordic region, are an object of study, while in the CTCC the Norwegian national research competence in chemistry is being developed (CASTL 2011; CTCC 2011).

Further proofs of UiT's strong research position as forerunner of innovation in the northern region are three centres of research-based innovation, established within a programme of the Research Council of Norway. *MabCent – Centre for Marine Bioactives and Drug Discovery* – conducts research into bioactive compounds from Arctic and sub-Arctic marine organisms with potential for further research, innovation and commercialization (MabCent 2012). *TTL – Tromsø Telemedicine Laboratory* – aims to establish an interdisciplinary environment that will conduct research concerning the combination of human caring with new technologies, in order to provide better health care and reduce the increasing burden on the health care system (TTL 2012). Finally, *iAD – Information Access Disruptions* – is the only centre of its kind, led by a commercial company with UiT as its collaborator. Its research concerns developing information access technology (iAD 2012).

Conclusion and an outlook for the future

While the story of a regional university and its cooperation with its environment may be told in many different ways, this chapter has focused on three nodal points that have been fundamental for the discourse concerning the institutionalization, survival and renewal of the university since its foundation in 1968. The three nodal points have been: 1) spatial Keynesianism, 2) sustaining and expanding regional economic life, and 3) the special status of UiT as a role model of the regional university in Norway. While the nodal points are helpful in structuring the discourse and textual narratives, the developments have been driven by interplay among these three closely interwoven factors that made it possible for UiT to be established, survive and remain a vibrant learning institution in the rather challenging natural environment of the Arctic region.

What is interesting in the case of Tromsø, is that geographic determinism has been overcome and used as an asset for building up and profiling UiT in relation to other higher education institutions in the country. From a comparative international perspective, the Norwegian specificity of regional development – with North Norway as an emblematic case – boils down to its stronger determination by the harsh climatic and geographical conditions. In the past these conditions rendered large parts of Norway, similar to Sweden, Finland and Iceland, unattractive or limited for human life and activity. In recent decades this very fact has led to a specific approach to spatial Keynesianism that has provided a balance in regional development through creating better social and employment conditions in order to prevent inhabitants from moving away and perhaps to attract new ones to settle down in the Arctic. While an institution of higher education seemed an ideal tool to realize this goal, the political will of the central government, recognition of the economic needs of the area and provision of an alternative to the established institutions of higher education in the country have all been keys to success.

The three above-mentioned aspects have been responsible for building up the capacity of Tromsø and its university in the Norwegian national setting, though it is the most recent developments in Northern Europe and the High North that may be fundamental for their future. With the changes in the international environment after the Cold War, the High North witnessed the creation of Barents Euro-Arctic Region. The region was an area of military confrontation during the Cold War; hence the underlying premise for its creation was that close cooperation would secure long-term stability in political and other relations and reduce possible tensions. The initial meeting launching this initiative took place in Kirkenes in 1993, with participation of the ministers of foreign affairs or representatives of Denmark, Finland, Iceland, Norway, the Russian Federation, Sweden and the Commission of the European Communities. Observers from the USA, Canada, France, Germany, Japan, Poland and the UK attended the meeting as well, thus showing interest in this regional experiment of replacing the confrontation and division of the past with a new partnership in the spirit of new regionalism. The signed declaration launched cooperation on two levels: intergovernmental and inter-regional, with sustainable development as the overall objective. The intergovernmental cooperation, realized by Barents Euro-Arctic Council (BEAC), includes four countries: Finland, Norway, Russia, Sweden, while the inter-regional level, realized by Barents Regional Council (BRC), covers cooperation among thirteen different counties or their equivalents in the four countries in question. Being forums of political and, ever more often, NGO cooperation for Norwegian, Finnish, Swedish and Russian regions in the north, both organizations provide numerous windows of opportunity to make use of the relatively rich UiT's expertise and research potential for the whole Barents area. While initially the overall objective for the new regional cooperation framework was sustainable development, it has become more substantial since working groups on indigenous peoples, investments and economic cooperation, Arctic shelf development, maritime safety etc., started practical cooperation and initiated common projects (BEAC 2011).

Furthermore, a new opportunity for UiT may come up in the critically important energy supply business. It is no longer any secret that, with an ever-increasing priority given to the security of energy supply on the part of the EU, the High North might become the key region of strategic interest for Norwegian relations with other countries (Offerdal 2010). So far, the natural gas and crude oil from Norwegian shores above the sixty-second parallel north has not been extracted, but in the case of economic or political need the petroleum industry is prepared to operate in this area (Arbo and Hersoug 2010). Most important of all is the fact that Norway is a reliable and predictable producer and that it has already developed a tight network of cooperation in the European and regional framework (Gheorghe *et al.* 2011: 402–403). If the petroleum industry were to intensify its operations in North-Norwegian sea waters, it would surely mean new dynamics for the business of the university and for the epistemic community concentrated in and around the University of Tromsø. In the Norwegian setting the developments could be similar to the rapid advancement of Stavanger at the outset of the Norwegian oil and gas industry (cf. Westnes *et al.* 2009). If these developments materialize, with the long-lasting and proven history of innovation in the specific areas of knowledge pertaining to living in the Arctic, UiT may possibly become not only a centre of epistemic expertise within national borders but also, as was desired by the founding fathers of the university in Tromsø, a real gateway to the Arctic with respect to knowledge and innovation, and with the reference area stretching from Novaya Zemlya in Russia to Reykjavik in Iceland.

Notes

1 This project was completed with the support granted by Iceland, Liechtenstein and Norway by means of co-financing from the European Economic Area Financial Mechanism and the Norwegian Financial Mechanism as part of the Scholarship and Training Fund, which enabled the author's study visit to Tromsø in September 2011.

2 The scandal that arguably broke the camel's back happened in 1962 when it turned out that in a hospital in Vardø, in the northernmost corner of Norway, a Danish student of medicine acted as a MD, treated patients and performed operations for over a year without any authority being able to control him (Aune 1996: 81–87). It made the politicians in Oslo aware of the necessity to provide greater resources for education and employment of medical personnel in North Norway. Throughout the 1960s support for building a specialized clinic and hospital in Tromsø was mounting, with the foundation of the university hospital in 1968 as an ultimate result of this process. The author wishes to thank Bjørn Hersoug for bringing up this micro-scale, though quite emblematic, event as a reason for decisions made in regional policy in North Norway.

Bibliography

Arbo, P. (1996) 'Fra industriorientert til kunnskapsorientert modernisering: Om Nord-Norges plass i kunnskapssamfunnet', in E. O. Eriksen (ed.) *Det nye Nord-Norge: Avhengighet og modernisering i Nord*. Bergen: Fakbokforlaget.

Arbo, P. (2011) 'Universitetets regionale betydning', *Plan*, 02, 18–22.

Arbo, P. and Eskelinen, H. (2006) 'Regions with small comprehensive universities: The cases of Joensuu and Tromsø', in J. E. Nielson (ed.), *The Role of Universities in Regional Innovation Systems – A Nordic Perspective*. Copenhagen: Copenhagen Business School Press.

Arbo, P. and Fulsås, N. (2002) 'The University of Tromsø', in D. C. Nord and G. R. Weller (eds), *Higher Education across the Circumpolar North: A circle of learning*. New York: Palgrave Macmillan.

Arbo, P. and Hersoug, B. (eds) (2010) *Oljevirksomhetens inntog i nord – næringsutvikling, politikk og samfunn*. Oslo: Gyldendal Akademisk.

Arnesen, A. and Lundahl, L. (2006) 'Still social and democratic? Inclusive education policies in the Nordic welfare states', *Scandinavian Journal of Educational Research*, 50(3), 285–300.

Arter, D. (2001) 'Regionalization in the European peripheries: The cases of Northern Norway and Finnish Lapland', *Regional & Federal Studies*, 11(2), 94–114.

Aune, A. (1996) *Kongens mann i nord*. Oslo: Aschehoug.

BEAC (2011) Available online at www.beac.st/in_English/Barents_Euro-Arctic_Council/Calendar_of_events.iw3 (accessed 14 July 2011).

Benner, M. and Sandström, U. (2000) 'Institutionalizing the triple helix: Research funding and norms in the academic system', *Research Policy*, 29(2), 291–301.

Bertelsen, T. I. (1991) *Kampen for et medisinsk fakultet i Tromsø*. Bergen: Universitetet i Bergen.

Bie, K. N. (1981) *Creating a new university – the establishment and the development of the University of Tromsø*. Oslo: Institute for Studies in Research and Higher Education.

Brenner, N. (2004) *New State Spaces: Urban governance and the rescaling of statehood*. Oxford: Oxford University Press.

Brox, O. (1966) *Hva skjer i Nord-Norge? En studie i norsk utkantpolitikk*. Oslo: Pax Forlag.

CASTL (2011) Available online at http://uit.no/castl (accessed 30 July 2011).

Christiansen, N. F. and Åmark, K. (2006) 'Conclusions', in N. F. Christiansen, K. Petersen, N. Edling and P. Haave (eds), *The Nordic Model of Welfare: An historical reappraisal*. Copenhagen: Museum Tusculanum Press.

CTCC (2011) Available online at www.ctcc.uit.no (accessed 30 July 2011).

DBH (2011) Database for statistikk om høgre utdanning. Available online at http://dbh.nsd.uib.no/ (accessed 7 September 2011).

Eriksen, E. O. (ed.) (1996) *Det nye Nord-Norge: Avhengighet og modernisering i Nord*. Bergen: Fakbokforlaget.

Etzkowitz, H. (2002) *The Triple Helix of University–Industry–Government: Implications for Policy and Evaluation*, Working paper 2002–11. Stockholm: Swedish Institute for Studies in Education and Research.

European Commission (2003) *The Role of the Universities in the Europe of Knowledge*. Brussels (=Communication from the Commission, COM (2003) 58).

Florax, R. (1992) *University: A regional booster*. Aldershot: Avebury.

Frølich, N. (2006) 'Still academic and national: – Internationalisation in Norwegian research and higher education', *Higher Education*, 52(3), 405–20.

Fulsås, N. (1993) *Universitetet i Tromsø 25 år*. Tromsø: Universitetet i Tromsø.

Fulsås, N. (2002) *The University of Tromsø: A Short Retrospect*. Edited by E. B. Jensen, Tromsø: University of Tromsø.

Gheorghe, A., Muresan, L., Celac, S. *et al.* (2011) 'An energy security strategy for Romania: Promoting energy efficiency and renewable energy sources', in A. Gheorghe and L. Muresan (eds), *Energy Security: International and Local Issues, Theoretical Perspectives, and Critical Energy Infrastructures*. Dordrecht: Springer.

Gornitzka, Å. (1999) 'Governmental policies and organisational change in higher education', *Higher Education*, 38(1), 5–31.

Harding, A., Scott, A. and Laskey, S. (eds) (2007) *Bright Satanic Mills: Universities, regional development and the knowledge economy*. Aldershot: Ashgate.

Hjort, F. (1973) *Universitetet i Tromsø: Hvorfor, hvordan og hvorhen*. Oslo: Polytekniskforening.

Hjort, P. F. (1976) *Universitetet i samfunnet: Artikler og foredrag om Universitet i Tromsø*. Tromsø: Universitetsforlaget.

Huggins, R., Jones, M. and Upton, S. (2008) 'Universities as drivers of knowledge-based regional development: A triple helix analysis of Wales', *International Journal of Innovation and Regional Development*, 1(1), 24–47.

Huisman, J., Norgård, J.D., Jørgen Gulddahl Rasmussen, J.G., and Stensaker, B. (2002) '"Alternative" universities revisited: A study of the distinctiveness of universities established in the spirit of 1968', *Tertiary Education and Management*, 8, 315–32.

Husén, T. (1998) 'The Swedish school reforms: Trends and issues', in A. Tjeldvoll (ed.), *Education and the Scandinavian Welfare State in the Year 2000: Equality, policy, and reform*, New York & London: Garland Publishing, Inc.

iAD (2012) Available online at www.iad-centre.no/ (accessed 28 January 2012).

Karlsson, C. and Zhang, W.-B. (2001) 'The role of universities in regional development: Endogenous human capital and growth in a two-region model', *The Annals of Regional Science*, 35(2), 179–97.

Kyvik, S. (2009) *The Dynamics of Change in the Organisational Field of Higher Education: Expansion and contraction*. Dordrecht: Springer.

Kyvik, S. and Ødegård, E. (1990) *Universitetene i Norden foran 90-tallet: Endringer i Styring og Finansiering av forskning*. København: Nordiska Ministerrådets Sekretariat.

Lloyd, T. (1956) 'The Norwegian–Soviet boundary: A study in political geography', *Norsk Geografisk Tidsskrift*, 15, 187–242.

MabCent (2012) Available online at www0.nfh.uit.no/mabcent/ (accessed 28 January 2012).

Musiał, K. (2010) 'A "glocal" university in Norden: Realising the third and fourth missions in the global age', in N. Götz, J. Hecker-Stampehl and S. M. Schröder (eds) *Vom alten Norden zum neuen Europa: Politische Kultur im Ostseeraum. Festschrift für Bernd Henningsen*. Berlin: Berliner Wissenschafts-Verlag.

NORUT (2011) Available online at www.norut.no/ (accessed 7 September 2011).

NOU/1991:24 'Organisering for helhet og mangfold i norsk forskning'. Available online at www.regjeringen.no/upload/kilde/odn/tmp/2002/0034/ddd/pdfv/154771-nou1991-24.pdf.

OECD (1999) *The Response of Higher Education Institutions to Regional Needs*. Paris: OECD.

OECD (2001) *Cities and Regions in the New Learning Economy*. Paris: OECD.

OECD (2007) *Higher Education and Regions: Globally competitive, locally engaged*. Paris: OECD.

Offerdal, K. (2010) 'Arctic energy in EU policy: Arbitrary interest in the Norwegian High North', *Arctic*, 63(1), 30–42.

Østerud, Ø. (ed.) (2007) *Norway in Transition: Transforming a stable democracy*. London and New York: Routledge.

Reiersen, J. E. (1982) 'Distriktspolitikk og regionale utviklingsprosesser', in N. Aarsæter, R. Nilsen and J. E. Reiersen (eds) *Folkemakt og regional utvikling*. Oslo: PaxForlag.

Rinne, R. (2010) 'The Nordic university model from a comparative and historical perspective', in J. Kauko, R. Rinne and H. Kynkäänniemi (eds) *Restructuring the Truth of Schooling: Essays on discursive practices in the sociology and politics of education*. Turku: Finnish Educational Research Association.

Rinne, R. and Koivula, J. (2005) 'The changing place of the university and a clash of values: The entrepreneurial university in the European knowledge society. A review of the literature', *Higher Education Management & Policy*, 17(3), 91–123.

Rokkan, S. and Valen, H. (1970) 'Regional contrasts in Norwegian politics: A review of data from official statistics and from sample surveys', in E. Allardt and S. Rokkan (eds) *Mass Politics: Studies in political sociology*. New York: The Free Press/London: Collier Macmillan.

SIVA (2011) Avilable online at www.siva.no/ (accessed 7 September 2011).

Stensaker, B. (2000) *Høyere utdanning i endring. Dokumentasjon og drøfting av kvalitetsutviklingstiltak ved seks norske universiteter og høgskoler 1989–1999*. Oslo: NIFU.

Stensaker, B. and Norgård, J. D. (2001) 'Innovation and isomorphism: A case-study of university identity struggle 1969–1999', *Higher Education*, 42(4), 473–93.

Teal, J. J. (1953) 'The rebirth of North Norway', *Foreign Affairs*, 32, 123–134.

Telhaug, A. O., Mediås, O. A. and Aasen, P. (2006) 'The Nordic model in education: Education as part of the political system in the last 50 years', *Scandinavian Journal of Educational Research*, 50(3), 245–83.

THiNC (2011) *Tromsø High North Cluster*. Available online at www2.uit.no/ikbViewer/page/ansatte/organisasjon/hjem?p_dimension_id=136656&p_menu=42374&p_lang=2 (accessed 7 September 2011).

TTL (2012) Available online at www.telemed.no/index.php?cat=77933 (accessed 28 January 2012).

UiT (2009a) *Strategidokument for Universitetet i Tromsø 2009–2013*. Available online at www2.uit.no/ikbViewer/Content/143103/Strategi.pdf (accessed 7 September 2011).

UiT (2009b) *Handlingsplan for Universitetet i Tromsø 2009–2013*. Available online at www2.uit.no/ikbViewer/Content/163115/Handlingsplan_UiT_2009_endelig.pdf (accessed 07 September 2011).

Valen, H. and Martinussen, W. (1972) *Velgere og politiske frontlinjer*. Oslo: Gyldendal.

Westnes, Petter, Hatakenaka, Sachi, Gjelsvik, Martin and Lester, Richard K. (2009) 'Role of universities in strengthening local capabilities for innovation: A comparative case study', *Higher Education Policy*, 22, 483–503.

Wæraas, A. and Solbakk, M. N. (2009) 'Defining the essence of a university: Lessons from higher education branding', *Higher Education*, 57(4), 449–62.

Index